Volcanoes of Europe

Also available from Dunedin Academic Press

Dougal Jerram: *Introducing Volcanology* (2011)

Alwyn Scarth: *Vesuvius - A Biography* (2009)

See www.dunedinacademicpress.co.uk for all our publications

Volcanoes of Europe
Second edition

Dougal Jerram

*Centre for Earth Evolution and Dynamics, University of Oslo
and DougalEARTH Ltd*

Alwyn Scarth

formerly University of Dundee

Jean-Claude Tanguy

Institut de Physique du Globe de Paris

DUNEDIN

EDINBURGH ◆ LONDON

Published by
Dunedin Academic Press
Head Office: Hudson House, 8 Albany Street, Edinburgh EH1 3QB
London Office: 352 Cromwell Tower, Barbican, London EC2Y 8NB

© 2017 Dougal Jerram, Alwyn Scarth and Jean-Claude Tanguy

ISBNs:
9781780460543 (Hardback)
9781780460420 (Paperback)
9781780465630 (ePub)
9781780465647 (Kindle)

First edition published by Terra Publishing (now an imprint of Dunedin Academic Press) 2001

British Library Cataloguing in Publication Data
A catalogue record for this book is available from the British Library

Design and prepress production by Makar Publishing Production, Edinburgh
Printed in Poland by Hussar Books

Contents

Preface

Europe, despite its many changes through history, has been home to volcanoes for many millions of years. These volcanoes pay no attention to human foibles such as historical periods, political boundaries (and how they change) and scientific definitions. Thus, the title *Volcanoes of Europe* disguises several kinds of arbitrary choices. We have included, for instance, the Canary Islands and the mid-Atlantic islands of Jan Mayen, Iceland, Svalbard, and the Azores within the European umbrella, although two of the Azores and half of Iceland belong to the North American plate, and the Canary Islands belong to the African plate. On the other hand, we do not describe the volcanoes of Turkey and the Caucasus, which many would, no doubt, call European.

It is altogether more difficult to define those volcanoes that are active, dormant, or extinct. Volcanoes do not always display the secrets of their past, nor do they always reveal their future intentions. Several times, even in the course of the twentieth century, expert volcanologists have been puzzled – not to say surprised – when certain volcanoes have suddenly burst into life after a long period of calm. Volcanoes are generally considered active if they have erupted in the last 10,000 years, though such a value may include volcanoes that are effectively extinct. It is also valuable to consider the older records of many of the volcanic areas as they help reveal how these spectacular landscapes develop and grow.

The notion of historical time is also extremely flexible, and historical records count for little within the defined span of 10,000 years. Even within the limited European context, the period during which eruptions could actually be recorded has varied greatly from place to place. Probably no volcano on Earth has a longer recorded history than Etna, where eye-witness accounts have recorded its eruptions, with admittedly varying degrees of fantasy, for thousands of years. However, the Italian volcanoes were in an exceptionally favoured position in the classical world. On the other hand, records in Iceland extend back only to the early centuries after the settlement in AD874, and no human being even settled in the Azores until 1439.

Beyond the historical context, accurate dates of eruptions are only just becoming available in many areas. The traditional methods of geological dating by fossils and stratigraphy are very hard to apply to volcanic edifices. The timespan is too short; the volcanic products preserve few animal or vegetal remains; and the erosion of valleys and their subsequent occupation by further lavas make large and active volcanoes a stratigrapher's nightmare. In many cases, too, the most recent eruptions have masked the products of their predecessors to such an extent that the story of the volcano can be difficult to read.

In recent decades, new techniques of absolute dating have done much to overcome these handicaps. Radiocarbon dates have been calibrated with greater precision, and volcanic rocks can be dated by thermoluminescence, argon dates (K–AR, Ar–Ar), and palaeomagnetic and archaeomagnetic studies, among other techniques. A whole range of these techniques is now being applied, especially to those more dangerous volcanoes whose tempestuous past must be discovered before their future furies can be predicted with accuracy. Nevertheless, the absolute dates of many European eruptions have yet to be established.

The availability of information is a further element that imposes its own limitations on any treatment of European volcanoes. In spite of the boom in volcanological research during the past few decades, some volcanoes are still imperfectly known. Thus, several Italian volcanoes are in intensive care, whereas those in the Azores, for instance, have undeservedly progressed little beyond the waiting list. Consequently, the balance and the treatment of active European volcanoes is, in part at least, influenced by the amount of the scientific literature that is available. The most recent of eruptions are being recorded at great length with a whole array of digital technologies from the ground, the air and up into space with satellites. Yet the older eruptions require the diligent studies of many scientists, and for the information to be made into the public domain. In this context, this contribution has used the wealth of information made available by the scientific and public communities that work on the volcanoes of Europe, as well as those gained by the authors in their volcanic careers.

Volcanoes hold a fascination that we have from a young age, and it is something that we would like to foster. The chief aim of this study is to stimulate a wide range of readers to encourage them to take an active, informed interest in some of the most sublime and fascinating volcanic features in the natural world and, especially, to go and see them. You can, in Europe, go and see an active volcano such as Stromboli, or explore ancient remains of these natural wonders. Aesthetic rewards will also enhance any scientific or knowledge-gaining pilgrimages, because the European volcanoes embellish landscapes beyond compare.

Authorship and acknowledgements

This updated colour version of *Volcanoes of Europe* has been put together by Dougal Jerram, expanding on the original content and scope of the first book, and with some new colour images provided by Jean-Claude. Updated colour images are used throughout, and new information is also included on the latest eruptions that have occurred. As with any of these ventures, debt is owed to many individuals and groups who have offered direct help, informative discussions over the years, and who have shared in a passion for earth science and particularly volcanoes. I will not be able to give credit to all here, but you know who you are. I must thank those who took part as *Meet the Scientists*: Matteo Lupi, Dave Pyle, Val Troll, Helen Robinson, Steffi Burchardt, Sigurður Gíslason, and Ben van Wyk de Vries, who also supplied materials. Specific help with parts and materials was also given by: Don Kaiser, John Howell, Angelo Cristaudo, Paraskevi Nomikou, Vikki Martin, Brian McMorrow, Josef Schalch, Sarah Gordee, Juan Carlos Carracedo, Alfredo Lainez Concepción, Breno Waichel, Paul Cole, Carlos Miguel, João Luís Gaspar, Abigail Barker, Frances Deegan, Morgan Jones, Sverre Planke, John Millett, Henrick Svensen, Reidar Trønnes, Torgeir Andersen, Trond Trosvik, Bæring Steinþórsson, Thórdís Högnadóttir, Jón Viðar Sigurðsson, Allan Treiman, Rolf Pedersen, Pierre Boivin, Cecile Olive-Garcia, Jörg Busch, Walter Müller, Universal Postal Union, TodoTenerife Team, Instituto Geográfico Nacional, Kappest/Vulkanpark, National Geopark Laacher See, US Geological Survey, NASA. Also, those people who made materials available to use through creative commons licences (credited along with the specific pictures) and from the Shutterstock archive. In the final construction of the book I give specific thanks to Helen Robinson for redrafting some original images, Francis Abbott for his efforts with Spain images, and Ben van Wyk de Vries for many efforts (both text and figures) within the chapter on France, as well as Anne Morton, David McLeod and Sue Butterworth for their great work in copy-editing, drafting and design and indexing. Final thanks to Anthony Kinahan at Dunedin for his continued support for the *Volcanoes of Europe* project. *Dougal Jerram (August 2016)*.

In the original B/W edition Jean-Claude Tanguy was primarily responsible for the sections on Etna, Vesuvius, Pantelleria and parts of the Aeolian Islands. Alwyn Scarth bore the responsibility for the remaining chapters; he gratefully acknowledges the invaluable assistance of Juan-Carlos Carracedo, Victor Hugo Forjaz, Harry Hine, Maxime Le Goff, and especially Anthony Newton.

The photographs without explicit credits were taken by the authors.

Chapter 1

An introduction to volcanology, igneous rocks and the volcanoes of Europe

There are a magnitude of varying processes and ways in which igneous rocks are formed and emplaced both as intrusions and as volcanic units. As we explore the volcanoes of Europe we will be exposed to the whole variety of rock types and volcano types that we have on our planet, and in many ways some of the most important and well-studied examples that we have are found in Europe (Fig. 1.1). Volcanism and its plumbing can be explored in a number of ways. The first way follows the 'present is the key to the past' ethos, in that we can observe volcanism in the present day, where it occurs, and how the resultant material is defined (e.g. Fig. 1.2A). Just like sedimentology and geomorphology, understanding the modern systems can help us a great deal

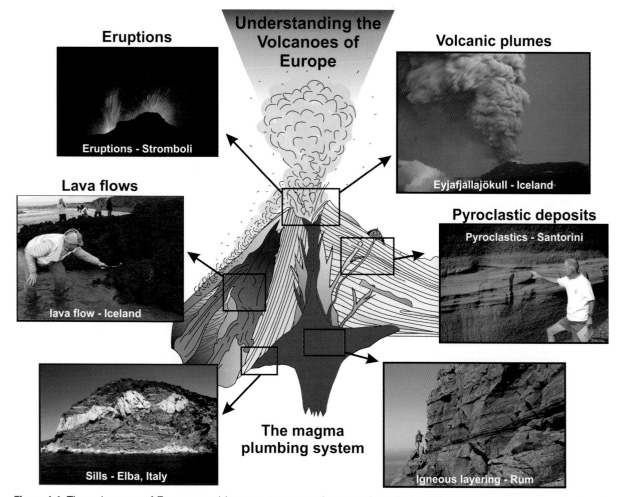

Eruptions

Eruptions - Stromboli

Understanding the Volcanoes of Europe

Volcanic plumes

Eyjafjallajökull - Iceland

Lava flows

lava flow - Iceland

Pyroclastic deposits

Pyroclastics - Santorini

The magma plumbing system

Sills - Elba, Italy

Igneous layering - Rum

Figure 1.1 The volcanoes of Europe provide a great opportunity to explore all the different types and styles of volcanic eruption and their plumbing systems (photo of Eyjafjallajökull courtesy of Jón Viðar Sigurðsson).

Figure 1.2 How we look at and record the volcanoes of Europe. **A)** Sampling at the lava front, Holuhraun eruption Iceland 2014. **B)** Artistic impression of Vesuvius (Vesuvius from Portici by Joseph Wright of Derby: 18th century painting). **C)** Inspecting volcanic deposits from Santorini.

in deciphering the past rock record and palaeo-environments. In Europe there is also a rich history of Man that goes back thousands of years, where volcanoes and their eruptions have been observed and recorded in script and in art (e.g. Fig. 1.2B). This anthropological volcanology becomes more important in Europe's recent past, when volcanoes such as Vesuvius are described erupting in great detail.

Finally, in outcrop we can use the normal range of field skills to measure and record ancient volcanic deposits, and to decipher the longer-term history of the volcanoes, where no other record of the eruptions is available (e.g. Fig. 1.2C). With the vast range of igneous rock types and the styles of deposits, we may be faced with everything from volcaniclastic sediments, lava flows or intrusions, to combinations of them all. In this case a basic understanding of the igneous rock types and styles of volcano will be useful. Additionally we can learn a great deal about the European volcanoes by understanding how rocks melt and where volcanoes occur on the planet. This in turn

will help us understand the distribution of volcanoes in Europe, and why there is such a wide and rich variation.

Igneous rock types

The first thing to consider when looking at volcanic systems and their deposits is the variety of igneous rocks possible. Igneous rocks are classified according different relative amounts of key minerals such as olivine, feldspar and quartz. Each mineral is made up of its constituent elements, and varying the quantity of each mineral type will produce a different rock composition. The volcanoes of Europe cover all the possible magmatic/volcanic types, so a familiarization with some of the terminology and classification is helpful when reading this book. In general, igneous rocks range in composition from those relatively rich in iron and magnesium (known as mafic) to rocks that are rich in silica and aluminium (known as acidic). Rocks can be classified by their grain size as well as by their chemistry (see Fig. 1.3). This is used

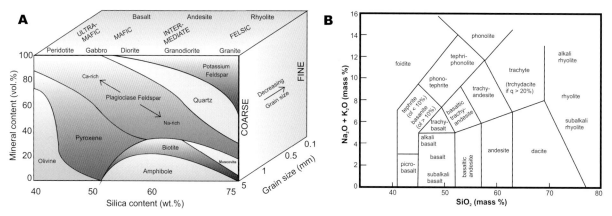

Figure 1.3 How we classify igneous rocks. **A)** Classification based on abundance of key minerals and crystal size. **B)** The Total Alkali–Silica plot used to classify igneous rocks from their geochemistry.

to help distinguish plutonic rocks (which have cooled slowly at depth), from shallow intrusions and volcanic rocks, which erupt at the Earth's surface and cool very quickly. With plutonic rocks, the slow crystallization leads to large crystals, whereas a rapid cooling results in very fine-grained crystals or even glass.

A simple classification scheme for igneous rocks is presented in Figure 1.3A, which combines the relative compositions of the rock types with the grain size to give a number of rock classification names. In some instances there may be some chemical analysis of the rocks, and where this is available the rock units can be classified according to the silica and alkali contents (e.g. Fig. 1.3B). Basic rocks rich in iron and magnesium range from coarse-grained gabbro through medium-grained dolerite to fine-grained basalt. Acidic rocks range from coarse-grained granites, through micro-granites to fine-grained rhyolite. Intermediate rocks include andesites and dacites, etc. (see Fig. 1.3). You may also come across rocks being termed 'basaltic' in composition, which can sometimes be used in a non-grain size sense, as well as the term 'granitic' to indicate silicic compositions.

Why rocks melt and where volcanoes occur

In order to understand at what temperature different rocks melt, and at what temperatures different crystals will start to solidify from molten rocks, a number of simple, yet highly revealing, melting experiments were used. A classic study in the early 1900s by Norman L. Bowen used melt experiments to show that minerals such as olivine, pyroxene and (calcium Ca-rich) plagioclase crystallize from hotter, more mafic compositions

(e.g. basalt), and minerals such as quartz, potassium feldspar, and micas from cooler acidic compositions (e.g. granite). In the Earth the rate at which the temperature increases with depth is known as the geothermal gradient, and commonly ranges from 25–30°C per kilometre of depth in most parts of the world. Exceptions to this are areas of thin crust/hot spots with very high geothermal gradients (30–50°C/km) and thickened crust in subduction settings (5–10°C). If we were to consider the geothermal gradient alone, we would soon reach the temperature to melt rocks. This is where the other factor of pressure comes in. If you try to melt rocks at high pressures, you need even higher temperatures than normal to do so. This simple relationship means that, for most circumstances on Earth, the pressure/temperature relationships are such that the rocks are below their melting point, and not hot enough to melt.

So how do we get rocks to melt? To help us understand this, we can consider a simple plot of an average geothermal gradient as shown in Figure 1.4. Also contained in this plot is a line known as the solidus. This marks the temperature/pressure relationship with depth at which melting will occur. At normal conditions the geotherm does not cross the solidus line, and so no melting will occur. However, if water is added to rocks (as happens at subduction zones) their melting temperature drops. This has the effect of moving the solidus line towards the geotherm, and if they cross, the rocks will start to melt. The other way of getting rocks to melt is by removing pressure from them. If we can quickly remove the pressure from a rock faster than it is allowed to cool, we will have the effect of raising the geotherm, as the rock will be hotter than expected at a shallower depth (see Fig. 1.4C). It can be shown that temperatures of ~700°C or greater are needed to melt rocks (this is at the low end for granitic compositions and in excess of 1100°C for basalts). Melting occurs at depths of 50–200km on average, and we need either to add fluids to the rocks or to release the pressure in order to get them to melt.

So where and how do we get volcanoes on earth? Well, this boils down mainly to plate tectonics and the convection of the Earth as it cools. The surface of the Earth is made of a number of rigid plates that are moved around on a more ductile, deeper mantle. The plates are made of the crust and the lithosphere, and can be continental or oceanic. As the Earth cools and convects, the plate boundaries are either pushed/rubbed against each other, or are pulled apart. The Earth cross-section on Figure 1.4 displays a schematic picture of various plate-tectonic scenarios involving both oceanic and continental plates.

Where plates pull apart, at constructive/divergent plate boundaries, we get upwelling of mantle and decompression resulting in melting, and rift volcanoes. Where plates subduct beneath one another (destructive/convergent plate boundaries) we introduce water and fluids into the hot mantle and cause melting where we lower the melting point, also resulting in volcanoes. Finally, where hotspots occur, marking where hot plumes of mantle material rise independently of the plates, we get decompression melting and volcanoes (see Fig. 1.4). Thus at various settings on the Earth we can generate

melt, and this explains why we have volcanoes on Earth. As we explore the volcanoes of Europe, we shall see how different plate-tectonic settings and hotspots are present, helping us to elucidate the origins of our European volcanoes.

Types of volcano

We can consider volcanoes firstly in terms of their morphology, and then in terms of the types and size of eruptive events (next section). This first step is important as volcanoes can have very similar morphologies but on quite different scales, for example from the smallest basalt lava flows to some of the largest basalt flows, known as flood basalts. The basic categories of volcano are presented in Figure 1.5, and we shall briefly consider the main morphological types below.

Fissure vents form as linear features on the Earth's surface and essentially represent lava erupting out of faults or cracks in the Earth's crust. Fissure vents can be found at constructive plate margins where the crust is being ripped apart under extension, e.g. Iceland and the Afar rift in Ethiopia, but are also common in hotspot volcanoes like Hawaii. In settings where the main lava type is predominantly basaltic, shield volcanoes are formed. These are large volcanoes that have a very broad relief, due mainly to the low viscosity of the lavas that make them up. Although they have a relatively

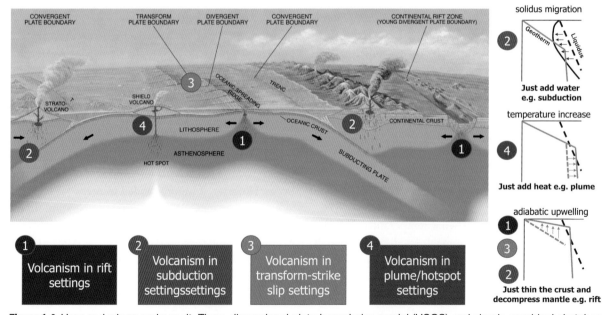

Figure 1.4 How and where rocks melt. Three-dimensional plate boundaries model (USGS) and simple graphical sketches of how to modify solidus or geotherm to promote melting. The numbers highlight the location of melting styles depicted in the graphs.

Volcano Type	Charactersitics	Examples	Simplified Diagram
Flood or Plateau Basalt	Very liquid lava; flows very widespread; emitted mainly from	North Atlantic Igneous Province, Columbia River, Siberian Traps, Paraná-Etendeka	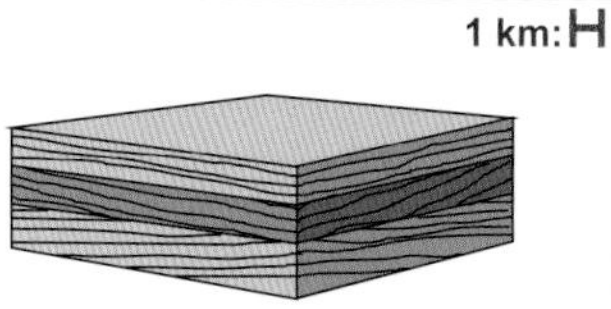 1 km:H
Shield Volcano	Liquid lava emitted from a central vent; large; sometimes has a collapse caldera	Skjaldbreiður, Iceland, Hawiian volcanoes,	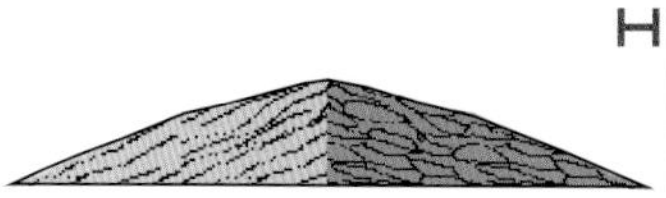 H
Cinder cone	Explosive liquid lava; small; emitted from a central vent; if continued long enough	Chaîne des Puys - France, Iceland,	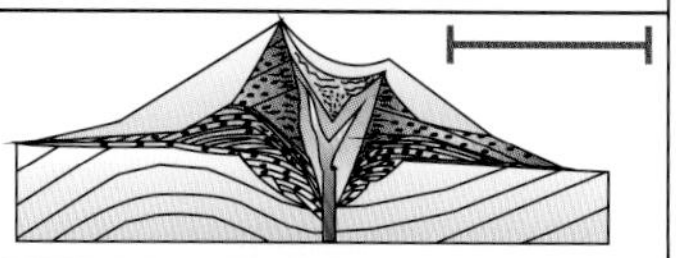
Composite or Stratovolcano	More viscous lavas, much explosive (pyroclastic) debris; large, emitted from	Vesuvius, Italy, Mount Rainier-USA, Colima-Mexico, Dmamavand-Iran Mount Fuji-Japan	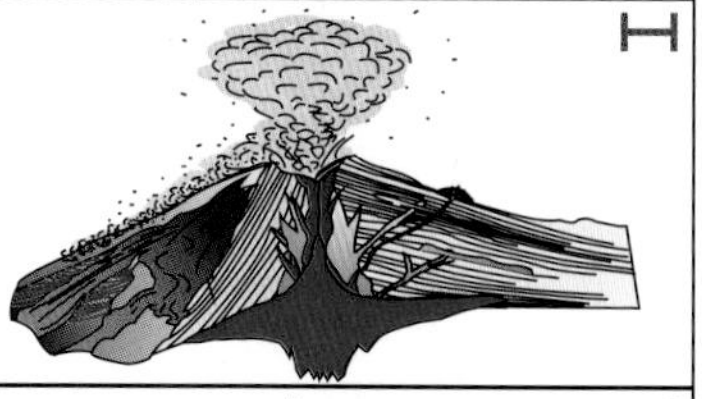 H
Volcanic Dome	Very viscous lava, relatively small; can be explosive; commonly occurs adjacent to/within craters of	Puy de Dôme, France Mt StHelens dome, Colima central	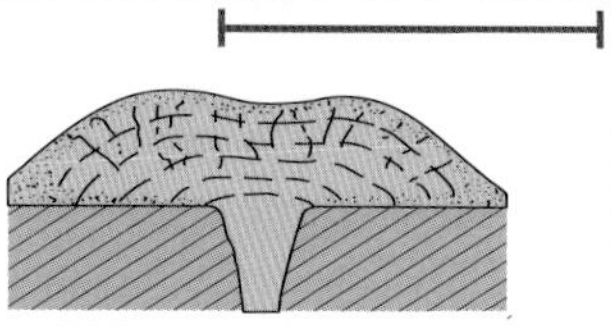
Caldera	Very large composite volcano collapsed after an explosive period; frequently associated	Campi Flegrei, Italy, Crater lake-USA, Long Valley-USA	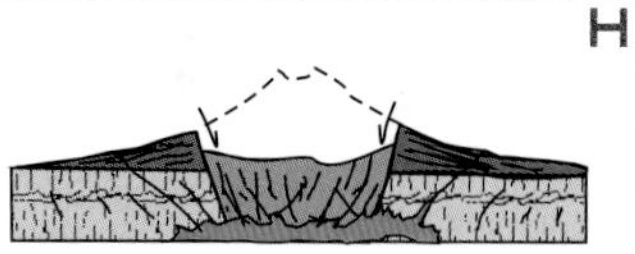 H

Increasing Violence
Increasing Viscosity

Figure 1.5 Types of volcano with examples (modified from USGS).

low profile, some examples can reach great heights due to their immense size (e.g. Mauna Loa, Hawaii 4169m (13,679ft). Cinder/scoria cones develop from small to moderate scale eruptions where pyroclastic scoria are erupted in fire-fountain and Strombolian type eruptions (see next section). These cones are very common and can be found as isolated or clustered vents in volcanic fields or along fissures.

Composite volcanoes (stratovolcanoes), rise up from shallow slopes at the base with steep-sided tops to the volcano, resulting in a cone-shaped mountain. They are a common form of volcano, often occurring in areas where more explosive silicic eruptions dominate. The sides of the volcanoes are littered with a mixture of both lava flows and pyroclastic debris, and commonly a number of valleys run off from the sides, which channel pyroclastic flows and debris flows from the volcano. Stratovolcanoes are mainly acidic (silicic) in composition, with rhyolites

and dacites, and hence their association with explosive eruptions, but they can show the full compositional spectrum including andesites and basalts. Lava domes are bulges of new lava that form in the craters of volcanoes. These often occur after large explosive eruptions and can be termed resurgent domes. Calderas/caldera volcanoes are a type of volcanic landform where the profile of the volcano, instead of being that of a classic cone or conical shape, reflects a circular/semi-circular depression on the surface. Such depressions or calderas are formed by the collapse of the land after the evacuation of magma from a shallow chamber that was present before the eruption.

Some volcanoes that are intricately involved with water are also significant. Submarine volcanoes occur when a volcano starts to erupt on the sea floor, and build up initially as underwater volcanoes. In some instances the build-up and growth of the volcanic edifice fills the water space and the volcano breaks through the sea surface with dramatic effects, known as an emergent volcano. Other water-influenced volcanoes include subglacial volcanoes, where an eruption occurs under ice. These are very complex in that they involve interactions between magma, water and ice, and are also constrained by the ice that covers them. Having introduced the types of volcano, we shall now look at the different styles of eruption and their relationship to the size of the eruptive event.

Types and scales of eruption

Types of eruption range from those that are not very explosive and have little or no eruption column to the largest type of explosive activity. An additional factor that can affect the style of eruption is the interaction with water or ice. If you have predominantly lava flows, explosive products in your volcanic basin will depend on the prominent type of eruption, and it should be noted that the style of eruption can change with time and also switch back and forth. A general classification of eruptions has developed, based on, and often named after, key volcanic events (Fig. 1.6). This captures the general scales of eruption based on observations of the eruption column and the explosiveness of the event. Below we shall briefly go through the different types of volcanic eruption; examples of some of these are also presented in Figure 1.7.

Surtseyan eruptions occur when a submarine volcano erupts through the waves and starts to produce a new volcanic island, also known as an emergent volcano. This style of eruption is named after the example that occurred off the coast of Iceland in 1963, leading to the formation of the Island of Surtsey (e.g. Fig. 1.7A). More generally they are known as hydrovolcanic (also phreatomagmatic) eruptions, signifying the interaction of lava with water.

Effusive eruptions are the very simplest of volcanic eruption on land, where lava effuses out of a vent forming a lava flow, and little or no explosive activity is involved. Such eruptions can occur commonly at lava lakes when they fill up and spill over, but it can often be difficult to develop a predominantly effusive eruption without some sort of more explosive event at its start. Hawaiian eruptions are more commonly seen in basaltic-dominated volcanoes. They are intimately associated with effusive eruptions, and this style of eruption can often feed much larger lava flows through time. Lava fountains that reach a few hundred metres into the air are common, and examples of this style of eruption along fissures can sometimes result in a wall of fire, which collapses to the ground to feed lava flows. A classic example of this type of eruption is the recent Icelandic eruptions of 2014 and 2015 (e.g. Fig. 1.7B).

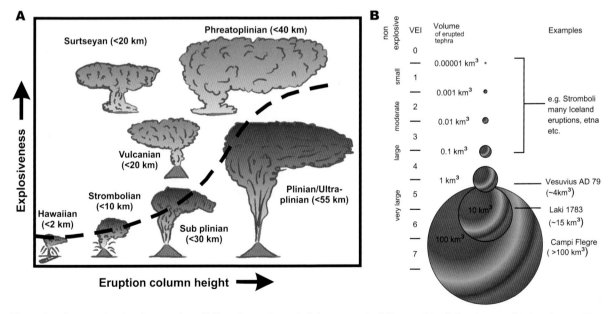

Figure 1.6 Scales of volcanic eruption. **A)** Eruption column height vs. explosivity used to define types of volcanic eruption (note the addition of water helps promote explosivity). **B)** The Volcanic Explosivity Index in graphical form with some European eruptions listed (adapted from USGS).

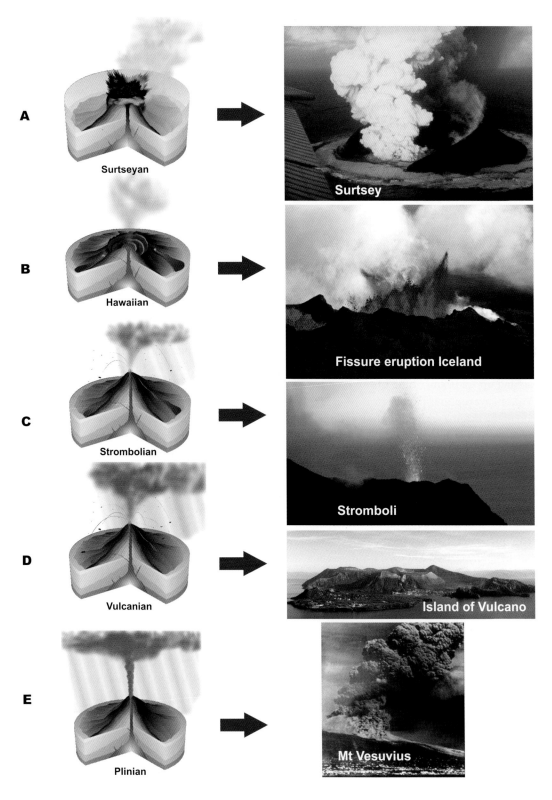

Figure 1.7 Examples of different eruption types and volcanoes with European examples. (3D images by Ivan d'Hostingue (aka Sémhur), inspired by the document about volcanism from the web portal on the prevention of major risks from the French Minister of Ecology, Environment and Sustainable Development) (photo **A**, NOAA; photo **D** by Brisk_g; photo **E** courtesy of Jacques Cordier).

Strombolian eruptions, some would say, are the most picturesque of volcanic styles, named after the volcano forming the Island of Stromboli in the Mediterranean. A Strombolian eruption is driven by the bursting of gas bubbles within the magma. Strombolian activity is characterized by scoria and bombs of hot material being ejected as the bubble bursts. In extreme cases, columns can measure from hundreds of metres to a few kilometres in height, but more commonly a Strombolian eruption looks somewhat like an energetic Roman candle firework when viewed at night (e.g. Fig. 1.7B).

Vulcanian eruptions, named after the 1888–1890 eruptions on Vulcano Island (Fig. 1.7C) in the Mediterranean, are more explosive still. These eruptions result from more viscous magma (andesite/dacite), from which it is difficult for gases to escape. The pressure builds up and eventually the blocked top of the volcano gives way, with an initial set of cannon-like explosions sending bombs and blocks through the air. The deposits contain a much larger amount of ash than their Strombolian counterparts, with eruptive columns reaching up to 10km. Some of the high gas contents in Vulcanian eruptions have been attributed to the interaction of the magma with groundwater, known as hydrovolcanic eruptions. In August 1888 to 1890, the island of Vulcano erupted with a number of explosions sounding like cannons going off at irregular intervals – these were the iconic eruptions that gave their name to this type. Peléan eruptions (also termed *nuée ardente*) are named after the 1902 explosive eruption of Mount Pelée in Martinique, and are similar in scale to vulcanian eruptions, but characterized by hot, glowing clouds of pyroclastic flows. These are driven by the collapse of domes, formed in the volcano's crater, which oversteepen and collapse.

Plinian eruptions are the largest types of eruption seen on Earth. They are named from the famous AD79 eruption of Mt. Vesuvius, observed by Pliny the Elder and Pliny the Younger. These eruptions are very powerful and result in a large column of ash and ejecta and a cloud of ash that travels high up into the atmosphere (up to 55km) and stretches out horizontally where it reaches the stratosphere (e.g. Fig. 1.7D).

In order to place some measure of the relative sizes of explosive volcanic eruptions, a classification scheme called the Volcanic Explosivity Index (VEI) has been developed. This uses a number of criteria to estimate the overall size and force of a volcanic event. The VEI runs from 0 where eruption is not explosive through Small (1), moderate (2), moderate–large (3), large (4) and very large (5–8). These different sizes in turn are related to a type or style of eruption (see table 1.1, Fig. 1.6B). As you can see from this table, the different types of eruption discussed above can be placed into the VEI with relative descriptions of explosivity and the height of the eruption column (as in Fig. 1.6).

The largest eruptions – Large Igneous Provinces

The largest eruptions that have occurred on earth are not necessarily the most explosive ones, but are made up, commonly, of basalt rocks similar to those that are erupted on Iceland and Hawaii today. These volcanic events are known as flood basalts and have occurred at key points in the Earth's history. Some of the largest individual eruptions measure thousands to tens of thousands in cubic kilometres, and can cover vast areas of land. Through geological time the Earth has experienced a number of large outpourings of lavas known as Large Igneous Provinces (LIPs), which have been mapped out to show their distribution (e.g. Fig 1.8). These LIPs or flood basalt provinces (a term used when they occur on land) have been associated with significant events in Earth's history, such as the break-up of supercontinents like Gondwanaland, and have been linked with extinction events. In the context of the volcanoes of Europe, the North Atlantic Igneous Province represents one of these large events that formed part of the break-up of Europe from North America. In this case the flood basalts are found outcropping in the UK and the Faroes Islands and are found extensively offshore UK and Norwegian margins. The ancient volcanoes that helped feed this massive province are exposed in some of the most classic examples of fossil volcanoes in the British Tertiary, and the volcanic activity in Iceland today is a legacy of this process, which started some 60 million years ago.

The term 'Super Volcano' has been used to describe a volcano that has the capability of erupting >1000km^3 of material in a single event (100–1000 times larger than historic eruptions). Clearly the eruptions associated with LIPs fit into this category, with many eruption events occurring with volumes in excess of 1000km^3, with the largest estimated at around 8000km^3. In the modern context six current volcanoes have been identified as possible supervolcanoes based on their previous activity: Yellowstone, Long Valley, and Valles Caldera, US; Lake Toba, Indonesia; Taupo Volcano, New Zealand; and Aira

Table 1.1 The Volcanic Explosivity Index with examples from European volcanoes (There is a discontinuity in the definition of the VEI between indices 1 and 2. The lower border of the volume of ejecta jumps by a factor of 100 from 10,000 to 1,000,000m3 while the factor is 10 between all higher indices).

VEI	Ejecta volume	Classification	Description	Plume	Frequency	Examples
0		Hawaiian	non-explosive	< 100 m	coninuous	Iceland fissure eruptions
1	> 10,000 m³	Hawaiian/Strombolian	gentle	- -	daily	Stromboli
2	> 1,000,000 m³	Strombolian/<u>Vulcanian</u>	explosive	1–5 km	weekly	Vulcano (1888)
3	> 10,000,000 m³	Vulcanian/<u>Peléan</u>	severe	3–15 km	yearly	Surtsey (1963)
4	> 0.1 km³	<u>Peléan</u>/<u>Plinian</u>	cataclysmic	10–25 km		Eyjafjallajökull (2010)
5	> 1 km³	<u>Plinian</u>	paroxysmal	> 25 km	⩾ 50 yrs	Vesuvius (AD 79)
6	> 10 km³	Plinian/Ultra-Plinian	colossal	> 25 km	⩾ 100 yrs	Laacher See (~12,900 BC)
7	> 100 km³	Plinian/Ultra-Plinian	super-colossal	> 25 km	⩾ 1000 yrs	Campi Flegrei (~39 ka)

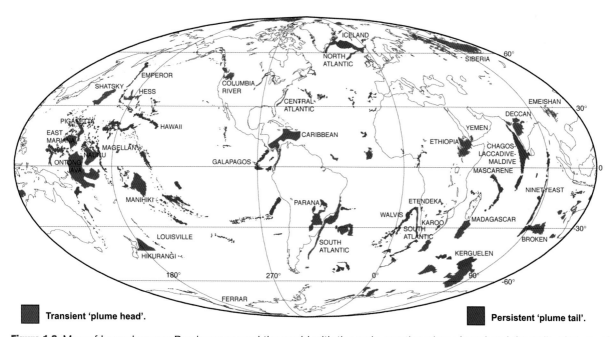

■ Transient 'plume head'.

■ Persistent 'plume tail'.

Figure 1.8 Map of Large Igneous Provinces around the world with the main eruption phase in red and the trails of plumes/hotspots in blue. The North Atlantic Igneous Province straddles the European plate and North American plate with its current expression of the Iceland plume still erupting today (image courtesy of Mike Coffin).

Caldera, Japan. Within Europe there is also one of these sleeping giants, Campi Flegrei, Italy, which in the past has erupted large volumes ~500km³ of material (e.g. the Campanian Ignimbrite ~39ka).

The plumbing systems of volcanoes

With the most modern volcanic systems we are mainly dealing with the eruptive products, e.g. lava flows, tephra, etc. Yet even in the modern systems it is useful to have an insight into the possible pathways or plumbing systems beneath our volcanoes. In some instances these can be excavated in caldera-forming eruptions such as that at Santorini, or they are exposed over time through erosion, which can be either very fast or over hundreds of thousands, if not millions, of years. A variety of types and geometries of intrusion can be found that feed the volcanic system (e.g. Fig. 1.9). These are often simply split into sheet-like intrusions, forming sills and dykes (e.g. Fig 1.10), and those more irregular and often low-aspect-ratio (large width vs. height) laccoliths, plugs and volcanic centres. It is important to have a grasp of these volcanic feeding systems, as they can actually be monitored in real time beneath active volcanoes (for example, the recent activity in Iceland 2014–2015).

More commonly, we need some erosion to explore the magma plumbing at depth. In dynamic environments such as Iceland, relatively young volcanoes can be excavated by ice and floods and their insides exposed. In other areas the fragility of the volcanic edifice can cause collapse, exposing material. Some of the most classically studied examples of the insides of volcanic systems are somewhat older and carved by glaciers and deeper erosion. In this context we can consider the volcanic centres in the British Palaeogene Province (formally the British Tertiary), as these eroded volcanoes formed a vital linchpin in connecting observations made at the modern systems in Europe with what must have been happening at depth.

Understanding the volcanoes of Europe

The distribution of the volcanoes of Europe is perhaps more diverse than on any other continent because of the complications caused by the collision between the Eurasian and the African plates. Most of the volcanoes occur on the margins of the European continent: in the Mid-Atlantic Ridge, the Canary Islands, southern Italy, and the Aegean Sea. The remaining areas of volcanic activity are broadly associated with old systems in France and Germany. So we have volcanoes on constructive margins related to the growth of the Eurasian plate along the Mid-Atlantic Ridge; others related to collision and subduction linked to the clash between Europe and Africa; and others that have no plate boundary connection and seem to relate to hotspots (e.g. Fig. 1.11). The old structural grain within Europe has a role to play in some of the locations of volcanoes, as well as the complex plate organizations and relationships in the Mediterranean, with small terrains and mini-plates complicating a simple plate-tectonic explanation.

The volcanoes on the Mid-Atlantic Ridge are clearly linked to the growing edge of the Eurasian plate. Eruptions are largely submarine and continuous along the whole length of the Ridge, which has been explored by remote geophysics and with submarine-based sampling. These eruptions occur chiefly from multitudes of fissures that produce the basalts that make the world's oceanic crust. In Jan Mayen, Iceland and the Azores, the crest of the Mid-Atlantic Ridge and part of its flanks have been built up above the waves, so that this vital volcanic activity can be inspected and analysed at close range. The emissions on these islands are not wholly basaltic, and they have also included eruptions of more evolved magmas after reservoirs have developed, and their complex relationships with the mid-Atlantic ridge and hotspot activity have, in some sense or another, helped in their emergence from the ocean. The Canary Islands, which lie on the African plate, seem to have been generated by a complex hotspot system that does not interact directly with the plate boundaries. Practically all the

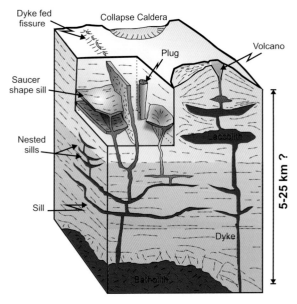

Figure 1.9 Schematic to show the types and styles of plumbing system beneath volcanoes.

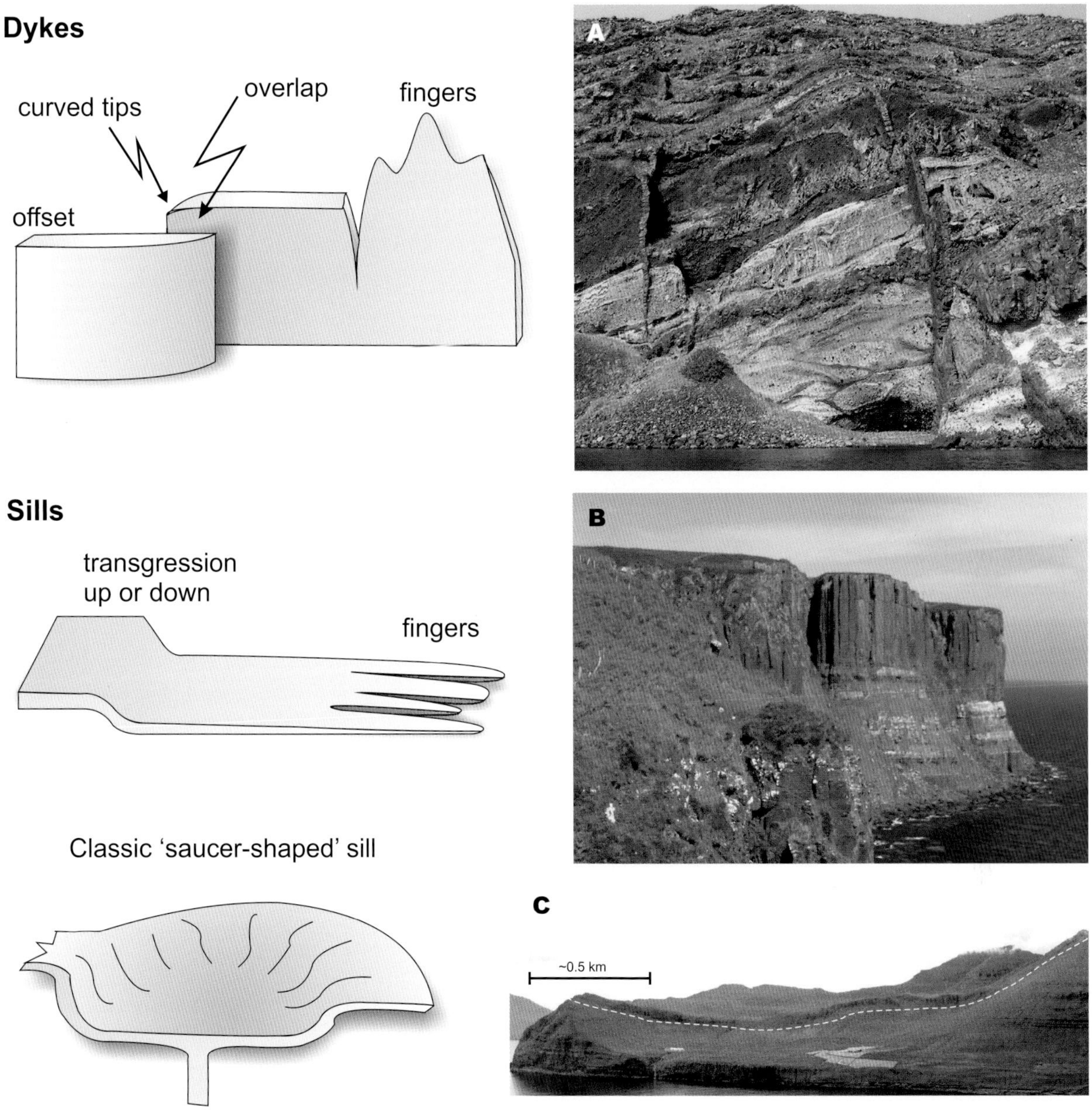

Dykes

curved tips

overlap

fingers

offset

Sills

transgression
up or down

fingers

Classic 'saucer-shaped' sill

A

B

C

~0.5 km

Figure 1.10 Sills and Dyke intrusions with examples. **A)** Santoríni, Greece. **B)** Kilt rock, Isle of Skye, Scotland. **C)** Streymooy Sill, Faroes.

volcanic islands in the North Atlantic Ocean represent considerable accumulations of volcanic material. Their bases often lie more than 2000m deep on the sea floor, and several volcanic peaks rise more than 2000m above the waves. Thus, Beerenberg in Jan Mayen, Öraefajökull in Iceland, Pico in the Azores, and Teide in the Canary Islands, form some of the most prominent mountains in the North Atlantic Ocean – and are, at least, on a par with any volcanic centres in the rest of Europe.

The volcanoes in Italy and Greece are closely linked to the prolonged collision of the African and Eurasian plates, during which several microplates/terrains have detached themselves from their parent masses and pursued varying and independent courses. This adds to the complexity of volcanism in the Mediterranean. At the same time, the edges of the microplates and the Eurasian plate were smashed, fractured and crumpled as the African plate advanced broadly northwards. Thus, continental sediments carried on the plates were thrust up and contorted to form the Atlas Mountains, the Alps, the Apennines, and the chains of the Balkans, Greece, and Turkey; and subduction/collision-related

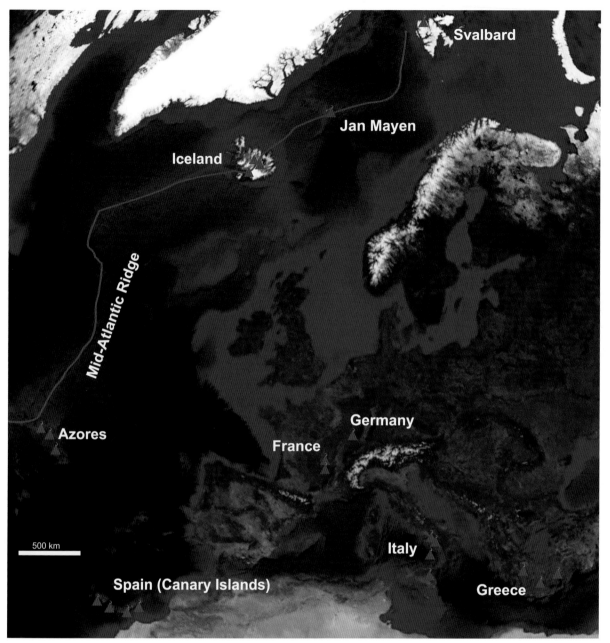

Figure 1.11 Summary map of the volcanoes of Europe. (Map based on NASA World Wind.)

magma made its way to the land surface up major faults that transect the Earth's crust along these sutures. Etna and Vesuvius may have formed in this way, as complex volcanoes straddled between subduction and collision. Parts of the forward edges of the African plate were also subducted beneath the Eurasian plate and the adjacent microplates, in a simpler sense. This subduction caused the eruptions that formed the Aeolian Islands and the volcanic islands in the Hellenic Arc in the Aegean Sea. Subduction is thus probably responsible for the most

violent European eruption during the past 4000 years, on the Greek island of Santoríni, and for the world's most diligent volcanic performer in modern times, Stromboli. But these Mediterranean volcanoes are as varied as the tectonic conditions that have given them birth. Some, such as Etna and Stromboli, have long histories of moderate and mainly basaltic eruptions; others, such as the Fossa cone at Vulcano, have erupted tuffs in more vigorous outbursts; yet others, such as Santoríni and Vesuvius, have erupted huge volumes of fragments (pyroclasts) during episodes of great violence that buried whole cities. In the case of Campe Flegre we have a system associated with tectonics, but maybe also a more localized hotspot activity, which is capable of very large, >500km³ eruptions, as demonstrated with the Campanian ignimbrite (~39ka).

The third main group of European volcanoes formed broadly in relation to the discontinuous rifts that traverse the continent from Oslo, in the north, to the Rhine Rift Valley and on to the Limagnes of central France. Eruptions have given rise to many cones and maars, in both the Eifel Massif in Germany and the Chain of Puys in central France, and the association of these with complex heat anomalies under these parts of Europe has been the subject of debate in recent years.

The Mediterranean volcanoes

The volcanoes and the major fold structures of the Mediterranean area have been caused fundamentally by the collision between the Eurasian and African plates. However, this generalization hides a great complexity of events that perhaps has no equal anywhere else on Earth. Although the Mediterranean area has been intensively studied, much work still needs to be done before the intricacies of the scenario of its development can be truly unravelled. It is thus difficult to explain the distribution and the causes of the Mediterranean volcanoes without making generalizations that may prove to be misleading, inadequate or even inaccurate as research progresses.

The collision takes place between two plates carrying continents that are themselves directly involved in the impact, which has shattered their edges, formed microplates that have moved in different directions, and crumpled and faulted the rocks for millions of years. Thus, collision has not only brought about subduction and deep faults that transect the whole crust, but also areas of crustal extension. As a result, individual Mediterranean volcanoes can have several different causes or, indeed, combinations of causes, which in some instances straddle our classic view of volcanic settings (e.g. Fig. 1.4). In broad terms, subduction seems to be responsible for the Greek volcanoes on the Hellenic arc in the Aegean Sea, and for the volcanoes in the Aeolian Islands, whereas Etna and Vesuvius perhaps owe their growth to eruptions at the intersection of deep major faults and complex subduction. But the relationships are far from simple, and the specialists rarely agree about the exact details of the course of events.

With this varied tectonic background, it is not surprising that the Mediterranean volcanoes have displayed virtually the complete range of eruptive styles. Thus, Vesuvius has been extremely violent, often erupting Plinian columns and pyroclastic density currents, but was largely effusive from 1631 to 1944; Etna has been chiefly effusive, but had some violent outbursts about 2000 years ago; Stromboli has been mildly explosive for many centuries; and Santoríni has produced only moderate eruptions since its great explosion in the Bronze Age. Thousands of smaller, yet not less fascinating, eruptions have formed cinder cones, lava flows and domes, and fumaroles and mudpots are commonly an indication of magma simmering close to the surface. Italy is the zone of greatest tectonic complexity, and it has also been the forum of the greatest, most varied and most closely studied volcanic activity in Europe.

The Mid-Atlantic Ridge

The Mid-Atlantic Ridge is the most clearly defined, and probably the best known, of all the mid-ocean ridges. It forms a sinuous curve bisecting the Atlantic Ocean from the Arctic to Antarctica, in a continuous chain of volcanic accumulations rising 2km or more from the ocean floor. A longitudinal rift runs along its crest, which marks the site of continual volcanic eruptions, where oceanic crust is generated as the North American and Eurasian plates diverge (e.g. Fig. 1.11). The main source of these eruptions is the multitude of fissures and dykes that run parallel to the trend of the crest. They have provided the basaltic pillow lavas and the black smokers that are characteristic of this environment. Generally speaking, the youngest volcanic rocks occur at the central parts of the ridge, whereas increasingly older rocks are found in roughly parallel strips further and further from the crest.

The ridge is mostly submerged, and it is only in exceptional circumstances that volcanic eruptions are so frequent as to have built it above sea level. The most common explanation for these exceptional conditions is that a mantle hotspot/plume lies beneath the mid-ocean ridge, and this seems to be the most likely reason for the emergence of Jan Mayen, Iceland, and the Azores as some of the culminating points on the Mid-Atlantic Ridge. Jan Mayen lies on the Eurasian flank of the ridge; Iceland is transected by the ridge so that its western part belongs to the North American plate and the eastern part to the Eurasian plate; and the ridge divides the Azores. Conversely, the Canary Islands are apparently not related to the Mid-Atlantic Ridge, and seem to have developed chiefly in response to one or more hotspots beneath the African plate.

The study of the islands along the Mid-Atlantic Ridge also shows significant variations from the simplified convergent margin pattern. Major offsets develop in the trend of the Mid-Atlantic Ridge that are associated with notable transform faults and fracture zones more or less at right angles to the trend of its crest. One of their broad effects is to create further fissures up which magma can then rise. They also tend to facilitate activity on the flanks of the ridge, where eruptions can build up and widen the ridge itself. Thus, the Azores rise from a broad, submerged platform on the flanks of the ridge. The rocks on these flanks usually increase in age with their distance from its crest, but the materials emitted during the flank eruptions are much younger than the rocks upon which they lie, and they do not usually increase in age with their distance from the crest of the ridge.

The morphology of the Mid-Atlantic Ridge is further complicated near the Azores by the development of a triple junction. Rifting occurs not only between the Eurasian and North American plates but also between the Eurasian and African plates. A zone of secondary spreading seems to have developed, and may have formed a microplate supporting the central and eastern Azores.

The fourth rather abnormal feature of the ridge is the eruptions from central clusters of vents, which are often related to flank volcanism. In Iceland, for instance, they often form large basaltic shields and pahoehoe surfaces, but sometimes more explosive eruptions take place if the magma has undergone some evolution in a reservoir. In these conditions, the prevalent basalts are replaced by intermediate lavas such as andesites or even rhyolites. At the same time, lava flows become less numerous and fragments become increasingly important as a stratovolcano is constructed. In this way, for instance, Hekla and Oraefajökull have grown up in Iceland, and Beerenberg

in Jan Mayen. In the Azores, most of the stratovolcanoes have also undergone a markedly explosive phase that led to the formation of large calderas on their summits. Iceland, too, has more than a dozen calderas, some of which, like Grímsvötn, are hidden beneath ice caps. Clearly the interaction of hotspots with the ridge leads to more complex crust and subsequently more complex volcanic systems.

Iceland is by far the largest emerged zone of any mid-ocean ridge in the world, and it is the best studied of the three European components of the Mid-Atlantic Ridge. The growth of the much smaller and less complex island of Jan Mayen has also been broadly elucidated. In the Azores, where the mixture of eruptions is greater, several important detailed studies in recent years have begun to clarify the picture of their development.

Summary

With some of the most iconic volcanoes in the world and some of the most intriguing historical examples of the struggle between volcanoes and Man, Europe's fiery past and its explosive present make for a fascinating subject. In the *Volcanoes of Europe*, we will explore the main active volcanic areas and some of the most recently active areas in Europe, as well as taking a look into the ancient volcanic centres that helped forge the study of volcanology and igneous geology. The structure of this book will look at the regions of interest from Italy (chapter 2), Greece, Spain, Portugal, Iceland, Norway, France, and finally Germany. Where possible we will touch on examples of Man's interaction with, and recording of, some of the most infamous eruptions, and we will find out from modern scientists about some of the exciting work that is going on at the volcanoes of Europe in our 'meet the scientist' sections. A glossary is provided to help you navigate through some of the terminology, and a list of the most well-documented eruptions in Europe's history is also provided.

Chapter 2

Italy

Introduction

Vesuvius, Etna, Stromboli and Vulcano are among the most famous active volcanoes on Earth. It is the Italian volcanoes, more than any others, that have given the Western world its views, fears, fascination and preconceived notions about volcanic activity for over 2000 years. Entangled with myths, deities, devils and saints, their varied and often spectacular eruptions have always inspired terror and fascination. The continual eruptions of Stromboli fully justify its nickname, 'Lighthouse of the Mediterranean'; Etna probably served as a model for the Polyphemus story in the *Odyssey*, and its emissions have often been televized in the past few decades. The awesome powers of Vesuvian outbursts were immortalized by Pliny the Younger in AD79 and, ever since, have forced many a Neapolitan sinner to the confessional. Vulcano itself, the forge of Vulcan, has stamped its name on practically every European language. The renown of the Italian volcanoes has sprung from centuries of intimate contact and observation by a large population that has clustered, in spite of all the dangers, on the rich soils of their flanks. The Seven Hills of Rome, too, were carved from old volcanic products, whereas Pompeii and Herculaneum were later buried by others.

The volcanoes of Italy take a whole gamut of forms, from calderas and stratovolcanoes to cinder cones, domes and mudpots. Their eruptions have ranged from vast Plinian outbursts with pyroclastic density currents, to lava flows and solfataras. Their products similarly cover a wide volcanic spectrum, from a predominance of basalts in Sicily, to the concentrations of potassium-rich rocks – trachybasalts, tephrites, trachytes, latites and phonolites – as well as rhyolites, in peninsular Italy. The active volcanoes have themselves strikingly different personalities and behaviour. Etna, Vesuvius, Stromboli and Vulcano are distinctive members of the volcanic family and, as a result, their activity has often been used to establish archetypes for a basic classification of styles of volcanic eruption. Long familiarity and study has also given Italy many type localities, and Italian is a major source of technical terms. Strombolian, Vulcanian, lapilli, scoria,

fumarole, solfatara, latite, atrio, not to mention volcano, provide just a few examples out of many. It is not surprising, therefore, that Italy has become the scene of some of the most vigorous contemporary volcanic research from several European countries. The progress of science has generated a vast bibliography. But, although the histories of many active and recently extinct Italian volcanoes have been elucidated, many intriguing problems about them have yet to be resolved, not least of which are those concerned with the very causes of Italian volcanic activity.

The volcanoes of Italy stretch in a broad but discontinuous band along the western flanks of the Apennines from Tuscany to Campania alongside the Bay of Naples and on, through the seamounts of the Tyrrhenian Sea, to the Aeolian Islands and eastern Sicily, following broadly the subduction trend that sweeps around Italy (Fig. 2.1). In addition, isolated outposts occur in Sardinia,

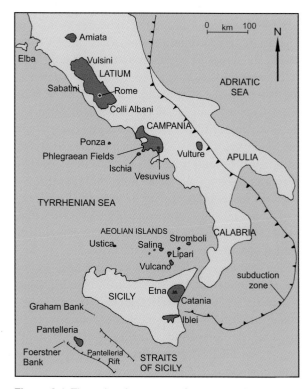

Figure 2.1 The volcanic zones and geo-tectonic setting of Southern Italy.

the Euganian Hills in the Po Valley, in Monte Vulture in Basilicata, and in the island of Pantelleria in the Straits of Sicily. In a very general way, volcanic activity started in the north and spread irregularly southwards. Thus, the northern areas seem to be wholly extinct, whereas many of the southern areas are clearly still active. The volcanism that is only recently extinct or is still active at the present time falls into five main zones: Tuscany, Latium, Campania, the Aeolian Islands and Sicily.

Most of the volcanic areas of Tuscany are small, scattered and eroded, and were formed between 5 million and 2.3 million years ago. However, Monte Amiata, much the youngest and largest volcanic area in Tuscany, dates from less than 400,000 years ago. Although Tuscany has no active volcanoes at present, the magmas still warm the waters circulating in the upper layers of the crust. This heat is harnessed at Larderello and Monte Amiata to such an extent that Tuscany is one of the most important sources of geothermal energy in mainland Europe.

The volcanoes of Latium, or Lazio, stretch in a broad band, 100km long, on the western margins of the Apennines and form the largest outcrop of volcanic rocks in Italy. Compared with Tuscany, the volcanoes are larger, more varied and more recent. They form distinct stratovolcanoes, calderas and vast sheets of pumice, as well as maars, domes and cinder cones. The eruptions are often associated with the formation of lakes such as Bolsena, Vico, Bracciano, Albano and Nemi, which embellish the rolling outlines of the Latium countryside. But the chief volcanic legacy in Latium is the series of hills dominating its skyline: the Monti Vulsini, Montefascione, Vico, Monti Cimini, Monti Sabatini to the north of Rome, and the Colli Albani (Alban Hills) to the south. For example, the Monti Vulsini were active from 400,000 to 60,000 years ago, the Monti Sabatini from 500,000 to 100,000 years ago, and the Colli Albani from 580,000 to 19,000 years ago. The eruptions produced rocks of a high potassic content, and most are rich in leucite. Leucitites, phonolites and tephrites are perhaps the most common, with smaller quantities of latites and trachytes.

The volcanoes of Campania are concentrated around the northern shores of the Bay of Naples, although Monte Vulture (in Basilicata) and Roccamonfina (on the border of southern Latium) also belong to this group. Like those in Latium, the Campanian volcanoes are associated with crustal stretching and the development of deep fractures delimiting elongated fault troughs and fault blocks. Some Campanian volcanoes have erupted during historical times, notably in Vesuvius, and the Phlegraean

Fields and Ischia. The areas on the fringe are older. The Ponza Islands were formed between 1 and 2 million years ago. The imposing mass of Roccamonfina began to erupt its leucitites, tephrites and phonolites about 1.2 million years ago, and its last sign of activity occurred 25,000 years ago. In the east rises the equally large mass of Monte Vulture, which erupted leucitites, tephrites, phonolites and trachytes between about 800,000 and 400,000 years ago.

The Phlegraean Fields form the core of Campania. Here, many vents are concentrated, from which chiefly trachytes and phonolites were emitted, often in notably explosive eruptions that produced extensive ashflows and blankets of pumice such as the Campanian ignimbrite. Similar eruptions have also marked the offshore islands of Procida, Nisida and Ischia. Hydrovolcanic eruptions, solfataras and fumaroles brought the activity into historical times. In the 1980s, vertical movements in the Pozzuoli area were substantial enough to give rise to fears of a renewed eruption, which has yet to materialize.

The glory of Campania is, of course Vesuvius – perhaps the only volcano in the world that needs no introduction. At the risk of provoking the most dangerous volcano in Europe, suffice it to say that it has been dormant since 1944, and that it usually erupts varieties of leucite–tephrite and phonolite.

The Aeolian Islands are probably a volcanic island arc, caused by the subduction of the African plate or the Ionian microplate. They are mostly less than a million years old and have erupted calcalkaline rocks, and latterly more potassic products, culminating in the shoshonitic lavas now emitted by Stromboli and Vulcano. The islands and their volcanoes are quite varied in age, appearance and style of activity. Several of them seem to be extinct, Lipari may be still active, Vulcano erupts infrequently and Stromboli has been one of the most continual volcanic performers of modern times.

Etna and the Monti Iblei constitute the contrasting volcanic areas on the island of Sicily, and offshore eruptions formed Pantelleria and Linosa to the south and Ustica to the northwest. Ustica in the Tyrrhenian Sea may be related to the Aeolian subduction zone. Pantelleria and Linosa erupted in submarine fault troughs. In 1831, they gained an ephemeral companion when Graham Island rose above the waves in a Surtseyan eruption. The Monti Iblei also erupted just below sea level. Their tholeiitic and alkali basalts erupted on a limestone platform on the outer margin of the African plate, where it had been broken by major faults.

Etna, by its size and origin, is in a class of its own in relation to the other Italian volcanoes. It seems to have formed where at least three major fractures intersect and usually allow magma to rise rapidly and relatively unhindered to the surface. It is a spectacular cone, more than 3000m high and covering an area of 1200km^2. Its eruptions began with tholeiitic basalts and continued with alkaline and more evolved lavas. Although Etna has experienced two caldera-forming eruptions in the past few thousand years, its predominant activity has usually been persistently effusive and mildly explosive. Its frequent eruptions have produced many lava flows and satellite cones.

CAMPANIA
Vesuvius

Vesuvius is irresistible. It is the focus of one of the world's most beautiful landscapes, sweeping up from the Bay of Naples in a sleek cone of exquisite proportions in front of the protecting arm of Monte Somma. The view from Naples has been painted and photographed so often that Vesuvius has been instantly recognizable for centuries (Fig. 2.2). It is by far the most violent volcano that has erupted in Europe in historical times, and its outbursts have been observed for nearly 2000 years, because they repeatedly destroyed the crops, property and lives of those attracted to the fertile soils around it. Vesuvius also has great methodological importance. In his two letters to the historian Tacitus, Pliny the Younger described the great cataclysm of AD79 with such accuracy and vivid clarity that they constitute the first scientific account of any eruption in the world, and this outburst rightly became the prototype of Plinian eruptions.

As if the aesthetic quality of the site, the exceptional length of its historical records, and the violence of its major eruptions were not enough, Vesuvius has also played an equally outstanding archaeological role. It is the most famous volcano in the world because it buried Pompeii and Herculaneum. As it looms over the City of Naples today, it represents one of the most dangerous volcanoes in Europe in terms of risks to human life. Having erupted with small force in 1944, its silence since has fuelled speculation that the next eruptions might be as big as those in 1631, or even worse.

Today, the main cone, or Gran Cono, of Vesuvius, rising to 1281m, is enclosed on the north and east by the semicircular rampart of Monte Somma, which is 1132m high. Between the two lies the Valle dell'Inferno, which extends westwards to valleys of the Atrio del Cavallo and the Fosso della Vetrana, which has been invaded by many lava flows in historical times. Monte Somma is the remains of the old Somma stratovolcano, which

Figure 2.2 Vesuvius from Naples, with the ridge of Monte Somma on the left, and extensive settlements on the right, which would be in danger of destruction during a sizeable new eruption.

covers an area of about 480km². The crest of Somma probably collapsed piecemeal during a series of violent eruptions that began about 18,000 years ago. They left behind an asymmetrical caldera, whose northern and northeastern rims now form Monte Somma (Fig. 2.3). Vesuvius grew up within this caldera. The much lower southern and southwestern sectors of the Somma caldera have been all but completely covered by eruptions of Vesuvius, although at the start of the twentieth century they still formed a shoulder on the flanks sloping towards the Bay of Naples. Thus, the southern, highly populated, sectors of the volcano have been wide open to the dangerous effects of virtually all the eruptions of Vesuvius for several thousand years. Both Vesuvius and Somma have had a strong central focus of activity and, although lateral fissures have erupted lava flows since 1631, few satellite cones appear to have formed. On the lower southern slopes, the prehistoric Camaldoli della Torre now forms a cone, 80m high, that is now almost completely swamped by lava. The two cones of Il Viulo and Fossa Monaca erupted nearby in about AD1000, and another followed in 1760. The lateral vents that developed in the southwest in 1794 and 1861 were no more than wide fissures, and they are already disappearing beneath the vegetation, but higher vents extruded the cupolas of the Colle Margherita in 1891 and the Colle Umberto in 1895.

The base of Somma–Vesuvius lies below sea level, so the volcano is larger than it seems. Nevertheless, Vesuvius itself is a relatively small stratovolcano, in spite of its strong focus and its long history of copious output. If the testimony of the many landscape painters is to be believed, the almost persistent activity from 1631 to 1944 produced remarkably few changes beyond the crater. The volcano reached its maximum recorded

altitude of 1335m on 22 May 1905, after six months of persistent explosive and effusive activity that had itself formed part of the continual agitation lasting since December 1875. The eruption in April 1906 decapitated this summit, and a subsequent landslide from the crater rim reduced it to 1186m, before it reached its present height of 1281m after the eruption in 1944.

Historical records of eruptions

Vesuvius has been accessible and easily visible in the midst of a well-populated area for more than 3000 years, and its behaviour has been scrutinized by literate observers for longer than any other volcano in Europe except Etna.

Vesuvius was calm for a long period before AD79, although there may have been an eruption in the eighth century BC and another, perhaps, in 217BC. In sharp contrast, hour by hour events of the eruption of AD79 have now been reconstructed. Thereafter, information is vague, although some kind of activity, probably persistent, occurred between AD80 and 120, 170 and 235 (especially in 203), and in the late fourth century. The details of the Pollena eruption that occurred in AD472 have been deduced from documentary and geological evidence, as well as from a broad range of radiocarbon dates. Activity was also noted, or claimed, in AD536, 685, 787, 968, 991, 999, 1006–1007, 1037, 1139, and possibly about 1350 and 1500. But it is not clear what erupted, or whether these dates represent separate eruptions that were part of longer periods of activity. Nevertheless, current archaeomagnetic studies indicate that, between about 750 and 1150, many copious lava flows erupted, which now form most of the coast between Portici and Torre Annunziata.

Vesuvius fell into repose well before the great eruption in 1631 initiated the revival of the pulsating activity that then lasted until 1944. After 1631, too, the more accurate, scientifically based analyses can often be compared with paintings and literature to provide a much more extensive picture. There were notable eruptions in 1707, 1737, 1754, and especially 1760, when vents opened low down on the southern slopes of the cone. Lava flows spread in several directions in 1767 and an immense lava fountain marked the eruptive climax in 1779. Further notable eruptions occurred in 1794, 1822, 1839, 1850, 1855, 1858, 1861 and 1872. The period between 1875 and 1906 was the most complex episode

Figure 2.3 The ridge of Monte Somma and the cone of Vesuvius from the air.

of persistent activity in modern times, although it continued on a slightly lesser scale until the eruption in 1944 brought this phase to an end.

The growth of Vesuvius

Vesuvius rises at the eastern end of the Campanian volcanic province in one of the most fractured and unstable zones of the Italian Peninsula. The volcanic activity arises where it is caught between the westward encroachment and subduction of the Apulian and Adriatic microplates on the one hand, and the eastward movements opening up the Tyrrhenian Sea basin on the other. Recent activity of Somma–Vesuvius has also been helped by reactivation of major regional faults along a fault trough running from northeast to southwest.

Although volcanic products from its southern basement have been dated to 0.3–0.5 million years ago, most of the Somma stratovolcano, which had a total volume of 100–150km³, grew up after the expulsion of the Campanian ignimbrite, about 37,000 years ago. It seems that repeated explosive and effusive eruptions of potassic lavas built Somma into a cone rising perhaps to 1800m above sea level. Then came the great eruptions and asymmetrical caldera collapses that decapitated Somma and left the caldera rim much higher on the north than on the south (see Fig. 2.4). The date of the formation of the caldera, and consequently the date of birth of Vesuvius within it, is one of the main structural and morphological problems regarding the volcano. It was often believed that the caldera was formed in AD79, but there is evidence to suggest that Vesuvius could be older. Most of the lavas found on the northern slopes of Somma are more than 18,000 years old. Therefore, the caldera and Monte Somma must have started to form at that date and prevented lavas from reaching beyond the ridge. Subsequent eruptions have remodelled the caldera and could have built up a cone that can be considered to be the direct ancestor of modern Vesuvius. Thus, the cataclysm of AD79 was only the last of the major outbursts that have altered the form of the caldera. Vesuvius has undergone several major cycles, usually lasting for a few millennia. These major eruptive cycles were all started by vast Plinian outbursts that are indicated by dated phonolitic pumice layers. They had all been preceded by long periods of repose that are revealed by the presence of ancient soils. The eruption of the basal pumice (*pomici di base*) initiated the caldera and the first cycle 18,000 years ago. The eruption of the Mercato pumice marked the start of the second cycle

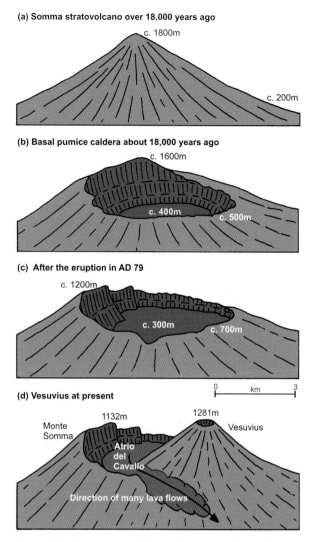

Figure 2.4 The growth of Monte Somma and Vesuvius.

about 8,000 years ago, and the Avellino pumice, dated to 3800 years ago, began the third cycle. The Pompeii pumice, erupted in AD79, initiated the latest cycle.

Periods of semi-persistent activity

Small-scale pulsating activity was typical of the behaviour of Vesuvius between 1631 and 1944, and especially between 1694 and 1872. The pulsations were similar but more prolonged after 1875. Phases of calm were followed by persistent mild activity, then by larger eruptions with bigger lava flows, and lastly by a short, so-called final eruption that could almost reach sub-Plinian proportions. The flows were produced by dark, rather primitive magmas, composed mainly of the basanites and tephrites that rose fairly rapidly from the depths. These eruptions completely emptied the vent, which then became blocked when its walls collapsed and withstood with increasing efficiency the magmatic pressure from below. Thus, a phase of calm, expelling only fumaroles, always followed one of these final eruptions.

Phases of calm on the volcano would be followed by a phase of mild continuous activity, which included Strombolian explosions and sometimes lava fountaining, which formed a small spatter and cinder cone, as well as minor lava flows inside the crater. These increased to intermediate eruptions that were Strombolian on a grander scale, and they were often superimposed upon the periods of persistent activity. During these phases, more lava, with a higher gas content, was expelled at a faster rate and built up a large cinder cone, and also flooded the Atriodel Cavallo and beyond with tephritic–leucititic lava flows. Forty-four such phases have been identified between 1631 and 1944.

The final eruptions were more vigorous, much shorter and less frequent than any other form of activity between 1631 and 1944. The 23 such phases each lasted only from one to two weeks. They often began by hydrovolcanic explosions, followed by an eruption of lava fountains and fine fragments that formed an eruptive column that rose into the stratosphere and showered down ash and pumice, and often obliterated the Sun more than 100km away. Vast fluid lava flows emerged from both the central crater and lateral fissures. Although the final eruptions are several orders of grandeur down the eruptive scale from the Plinian eruption of AD79, they have been sufficiently imposing to impress painters such as Tomasso Ruiz, poets such as Lamartine in 1822, scientists such as Poullet Scrope in 1822 and Perret in 1906, as well as the Allied troops in 1944. In 1794, for example, lava surged from a fissure that opened just above Torre del Greco, destroyed three-quarters of the town within a few hours, and trapped and killed 15 people. In 1872, flows spread over the northwestern slopes of the Gran Cono from a fissure that suddenly opened and fired out volcanic bombs that caused 12 deaths. These final eruptions bring the cycles of persistent activity to a close, because they are invariably followed by a period of calm.

The eruption in 1906

The final eruption of 1906 lasted an unusually long time, from 4 April to 23 April, second only to the 23-day final eruption in 1794. It had an eminent eye-witness, Frank Perret, who stayed at the Vesuvian Observatory with the director, Matteucci, throughout its climax. The eruption began when moderate explosions intensified

at the crater, and lava flows emerged from new vents on the upper southern slopes of the main cone. On 7 April, as tremors increased in intensity, lava fountains soared 3km skywards, and torrents of lava gushed from a fissure and partly destroyed Boscotrecase on the lower southeastern flanks of Vesuvius. Early on 8 April, the eruption column shot obliquely to the northeast and showered the lower flanks of Monte Somma with so many lapilli and bombs that many roofs caved in at San Giuseppe and especially at Ottaviano, where 100 people were killed when the church roof collapsed on them. Then, a sub-Plinian column rose 13km into the air as if it had been generated by a giant steam engine. This lasted for 18 hours and eventually decapitated the summit of the cone. From 9 April until the eruption ended on 23 April, large amounts of fine, often pale, ash were expelled. Heavy rain also caused mudflows that severely damaged Ottaviano. Beneath the falling ash in Naples, processions of panicstricken inhabitants filled the streets invoking the intercession of the Saints. This eruption killed at least 218 people. Molten lava trapped and killed three old men at Boscotrecase, and the rest died when roofs collapsed. Calm duly returned and remained for seven years. Vesuvius had lost 115m from its summit and the crater was now an abyss 700m across and more than 600m deep.

The eruption in 1944

When Vesuvius resumed activity on 5 July 1913, it began a long period of persistent activity that filled the crater formed in 1906 to such an extent that the summit of the inner cone that nested within it could be seen from Naples rising above the main crater rim. The final eruption of 1944, which was witnessed by Giuseppe Imbò, lasted from 18 to 29 March, as the Allied armies advanced through Campania (see wartime images eruption in 1944, Fig. 2.5). From 13 March, oscillations of the magmatic column destroyed the inner cone, which collapsed into the main vent. At 16.30 on 18 March, lavas emerged from the crater, channelled westwards along the Atrio del Cavallo, and caused the greatest damage during the eruption when they engulfed San Sebastiano and Massa within the next three days. On 20–21 March, eight lava fountains, each lasting 20–40 minutes, gushed 2000m above the crater, with smaller lava fragments reaching as high as 4km. They were followed on 22 March by violent hydrovolcanic explosions that ejected ash-laden clouds and small pyroclastic density currents. The wind blew great quantities of ash eastwards, where,

Figure 2.5 Wartime images of the 1944 eruption of Vesiuvius. **A&B)** Vesuvius erupting from the air. **C)** Crew member cleaning ash off the wing of a B-25 from the 320th Bomb Group. **D)** B-25s from the 447th Squadron of the 321st Bombardment Group passing the erupting volcano. **E)** A steaming lava flow engulfs buildings on the flanks of Vesuvius. (Images from the US National Archives and Records Administration, courtesy of Don Kaiser).

for instance, 80cm accumulated 5km away at Terzigno. Gradually the ash explosions waned, and they ceased altogether on 29 March. The eruption of 1944 removed most of the material that had filled the crater during the previous 31 years of persistent emissions. The new crater had shifted 200m to the south and was only one-third of the size of its predecessor. It was 300m deep, 580m long and 480m wide, and has been the scene of variable mild fumarole activity throughout the ensuing period of repose.

Repose periods
Since 1631, repose periods have usually lasted for months or years after the final eruptions. Half of these calm phases lasted only between one and three years,

and none lasted for longer than seven years. However, the calm since 1944 has now lasted so long that it far exceeds any period in modern times, and may signal a complete change in the rhythm of activity of Vesuvius. If so, it may indicate either the close of the period that began in 1631, or even the end of the major Pompeian cycle that started in AD79. In the first case, the calm will last only decades or a few centuries, as happened, for instance, before the sub-Plinian eruptions in AD472 and 1631; Vesuvius will, therefore, most probably erupt with sub-Plinian violence, perhaps between 50 and 250 years hence. In the second case, if the repose since 1944 marks the last phase of the Pompeian cycle, then complete calm should reign at Vesuvius for at least several centuries before a Plinian eruption initiates a new major cycle.

Sub-Plinian eruptions

Sub-Plinian eruptions are smaller versions of Plinian eruptions, but are just as dangerous to those living nearby. They seem to be set in motion by the upsurge of unusual volumes of fresh magma from the depths, bringing a long period of repose to an end. The eruptions in AD472 and 1631, for example, were preceded by at least several dormant decades. These sub-Plinian eruptions consisted of violent explosions of pumice and pyroclastic density currents, and they discharged fairly large volumes of 0.3–1km³ of material from intermediate or evolved magmas, usually ranging from tephrites to phonolites. In some cases, the sub-Plinian eruptions seemed to clear the vent for a long period of smaller scale semi-persistent activity.

The Pollena eruption in AD472

The sub-Plinian Pollena eruption that occurred in AD472 deposited more than 30m of pyroclastic density current fragments at Pollena on the northwestern slopes of Monte Somma, and 8cm of ash on Constantinople, 1200km away. On 5–6 November, 'Vesuvius … vomited up its completely consumed inner parts and turned day into night, covering the whole surface of Europe with fine dust. Every year, on 6 November, the people of Constantinople celebrate the memory of those terrifying ashes'. Cassiodorus also clearly described the pyroclastic density currents as 'rivers of dust and sterile sand that a raging blast had raised into liquid flows'. The Pollena deposits indicate a large eruption with a high eruptive column, with pyroclastic density currents on the northwest and pumice falls that accumulated thickly, especially to the east-northeast of Vesuvius. As the eruption progressed, evolved phonolitic magmas were expelled first, and they were followed by phonotephritic magmas from the lower reaches of the reservoir. Eventually, hydrovolcanic activity brought the eruption to a close. Some 0.32km³ of fragments were ejected in the course of about a week. It was probably the most violent of all the outbursts of Vesuvius after AD79.

The eruption of 1631

The eruption of 1631 is of great importance in the history of Vesuvius because it not only initiated the semi-persistent activity that lasted until 1944, but also because it was the most violent eruption since AD472, and it ended the long interlude of Renaissance calm. Vesuvius had been shaken by some earthquakes during the preceding five months, but they became more frequent on 10 December, when the ground rumbled and wells dried up. The shocks reached alarming proportions during the night of 15–16 December. The eruption began with a violent explosion at 07.00 on 16 December 1631. A subPlinian column soared to the stratosphere for 11 hours. The volcano then roared throughout the night. At 10.00 the following morning, pyroclastic density currents burst from the cone, swept to the sea between Portici and Torre Annunziata, and more or less engulfed Torre del Greco, Pugliano, Portici, as well as Resina, which had been built, although no one yet knew it, over part of Herculaneum (Fig. 2.6). These pyroclastic density currents (also termed pyroclastic flows or surges) killed most of the victims of the eruption; and at least 20,000 people fled in terror to Naples, where they were housed in the leper colony to prevent the spread of disease. Soon, violent explosions of ash also blanketed the volcano. Even at Naples, 12km away to the west, the ash accumulated to a depth of 30cm in total darkness that lasted for over two days. Torrential rain mixed with the ash on Monte Somma and generated mudflows that swamped Pollena, Massa and Ottaviano, and devastated the plain of Nola. After about a week, the power of the eruption waned considerably, but carried on into January 1632.

Contrary to long-established opinion, no lava flows seem to have been expelled during this eruption; the supposed fast-moving lava flows were, in fact, pyroclastic density currents. The lavas that have been attributed to this eruption really erupted between about 750 and 1139. In 1631, the initial eruptions discharged

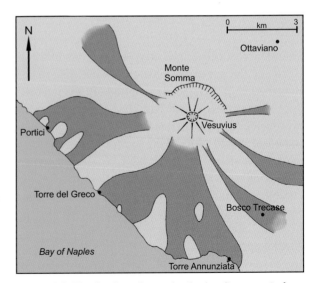

Figure 2.6 Distribution of pyroclastic density currents from Vesuvius in December 1631.

phonolitic leucitites from the upper parts of the evolving magma reservoir, and tephritic leucitites were expelled from its lower reaches as the eruption concluded. At the end of the eruption, at least 4000 people, and perhaps as many as 10,000, had been killed. Vesuvius had lost 470m from its crest, the sub-Plinian gas blast had widened the crater to 1600m across; and the whole area around the volcano had been devastated. Calm then returned until the first phase of semipersistent effusive activity in modern times began on 15 April 1638.

The large-scale Plinian eruptions

The Plinian outbursts of Vesuvius were eruptions of enormous proportions, and their deposits all reveal a similar pattern. They began with explosions of trachytic or phonolitic ash and pumice that rose over 20km high in an eruptive column and spread out into a typical umbrella pine tree shape. The fragments were first white and then grey. As the column collapsed from time to time, pyroclastic density currents – the really lethal elements of the eruption – rushed down the volcano. At length, when most of the contents of the reservoir had been ejected, the summit of the volcano collapsed into a caldera, which every subsequent outburst then widened. In all, up to 5–10km³ of fragments, covering about 500km² in area, were probably expelled in periods of two or three days. Thus, these Plinian eruptions occupied phases totalling scarcely 15 days in 18,000 years, but enough material was discharged at sufficient speed to bury several towns completely when the occasion presented itself in AD79.

The cycle that began with the eruption of the Avellino pumice about 3800 years ago included at least six sub-Plinian eruptions, the last of which might have been the doubtful eruption in 217BC. However, scholars such as Diodorus Siculus, Vitruvius and Strabo recognized that Vesuvius was a volcano because of the burnt stones strewn across its summit. Nevertheless, the main perceived hazard in the district was undoubtedly the earthquakes that often shook Campania. Most of them caused no more than a shudder in the ground, but the earthquake on 5 February AD62 had caused considerable damage in and around Pompeii, which had not been entirely repaired 17 years later.

The eruption in AD79

In AD79, Pompeii was probably one of the largest towns in Campania, with a population of about 20,000, 10km southeast of Vesuvius. Herculaneum was a smaller

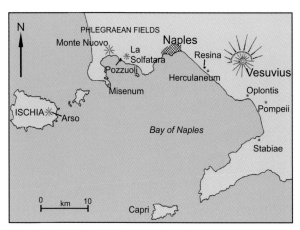

Figure 2.7 Vesuvius and the Phlegraean Fields, with Monte Nuovo, Arso and La Solfatara.

seaside town of about 5000 people, 7km due west of Vesuvius. Naples, which was probably then, as now, the chief centre, and Stabiae, Oplontis and the naval base of Misenum in the west, were the other major settlements on the Bay of Naples (Fig. 2.7). All were built on recent volcanic rocks, and Pompeii, Herculaneum and Oplontis were constructed on rocks expelled by Vesuvius within the previous 10,000 years. Being on the southern side of the volcano, they would thus be vulnerable if Vesuvius were ever to revive, and both towns lived with the volcano as their backdrop (Fig. 2.8).

In the summer of AD79, the 17-year-old Pliny the Younger was staying with his mother and her brother, Pliny the Elder, at Misenum, 32km west of Vesuvius. His first letter to Tacitus describes his uncle's journey to his death at Stabiae, and the second his own experiences at Misenum. There is no mention of Pompeii or Herculaneum in Pliny's narratives.

The tectonic earthquake that damaged much of Campania in AD62 could have been a precursor of the eruption 17 years later. Two features, recounted by Seneca (VI: 1 and 27), suggest that it could have been related to an increase in magmatic pressure beneath the volcano: the earthquake was centred on the area around Vesuvius, and 600 sheep were poisoned in the Pompeii region, 'when, at ground level, they inhaled some baleful air', probably carbon dioxide.

Earth tremors of increasing frequency began several days before and gave premonitory but unheeded signs of the impending eruption. Early on the morning of 24 August, the initial hydrovolcanic explosions cleared the conduit and distributed fine ash, especially to the north and east.

Figure 2.8 Living under the backdrop of Vesuvius: **A)** the ruins of Pompeii with Vesuvius behind; **B)** the excavations at Herculaneum, with an eerily lit Vesuvius on the skyline.

Just before 13.00 on 24 August, the Plinian climax began and its eruptive column of gas and mainly phonolitic fragments rose from the upper layers of the magma reservoir and soon reached a height of 27km, as the wind winnowed out the ash and white pumice, raining them down in a southeasterly direction. Pompeii was thus situated in the zone of maximum accumulation. Throughout the next seven hours, a thickness of about 1.40m of ash, pumice and fist-size lithic fragments fell on Pompeii. Roofs collapsed and, by the evening, most of the people had probably panicked, fled, and saved their lives.

From about 20.00 on 24 August, darker tephriphonolitic magma exploded from the middle zones of the magma reservoir. The Plinian column rose to a height of 33km. The debris expelled became coarser and changed to a greenish grey. By midnight, a thick blanket covered an area southeast of Vesuvius from Terzigno to Oplontis. On top of the white pumice, 1.3m of grey pumice eventually fell on Pompeii that night. However, the western areas around Misenum remained unaffected on 24 August, and Herculaneum, much closer to Vesuvius, had been spared all but showers of fine ash.

At 01.00 on 25 August came the major change in the Plinian eruption. Magma that was rather poorer in explosive gases began to be emitted. Thus, the column could not always sustain its upward impetus, and its lower parts collapsed from time to time, crumbling like a pillar of fire, to form pyroclastic density currents. The first pyroclastic density current was funnelled westwards by the Atrio del Cavallo at a speed of at least 100km an hour and reached Herculaneum in less than four minutes. Its remaining inhabitants had time only to flee

to the shore, where they died choking in the swirling cloud. Hundreds of victims were discovered there in the excavations in 1982 (Fig. 2.9). As fewer than a dozen victims had previously been discovered in Herculaneum, it was assumed that the townspeople had been able to flee to safety, and thus that the town must have been among the last casualties of the eruption. Herculaneum, in fact, was the first major victim of Vesuvius in AD79.

Meanwhile, as the grey pumice continued to fall, a second pyroclastic density current at about 02.00 completed the burial of Herculaneum. A third pyroclastic density current, at about 06.30, buried Oplontis and spread as far as the northern walls of Pompeii, and may have prompted many other citizens to leave. A great increase in tremors about that time, marking perhaps the beginning of caldera collapse, can only have reinforced their fears. Pompeii was already blanketed in 2.4m of ash, pumice and even larger rock fragments, and they were still falling in the stifling darkness. At 07.30 on 25 August, a fourth pyroclastic density current completely overwhelmed Pompeii and all the 2000 people remaining in the city. Some were killed by falling columns or tiles and bricks ripped from the buildings, others asphyxiated by a mixture of hot ash and mucus inhaled with their last breaths, but many were baked by the pyroclastic density currents where they finally stood.

The plaster casts made during the excavations reveal their terrible fate and now make gruesome archaeological exhibits (Fig. 2.10). Five minutes later, a fifth and even larger pyroclastic density current swept over Pompeii and carried onwards to the outskirts of Stabiae, 17km from the volcano. Earthquakes, probably related to further collapse of the caldera, reached a frightening

Figure 2.9 The boathouses at Herculaneum. **A)** The boathouse entrances at the original shoreline. **B)** A boat excavated from just in front of these boat houses. **C)** Bodies discovered hiding within the houses, many still showing resting position up against the walls. **D)** Blackened bone ends reveal the searing heat that ensued when the pyroclastic density current hit Herculaneum, and killed the people at the shore's edge.

pitch throughout the region. About 08.00 on 25 August, the sixth and largest pyroclastic density current swept onwards to Stabiae and killed Pliny the Elder. Another branch of the same pyroclastic density current surged westwards across the waters of the Bay of Naples and halted, 32km away, within sight of Pliny the Younger and his mother above Misenum.

Vesuvius released other pyroclastic density currents throughout the morning of 25 August and still more pumice rained down incessantly. Then, as the day went on, the pyroclastic density currents diminished in vigour, the Plinian column reduced in height, and the discharge of the most gaseous and evolved phonolitic parts of the magma reservoir was completed. The final hydrovolcanic phase ensued as water entered the vent when the magma reservoir collapsed. Many lithic fragments, in a fine ash replete with accretionary lapilli, were scattered, 1m thick, all around the flanks of Vesuvius. Eventually,

glimmers of daylight returned to the more distant areas such as Misenum in the west, the tremors declined in number and size, and Pliny the Younger and his mother returned home.

When the daylight returned, all Campania was covered with a thick blanket of fine grey pumice in undulating dunes. About 300km² around Vesuvius was completely devastated and the farms, villages, towns and cities of the plain had vanished. Vesuvius had discharged 1km³ of white pumice, 2.6km³ of grey pumice, 0.37km³ of pyroclastic density current deposits, and 0.16km³ of hydrovolcanic materials, expressed as dense rock equivalent (DRE). The magma reservoir thus delivered 4.13km³ of dense rock, or about 10km³ of fragments, although these figures are often revised with new mapping. The deposits themselves reveal the layers of white and grey pumice, often delicately encasing complete pottery, and the upper parts show the

Figure 2.10 The moulds of bodies at Pompeii: **A–D** various views of the body moulds at Pompeii, some in original positions and some moved to show case areas; **E)** a view from one of Pompeii's victims with the cone of Vesuvius in the backdrop.

cross-bedded and ash-rich pyroclastic density current deposits (Fig. 2.11). It is in these upper deposits that most of Pompeii's victims were found.

Preventive techniques may well have improved when the next Plinian eruption of Vesuvius occurs. If such an eruption were to occur tomorrow in defiance of all trends, no known palliative technique could save the many towns on the southern flanks of Vesuvius, and the only effective defence would seem to be immediate and comprehensive evacuation.

PDC deposits

Detail of pyroclastic density current layers

Body mould

Wood mould

Grey pumice

White pumice (fall out from air)

Figure 2.11 Deposits from the eruption. The white and grey pumice from the fallout of material from the Plinian cloud, here encasing a large pottery vase. Above these are the cross-bedded ash and lapilli of the pyroclastic density currents that finally hit Pompeii. It is at this level that most of the bodies are found.

The Phlegraean Fields

The Phlegraean Fields form an array of cones, craters, calderas and occasional domes and lakes, covering a low-lying area of about 100km², due west of Naples (Fig. 2.12). They do not really live up to their name, which is derived from the Greek *phlegein* (to burn), although they

The eruption that killed Pliny the Elder

'At that time [24 August AD79] my uncle was at Misenum in command of the fleet. About one in the afternoon, my mother pointed out a cloud with an odd size and appearance that had just formed. From that distance it was not clear from which mountain the cloud was rising, although it was found afterwards to be Vesuvius. The cloud could best be described as more like an umbrella pine than any other tree, because it rose high up in a kind of trunk and then divided into branches. I imagine that this was because it was thrust up by the initial blast until its power weakened and it was left unsupported and spread out sideways under its own weight. Sometimes it looked light-coloured, sometimes it looked mottled and dirty with the earth and ash it had carried up. Like a true scholar, my uncle saw at once that it deserved closer study and ordered a boat to be prepared. He said that I could go with him, but I chose to continue my studies. Just as he was leaving the house, he was handed a message from Rectina, the wife of Tascus, whose home was at the foot of the mountain, and had no way of escape except by boat. She was terrified by the threatening danger and begged him to rescue her. He changed plan at once and what he had started in a spirit of scientific curiosity he ended as a hero. He ordered the large galleys to be launched and set sail. He steered bravely straight for the danger zone that everyone else was leaving in fear and haste, but still kept on noting his observations. The ash already falling became hotter and thicker as the ships approached the coast, and it was soon superseded by pumice and blackened burnt stones shattered by the fire. Suddenly the sea shallowed where the shore was obstructed and choked by debris from the mountain. He wondered whether to turn back, as the captain advised, but decided instead to go on. 'Fortune favours the brave,' he said, 'take me to Pomponianus.' Pomponianus lived at Stabiae across the Bay of Naples, which was not yet in danger, but would be threatened if it spread. Pomponianus had already put his belongings into a boat to escape as soon as the contrary onshore wind changed. This wind, of course, was fully in my uncle's favour and quickly brought his boat to Stabiae. My uncle calmed and encouraged his terrified friend and was cheerful, or at least pretended to be, which was just as brave.

Meanwhile, tall broad flames blazed from several places on Vesuvius and glared out through the darkness of the night. My uncle soothed the fears of his companions by saying that they were nothing more than fires left by the terrified peasants, or empty abandoned houses that were blazing. He went to bed and apparently fell asleep, for his loud, heavy breathing was heard by those passing his door. But, eventually, the courtyard outside began to fill with so much ash and pumice that, if he had stayed in his room, he would never have been able to get out. He was awakened and joined Pomponianus and his servants who had sat up all night. They wondered whether to stay indoors or go out into the open, because the buildings were now swaying back and forth and shaking with more violent tremors. Outside, there was the danger from the falling pumice, although it was only light and porous. After weighing up the risks, they chose the open country and tied pillows over their heads with cloths for protection.

It was daylight everywhere else by this time, but they were still enveloped in a darkness that was blacker and denser than any night, and they were forced to light their torches and lamps. My uncle went down to the shore to see if there was any chance of escape by sea, but the waves were still running far too high. He lay down to rest on a sheet and called for drinks of cold water. Then, suddenly, flames and a strong smell of sulphur, giving warning of yet more flames to come, forced the others to flee. He himself stood up, with the support of two slaves, and then he suddenly collapsed and died, because, I imagine, he was suffocated when the dense fumes choked him. When light returned on the third day after the last day that he had seen [on 26 August], his body was found intact and uninjured, still fully clothed and looking more like a man asleep than dead.' (Pliny VI: 16)

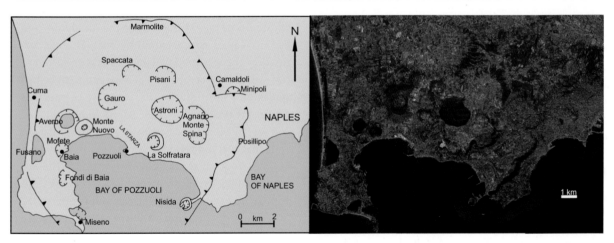

Figure 2.12 Location map of the Phlegraean Fields (satellite map from NASA Worldwind).

have always been notorious for the sinister fumes that the Greeks and Romans believed were the exhalations of the underworld. Odysseus (*Odyssey* XI) and Aeneas (*Aeneid* VI) both visited Hades from here and, at the Mare Morto, Charon was reputed to row the souls of the dead across the Styx to the underworld. Innumerable hot springs made Baia a leading spa; the sulphurous emanations from La Solfatara were known as the Forum Vulcani, and the old crater forming the harbour at Misenum was a major base of the Roman Fleet.

Many eruptions have taken place in the Phlegraean Fields in the past 12,000 years and it is possible that

some of them could have been witnessed by the Greek colonists, who first named and settled in the area in the eighth century BC. However, the formation of Monte Nuovo in AD1538 is the only eruption recorded in historical times, and radiocarbon dating suggests that it followed 3000 years of quiescence. On the other hand, the recent activity of the Phlegraean Fields has been concentrated on the emissions at La Solfatara and, more dangerously, on the bradyseismic movements that are caused by oscillations in the levels of the subterranean magma. They are now focused on Pozzuoli, but they have been common in the Phlegraean Fields at least since the ground first rumbled beneath the feet of Aeneas. Bradyseismic uplift increased markedly again during the two years before Monte Nuovo erupted in 1538. In modern times, uplift has occurred at Pozzuoli in 1969–72 and between June 1982 and December 1984. Disquiet was followed by panic, departure and evacuation, as the population, the administration and scientists prepared for a renewal of volcanic activity, which fortunately did not take place. But the incidence of volcanic hazards is high in this area, where some 1.5 million people live. Although any future eruption is most likely to form a cone like Monte Nuovo, it will most probably take place in the densely populated area centred on Pozzuoli. However, this volcanic area has a potentially more sinister past with much larger volcanic eruptions, as evidenced by the thick volcanic deposits that form the Campanian Ignimbrite and the Neapolitan Yellow Tuff.

The Campanian Ignimbrite

The arrival of fresh magma in the reservoir provoked perhaps the greatest eruption in the Mediterranean area during the past million years. The summit of the stratovolcano collapsed as the grey Campanian Ignimbrite was expelled. Its volume has been estimated at between 80km^3 and 150km^3 of magma, which represents up to 500km^3 of ignimbrites.

The materials were possibly expelled from a ring fracture fed by a subterranean system of cone sheets. The enormous momentum of the explosion spread the finer fragments in a rapidly expanding incandescent cloud, which formed the Campanian Ignimbrite. Its trachytic pumice and lithic fragments are embedded in an ashy matrix, and it thickens to as much as 60m in lower-lying areas. This enormous ashflow not only blanketed the whole of the Campanian Plain, swept across the Bay of Naples into the Tyrrhenian Sea and rode over hills 1000m high in the Sorrento peninsula, but covered 30,000km^2 and much of southern Italy as well. It has been dated most recently to 36,000–37,000 years ago. The cliffs on top of which Sorrento is mostly built today are made of the thick deposits from this colossal eruption (Fig. 2.13A). It is usually supposed that it burst out in a single unit in less than about a week. However, cores taken from the Tyrrhenian sea floor have revealed five separate trachytic ash layers, which suggest that the Campanian Ignimbrite was ejected in five distinct stages. The chief episodes occurred 36,000, 33,500 and 26,900 years ago, with smaller emissions about 38,700 and 24,000 years ago. If these correlations are substantiated, then the ignimbrite would have been erupted in brief individual episodes extending over an unusually long period. The irregular scalloped outline of the associated caldera, which could indicate piecemeal foundering, gives some support to the concept of five-stage eruptions.

The collapse of the caldera was the key event in the history of the Phlegraean Fields. Its floor foundered by at least 700m and formed a hollow 12–15km in diameter, making it one of the largest calderas in Europe. But it has only recently been properly identified, mainly because subsequent eruptions have masked both its nature and

Figure 2.13 The thick deposits from some of the largest eruptions. **A**) The wall of Campanian ignimbrite deposit on which much of modern Sorrento is built, the other side of the bay from the eruption. **B**) Thick Neapolitan yellow tuff deposits seen in and around Naples. Many of the buildings are made using this for building blocks (photo **A** by Norbert Nagel/Wikimedia Commons; photo **B** by Yiftah-s).

its limits. Part of its floor and its southern rim sank below the Bay of Naples, but remnants of its scalloped rim can be seen at the Monti di Procida in the south, at Cuma in the west, at Marmolite in the north, and especially at Camaldoli in the east. They demarcate the zone in which all subsequent volcanic activity took place in the Phlegraean Fields. Thereafter, the eruptions decreased in intensity and became increasingly concentrated on the centre of the caldera.

The Neapolitan Yellow Tuff

Soon, eruptions began to fill the sea-flooded caldera, and perhaps about 23km³ of magma erupted altogether, but the chronological sequence of the trachytic tuffs lying between layers of latitic and trachytic flows is hard to establish. However, this period concluded with a major event: the eruption of the Neapolitan yellow tuff, whose colour comes from the diagenetic alteration of trachyte erupted in shallow water (Fig. 2.13B). The outburst occurred about 12,000 years ago, and it has been linked to an important ash layer in the Tyrrhenian Sea that erupted about 12,300 years ago. In all, some 20–50km³ of these hydrovolcanic trachytic tuffs were expelled, either in a single eruption, or from multiple emissions with their main source in the east, or from vents arranged in small arcs around the inner rim of the caldera as well as in its centre. The arcs could well represent a persistence of ring fractures fed by cone sheets. The vents in the arc stretch from Miseno to Monticelli, and those belonging to the eastern arc extend along Posillipo Hill. The central vent formed the Gauro volcano, the largest within the caldera. These explosively hydrovolcanic or even hydroPlinian eruptions produced pumice, accretionary lapilli and ash. Sometimes the yellow tuff is stratified, as at Mofete and Miseno; sometimes it is chaotic, as at Camaldoli. This major eruption was accompanied by the formation of the Neapolitan Yellow Tuff caldera, which occupied the centre of the Campanian Ignimbrite caldera and foundered some 60m below sea level.

Activity during the past 12,000 years

Activity in the Phlegraean Fields during the past 12,000 years was concentrated in three periods. The first period lasted from 12,000 to 9500 years ago, when 34 mainly hydrovolcanic eruptions took place, in or near the sea, around the perimeter of the Neapolitan Yellow Tuff caldera. The resultant cones now form a circle stretching from Capo Miseno volcano, in the west, to Montagna Spaccata in the north and round to Nisida volcano in the east. A thousand years of quiescence then ensued, during which a palaeosol developed on the land areas within the caldera. The second period of activity was marked by six relatively mild eruptions that began 8600 years ago with the formation of the Fondi di Baia cone in the west and ended 8200 years ago with the eruption of the San Martino cone in the north. A long period of calm followed this brief eruptive episode and a major palaeosol developed widely over the land surface. Towards the end of this calm period, the central part of the caldera was uplifted by some 40m and formed the La Starza terrace, which now stretches northwestwards from Pozzuoli. The third and latest period of volcanic activity lasted from 4800 to 3800 years ago. It witnessed 16 explosive and 4 effusive eruptions, all of which took place in the northeastern sector of the Neapolitan Yellow Tuff caldera, except the explosion of the Averno volcano in the west. Most of these eruptions were trachytic. The main events of this episode included the eruptions of Agnano–Monte Spina volcano (which began the series 4800 years ago), the explosion of Averno volcano about 4500 years ago, the extrusion of the dome of Monte Olibano and the eruption of La Solfatara 3800 years ago, which were immediately followed by the explosions of the Astroni tuff ring and the Fossa Lupara, which brought the third period to a close. The volume of volcanic material expelled during the past 12,000 years amounts to no more than 4.5km³, which indicates how the magma reservoir has continued to shrink and cool. However, although the formation of Monte Nuovo in 1538 marks the only subsequent magmatic event, the Phlegraean Fields are far from extinct, and both hydrothermal emissions and bradyseismic movements have been common.

Bradyseismic movements at Pozzuoli

Pozzuoli is built over the hub of the Phlegraean Fields caldera in the zone most sensitive to recent bradyseismic movements, which are probably related to a resurgence within the caldera linked to thermodynamic oscillations in the magma reservoir situated at a depth of between four and five kilometres. Thus, although its molten parts are now only small, its fluctuations are easily propagated to the land surface. As the area around Pozzuoli has been quite densely populated since antiquity, even small movements have been noticed. In fact, bradyseismic movements can be traced back to when the terrace called La Starza was uplifted 5400 years ago near Pozzuoli.

Some of the bradyseismic movements of antiquity were registered on the columns of the old Roman market (Fig. 2.14), built about 100BC, which early archaeologists mistook for the Temple of Serapis. Soon afterwards, the pavement had to be built 2m higher, after sinking ground had taken the original floor below the waves. Subsequent fluctuations were recorded on its columns. When the Roman Empire fell, the building was abandoned, and the lower 3.6m of its columns were covered by earth and rubble. Apparently the area sank by about 5.8m during the tenth century and flooded the ruins. The submerged parts of the columns were then attacked by the stone-boring mollusc, *Lithodomus lithofagus*. Later, bradyseismic movements elevated the land, and the waters drained away to reveal the pitted surface of the pillars. The depredations of *Lithodomus* revealed the old depth of the sea water very accurately. The earth and rubble had protected the lower partsof the columns from the *Lithodomus* and, when it was removed, the smooth bases of the columns were also revealed, still intact. These upward movements, amounting to about 6m in all, probably occurred in the decades preceding the eruption of Monte Nuovo. After the eruption the ground sank again, perhaps by as much as four metres. All these displacements were, of course, independent of worldwide eustatic changes of sea level and contemporaneous Italian tectonic movements.

The first 60 years of the twentieth century witnessed an irregular overall sinking of about 60cm, punctuated by minor uplifts and periods of stability. This trend was reversed in mid-1969 by the onset of a phase of uplift that continued until mid-1972. It produced a maximum uplift of 1.70m, which was accompanied by tremors and caused considerable anxiety at Pozzuoli. During the next two years the ground sank by about 20cm, thus giving a net overall maximum rise of 1.50m from the levels of 1969. However, the feared earthquake or volcanic eruption fortunately did not occur. The crisis did highlight the need for civil defence plans, if evacuation were needed, as well as systematic monitoring of the whole area. It is astonishing that these facilities were not available in 1969, although a variety of political, social and scientific explanations may account for their absence.

These omissions were largely rectified during the ensuing period of relative calm. Then, further uplift occurred from August 1982 until December 1984. The greatest rise (1.85m) was registered at the harbour of Pozzuoli (see Fig. 2.15). Thus, central Pozzuoli has risen 3.35m since mid-1969. The area had an established network of seismic, levelling and tidal gauge stations,

Figure 2.14 The Roman market at Pozzuoli, known as the Temple of Serapis. After the destruction and flooding of the edifice, the lowest parts of the columns were protected by rubble; the parts exposed to the water were attacked by the stoneboring mollusc *Lithodomus lithofagus* and appear rough and eaten; the middle and upper parts were exposed to the air and remained relatively unaffected by atmospheric weathering (photo by Ferdinando Marfella – Creative Commons).

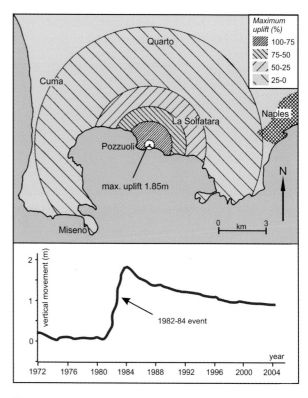

Figure 2.15 Bradyseismic uplift around Pozzuoli highlighting the 1982–4 event.

gravity and geothermal monitoring – all coordinated in a research centre at the Vesuvius Observatory. The ground deformation occurred throughout the Phlegraean Fields, but was most marked at the hub of the caldera at Pozzuoli. Analysis of the hydrothermal emissions, especially close to the hub at La Solfatara, showed an initial increase of activity, but this was not maintained, which suggested that the magma was not itself rising.

The earthquakes did not begin until November 1982, after uplift had been under way for some months. The initial magnitude 3.5 earthquake occurred at La Solfatara on 15 May 1983 and a magnitude 4.0 earthquake occurred at Pozzuoli on 4 October 1983. As a result, 40,000 people were evacuated from the centre of Pozzuoli. This was a sound safety measure, because many of the buildings in the town were damaged and some later collapsed during the swarms of small tremors between October 1983 and April 1984. In all, some 8000 buildings were damaged, and the harbour is now too shallow for large boats, and lower quays have been built alongside their now uselessly high predecessors. But the earthquake foci remained at the same depth of between three and four kilometres, which again suggested that the magma was not rising and, therefore, that a volcanic eruption was unlikely unless there was a radical change. The crisis ended when both the uplift and the earthquakes stopped in December 1984. In spite of the panic, the media exaggeration, and the evacuation, which with hindsight was perhaps not entirely necessary, the crisis was handled quite successfully. However, the crucial symptoms indicating an imminent eruption in these conditions are still not fully known. Pozzuoli is still in the forefront of volcanotectonic analysis. The vital need to solve the problems is emphasized by the great vulnerability of a population of 400,000 living in the potential danger zone.

La Solfatara

As its name indicates, La Solfatara is famous for its sulphurous emissions, for which it has become the type locality (Fig. 2.16). La Solfatara lies only 1.5km east of Pozzuoli, very close to the centre of the most recent volcanic and tectonic activity in the Phlegraean Fields, and it may be wrong to assume that these emissions represent the dying phases of activity on this vent. La Solfatara forms a horseshoe cone, wide open on its southwestern side. Its crater, the Piano Sterile, is 750m long and almost 600m broad; its almost rectangular outline seems to have been determined by two sets of faults, one trending from northwest to southeast, the other from northeast to southwest.

The latest episode of eruptions began in the district less than 3800 years ago with the extrusion of the trachytic dome of Monte Olibano. Immediately afterwards,

Figure 2.16 La Solfatara. **A**) Panoramic view of the crater towards the southeast. **B**) Sulphur-coloured fumaroles are popular with tourists and volcanologists. **C**) Bubbling mud pools. (Photo **A** by Patrick Massot.)

Meet the Scientist Matteo Lupi

Previously at the University of Pisa, Italy; University of Edinburgh, Scotland; and Bonn, Germany, since 2013 Matteo has been a researcher at ETH of Zurich, Switzerland, and is funded by the Marie Curie ETH program and SNF. His research focuses on fluid-driven seismicity and seismically-induced fluid flow in volcanic and hydrothermal environments and in tectonically active systems. Recently he has started looking into the ground deformation in the Phlegraean Fields.

How long have you been working on Phlegraean Fields (Campi Flegre)?
It all started in September 2013 when I was in a conference in Italy. There I was chatting with my MSc adviser, Prof. Rosi, who is currently the Director of the Seismic and Volcanic Risk Office for the Italian civil protection. There, I have been told that the uplift inside the Campi Flegrei caldera accelerated once again to rates as high as 2cm per month. I have always been attracted by volcanic hazard and risk mitigation, so this seemed a good occasion to start to look into the problem at the Campi Flegre.

What has been your most exciting discovery about it?
In general, the most exciting discovery of my scientific career is the mutual interaction between geological processes and how little it takes to activate giant geological systems. We have the tendency to think that volcanoes, earthquakes, black smokers, faults, and many other geological systems are only weakly affected by the surrounding geological environment. Yet, with the current work on the Campi Flegrei, it is becoming clearer how important is the mutual interaction between geological forces.

What inspired you to study volcanoes?
I was always fascinated by nature, and volcanoes are the pristine expression of nature's power. There is nothing like sitting on a volcanic peak and watching explosions (like in Stromboli, Italy) or being at 4400m above the ground and observing Colima's peak (Mexico) from the nearby observatory. Volcanoes are undoubtedly mystic places where nature is at its best and geology is in action!

How important do you think the volcanoes of Italy have been in our understanding of volcanology?
Italian volcanoes have played a major role for the development of volcanology. Their logistically easy access, the frequent eruptions of some of them, and the most well-described historical eruption provided by Pliny for the Vesuvius AD79 eruption strongly contributed to the development of volcanology. Italian volcanoes attract scientists from around the globe, and the Neapolitan area is the most urbanized volcanic region of the planet, representing a challenge for the development of risk mitigation plans for volcanic hazards. Italian volcanoes are definitely ideal natural laboratories where to investigate magma in motion.

eruptions from a new vent, situated only 300m to the north, shattered the northern half of the dome, and built a cone of ash and white pumice alongside it. This was the cone of La Solfatara. Violent explosions then opened up all its southwestern flank, and small pyroclastic density currents surged forth and covered almost 1km². In the final act of the eruption, a piston of viscous trachyte plugged the vent and intruded the northern wall of the crater. Thus, the Piano Sterile of La Solfatara is encircled by a wall composed of a plug in the north, an ash and pumice cone in the centre, and a broken dome in the south. The fumaroles have greatly altered all the surrounding rocks. Some of the fumaroles reach temperatures of 160°C and are composed of over 80 per cent of steam and water, but also exhale carbon dioxide, hydrogen sulphide and small amounts of nitrogen, hydrogen and carbon monoxide. The basic cause of all this fumarole activity is the recycling of atmospheric waters that are heated by the high temperatures prevailing just below the surface.

Fumarole activity has probably been more or less continuous at La Solfatara ever since the crater was formed. Various small vents emitting mud, or creating small hot muddy lakes in the Piano Sterile, were known to the

Romans and were also recorded as early as the fifteenth century, although activity seems to have declined since the beginning of the eighteenth century. Perhaps the greatest event recorded in the crater in historical times occurred in 1198. It has sometimes been described as a hydrovolcanic eruption, but was most probably a muddy geyser that rose about 10m into the air after a small earthquake. A similar event occurred in 23 July 1930, when mud clots were thrown 30m skywards.

The crater of La Solfatara forms a wide arena floored with hardened grey mud, pockmarked with bubbling muddy pools and hissing vents exhaling steam and sulphurous fumes. At present, several small active fumaroles cluster on the eastern half of the Piano Sterile and the adjacent crater walls. Many seem to be related to fractures following the regional trends: northwest–southeast and northeast–southwest. Each cluster has its individual characteristics: the Forum Vulcani is small but vigorous; the Bocca Grande is large and hottest of all. The Fangaia area, nearest the centre of the crater, has hot fumaroles and, as its name suggests, pools of hot mud that are usually 2m deep and cover up to 500m^2. Most of the fumaroles are encrusted with beautiful but fragile sulphur crystals, derived from the oxidation of hydrogen sulphide when it reaches the air. This sulphur then reacts with water to produce sulphuric acid, which contributes greatly to the chemical alteration of the rocks, which are then often bleached to form the hardened mud that makes little pale grey mounds between the new muddy pools that appear from time to time. In contrast, the northwestern part of the crater floor is calm enough to support a campsite and

bar for those who appreciate sulphurous aromas. The inactive areas are coated with greyish-white hardened mud that resonates when walked upon. Sometimes, however, the hardened crust that masks the boiling waters below is dangerously thin, and visitors are well advised to follow the designated safe paths.

Monte Nuovo

The formation of Monte Nuovo marked the latest, but almost certainly not the final, eruption in the Phlegraean Fields. It took place in the space of a week, between Sunday 29 September and Sunday 6 October 1538, 5km west of Pozzuoli, at the hamlet of Tripergole, which had been reputed for its hot springs since Roman times. The eruption formed a small squat cone, 130m high, on the La Starza terrace (Fig. 2.17).

Throughout the Middle Ages, the area had been generally subsiding, and the Roman Lago Averno had become an arm of the Bay of Naples. The precursory phase reversed this trend: the land began to rise, and increasingly frequent earthquakes started to shake the district around Pozzuoli, especially from 1537 onwards. Rapid but very localized uplift, accompanied by stronger earthquakes, took place on 27 and 28 September 1538. The Temple of Serapis at Pozzuoli apparently rose by some 5m, and a tract of the sea bed some 350m wide was exposed along the coast. By this time, the earthquakes had reduced much of Pozzuoli to ruins. The first and most important (hydrovolcanic) eruptive phase began on the evening of 29 September at 19.00 when first water, then gas, and finally magma gushed from a fracture that

Figure 2.17 Monte Nuovo over Lago Averno, with a contemporary view of the eruption in 1538 (photo by Yiftah-s).

The eruption of Monte Nuovo according to Scipione Miccio

'In the year 1538, the city of Pozzuoli and all the Terra di Lavoro were much tried by strong earthquakes. On the 27th of the month of September, they ceased neither by day nor night in that city. The plain that stretches from Lago Averno to Monte Barbaro arose by a certain amount, and opened up in many places. Water surged forth from these openings and, at the same time, the sea alongside that plain dried up over a distance of 200 paces (370m). As a result, the fish were left stranded high and dry, and fell prey to the local inhabitants.

On the 29th of the aforementioned month, about the second hour of the night [20.00], the ground opened near the lake. It exposed a most horrifying abyss, from which smoke, fire, stones and ashy mud then issued furiously. The opening made a noise like a roll of thunder, which was heard as far away as Naples. The fire coming out of the abyss ran close to the walls of the unhappy city [Pozzuoli]. The smoke was both black and white: the black part covered the land in darkness, and the white part resembled the whitest bombax [cotton]. Having risen into the air, this smoke eventually touched the sky. When the stones were thrown out, the all-devouring flames converted them into pumice. Some say that the largest of those that were thrown a great distance were the size of an ox. These stones rose as far as a crossbow shot into the air, and then fell back down, sometimes into the abyss, and sometimes onto its margins. In fact, the expulsion of many of them could not be seen because of the obscurity of the smoke. But they were clearly evident when they emerged from the smoking fog; and they had no little odour of foetid sulphur, just as the cannonballs of a bombardment can be seen when the smoke produced by the powder has dispersed.

The mud had the colour of ash, which was very liquid at first, but, little by little, it eventually became drier. It was expelled in such quantities, along with the aforementioned stones, that a mountain a mile high was created in less than 12 hours. Not only Pozzuoli and the neighbouring area, but also the city of Naples, were covered by this ash, where it spoilt much of the beauty of its palaces. Transported by the rage of the winds, it swirled and burned the green vegetation and trees through which it passed, and broke down many trees under its weight. Apart from this, many birds and various animals that were coated with this mud left themselves prey to man.

This eruption continued incessantly for two nights and two days. In fact, sometimes it was more powerful than others. When it was at its most powerful, a great din, noise and booming resonance could be heard in Naples – like the thundering noise when two armed enemy forces enter into conflict. The eruption stopped on the third day, when the mountain was exposed without its cover, causing no little amazement among all those who saw it. From the summit of the said mountain, a round concavity was seen stretching down [within it] to its feet. It was a quarter of a mile wide, and the stones that had fallen back into it could be seen boiling in its midst – like a great cauldron of water boils on top of the naked flames.

The citizens of Pozzuoli abandoned their homes and fled, with their wives and children – some by sea, and others over land. The Viceroy [Pedro de Toledo] at once took horse in the direction of the city, and halted on the Monte San Gennaro to observe the fearsome spectacle and the misery of the city, which was so completely covered with ash that any traces of the houses could scarcely be seen. These appalling ruins terrified the citizens of Pozzuoli, and they determined to abandon the city. But the viceroy, who was unwilling to consent to the desolation of such an ancient city that was of such use to the world, decreed that all the citizens should be repatriated and exempted from taxes for many years. To demonstrate his good faith, he himself built a palace with a fine strong tower, and erected public fountains and a terrace a mile long, with many gardens and springs. He reconstructed the road to Naples and widened the tunnel so that it could be traversed without lights. He built the San Francesco church at his own expense. He also had the satisfaction of completing his own palace, and of seeing that many Neapolitan gentlemen had built mansions there too. He also restored the hot baths as successfully as possible, and had the city walls rebuilt. And, to stimulate interest in the city, he decided to spend half the year in Pozzuoli, although ill health subsequently enabled him to stay there only in the spring'. Scipione Miccio, 1600. *Vita di Don Pedro da Toledo*. Archivio Storico Italiano IX (1846).

quickly progressed towards Tripergole. Soon, wet trachytic ash rained down as far as Naples, and black and white clouds rose about 4km high. The hydrovolcanic eruption produced a mixture of pumice, yellow tuffs and old lava fragments in a coarse ash matrix that piled up near the vent to build up the bulk of the cone of Monte Nuovo by the end of 30 September 1538. It emerged for the first time from its pall of thick black smoke on the next afternoon, 1 October. The fragments also again built up the isthmus separating Lago Averno from the Bay of Naples. During the calm interlude on 2 October, observers climbed the new cone and saw fuming lava bubbling in the crater. Then, when water was no longer invading the vent, mild Strombolian explosions covered the cone with about 3m of typical cinders. However, a short violent explosion on the afternoon of 3 October expelled a pyroclastic density current that rushed about 7km southwards into the Bay of Naples. It was followed by three days of calm. The final eruption was the shortest, most restricted in extent, and the most disastrous. It lasted little more than a few minutes on 6 October 1538. On that Sunday afternoon, the sudden explosion of a pyroclastic density current killed 24 people who were climbing up the southern flanks of the cone. The bodies of several victims were never recovered. It was the last fling of Monte Nuovo. The new mountain was built by both Surtseyan and Strombolian types of eruption; the former gave it its characteristic broad squat cone and wide crater 125m deep; the latter gave it its covering of dark cinders.

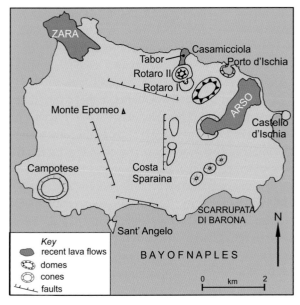

Figure 2.18 Key features and recent eruptions on Ischia.

Ischia

Rich green with its mantle of pines, olives and vineyards, Ischia stands guard over the entrance to the Bay of Naples and the Campanian volcanic region along its northern shores, and, with its predominantly trachytic eruptions, it resembles the Phlegraean Fields nearby. It lies on a major fault that can be traced through Procida, the Phlegraean Fields and on to the Apennines at Benevento. But the island has also developed a clear individuality, in which the landscape owes much to local volcanotectonic uplift. Ischia covers an area of about 45km² and stretches 9km from east to west and 6km from north to south (Fig. 2.18). It rises to 787m at Monte Epomeo, which marks the crest of the uplifted block dominating the island and, indeed, is higher than any volcanic summit in the adjacent Phlegraean Fields. It is thought to have been uplifted, faulted, and fractured by an increase of magmatic pressure in the shallow magma reservoir below. This mass of magma has not yet completely cooled. After the uplift began, eruptions took place along these fractures, so that perhaps 50 volcanic vents can be distinguished on the island. Eruptions occurred for many thousands of years, and they continued after the Greeks settled on the island in about 770BC, through the Roman period and into the Middle Ages, when the emission of the Arso lava flow in 1302 marked the latest episode of activity. Eruptions on Ischia were noted in about 470BC by Pindar and Strabo, in about 350BC by Strabo and Pliny the Elder, and again in 19BC by Pliny the Elder; however, it is difficult to link these references with specific vents on the island.

The island is still tectonically unstable; but, rather paradoxically in an island that owes much of its altitude to uplift, the net trend of movements since Roman times has been downwards. Ischia also suffered a major earthquake on 28 July 1883, which destroyed much of Casamicciola and cost over 2000 lives. This was only the latest and most lethal of several such earthquakes recorded during historical times. The island has many hot springs that support a tourist industry, notably at Casamicciola. These relatively recent volcanic, seismic and hydrothermal features suggest that Ischia may not have witnessed its final volcanic eruption.

Ischia began life as a submarine volcano. Its base lies at a depth of about 500m in the Tyrrhenian Sea, except where it faces the Phlegraean Fields, and it now forms a much-eroded volcanic complex that outcrops on the southeast of the island. When the complex had been built above sea level, explosive eruptions expelled predominantly trachytic fragments. As is shown by the Scarrupata di Barona cliff face on the southeast of the island, prolonged activity was punctuated by episodes of repose, when fossil soils had time to develop before eruptions resumed. Most of this basal complex was then eroded away, and the episode was brought to a sudden close more than 150,000 years ago by the collapse of a caldera, of which little direct morphological evidence remains. There then followed five more phases of its construction. A second phase may well have lasted from about 150,000 to 74,000 years ago, although activity seems to have been concentrated around 130,000 years ago. The third phase, lasting from about 55,000 to 30,000 years ago, was marked by the vast eruption of trachytic ignimbrite that deposited the Monte Epomeo green tuff at about 55,000 years ago. The fourth phase in the growth of Ischia lasted from about 28,000 to 18,000 years ago, with many eruptions occurring around the rising Monte Epomeo block. At this time, too, the Campotese volcano erupted in the southwest of the island and produced trachytic lava flows before ending with a great explosion 18,000 years ago, which breached the crater and expelled the Citara–Serrara tuffs.

The fifth, and current, phase of the growth of Ischia began about 10,000 years ago. It was marked by the increasingly rapid uplift of the Monte Epomeo fault block, accompanied from time to time by eruptions around its edges, especially in the lower area to the east of Monte Epomeo. Thus, at least 14 eruptions have taken place during the past 5000 years, and some of these have been dated. About 3800 years ago a violent Surtseyan eruption occurred 2km off the southeastern coast at the Secca di Ischia and deposited the Piano Liguori unit over much of southeastern Ischia. Between 3000 and 4500 years ago, trachytic lavas oozed out in a wall nearly 1km long and 200m high, which forms the Costa Sparaina. Other eruptions also took place in the far northwest and formed the pair of partly collapsed trachytic domes and lava flows comprising the Zara complex. This reaches a height of 200m and constitutes the northwestern peninsula of the island. At least one of the flows is older than the fourteenth century BC, because a Bronze Age settlement of that date was built upon it. About 930BC, an eruption probably took place from the Cannavale vent, in the east of the island. A Surtseyan eruption built the tuff ring surrounding the maar forming the Porto. It occurred possibly as late as the fourth century BC, because the tuffs lie on top of Greek ware from the fifth century BC. (A channel was cut through the tuff ring to the open sea in 1834 to make the present harbour.)

Perhaps the most varied eruptions in Ischia occurred on the short Rotaro fissure near the northeastern coast of the island, 1.5km south of Casamicciola. The vents erupted at irregular time intervals, at regular distances of 300m. The main cone, Monte Rotaro I, grew up first in the south as a small stratovolcano in about 600BC. Then a violent hydrovolcanic explosion, some 300m along the fissure to the north, destroyed the northern flank of Monte Rotaro and formed a crater 500m wide. During the Roman period, a trachytic dome, Rotaro II, rose to a height of 175m within this crater, almost filling it. Very soon afterwards, another 300m further north along the fissure, an explosion blew a hole, 40m deep, into the northern base of this dome. This explosion may have occurred as the dome was solidifying, because viscous alkali trachytes emerged from the hole and flowed out to reach the shore in a small promontory, the Punta della Scrofa, almost 1km away. Finally, a nearly solid piston of viscous lava pushed like toothpaste up from a vent some 300m north of the source of the flow. This piston, forming Monte Tabor, still carries a cap of older tuffs taken upwards as the piston rose. It now rises 76m and is 150m in diameter, which is probably scarcely wider than the original vent from which it emerged. The formation of Monte Tabor seems to have brought the eruptions on the northern part of this fissure to a close.

Additional explosive eruptions occurred about AD60, and sometime later the contiguous domes of Montagnone and Maschiatta probably extruded from the same vent. The little cones of Molara, Vateliero and Cava Nocelle erupted at 500m intervals along a northeast-trending fissure. The cinders from Vateliero cover a fossil soil containing Roman pottery from the second and third centuries AD, and these eruptions could perhaps be related activity dated to AD430. A further explosive eruption, which laid down the Fiaiano pumice, occurred from about AD670 to AD890 from the vent that was to produce the Arso eruption in 1302. The eruption of the Arso cone marked the latest episode of activity on Ischia. It began on Friday 18 January 1302 and lasted for about two months. It broke out on the eastern flanks of Monte Epomeo, probably on an extension of the Rotaro fissure. The eruption began with a violent hydrovolcanic explosion of pumice that choked the sea nearby, and ash spread in a sulphurous cloud all over Campania and even covered Avellino, 75km away, in a blanket like snow. It was soon followed by emissions of olivine trachytes that formed a crescent-shape ridge of spatter, 50m high and 450m wide, and by a lava flow that set fire to the vegetation and spread 2.7km to the sea in a broad band between 200m and 1km wide. Many terrified islanders are said to have fled to Procida and Capri, as well as to the mainland, and some commentators even claimed that many people and their animals had been killed.

Aeolian Islands

The Aeolian Islands, north of Sicily, are the most famous volcanic archipelago in the Mediterranean Sea (Fig. 2.19). Seven volcanic islands and some islets form an arc with a north–south trending pendant, making a T-shaped group about 100km across. Vulcano, Lìpari and Salina, forming the pendant, are the largest islands, but Stromboli is now by far the most active. The islands are only the summits of bulky volcanic accumulations, whose bases lie on the floor of the Tyrrhenian Sea at depths of 1000–2000m, and whose crests rise a further 500–900m above the waves. They are usually composed of one or more stratovolcanoes, with several calderas, smaller domes, cones or lava flows.

During a volcanic history stretching back more than a million years, the focus of activity has shifted from island to island. The Aeolian Islands have given the volcanic world the Strombolian and Vulcanian types of eruption, which for many years provided a useful shorthand means of describing two contrasting forms

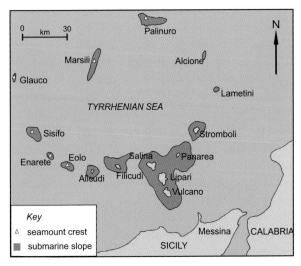

Figure 2.19 The Aeolian Islands and surrounding seamounts.

of activity. Stromboli is deservedly famous as the 'Lighthouse of the Mediterranean', and is the most diligent volcano on Earth at present. It was probably a latecomer to the Aeolian scene, but has apparently been in almost continual activity for at least 200 and perhaps for 2500 years. Vulcano, believed in antiquity to be the home of Vulcan (the god of fire), has several calderas and a beautiful tuff cone in the Fossa, as well as hot, bubbling fissures and mudpots. Its efforts seem to have been more episodic than Stromboli's, and it has been resting since the Fossa last erupted in 1888–90. Lìpari last had a major eruption at Monte Pilato, probably in AD729, and its fumaroles and seismic activity are lethargic, but it has the most varied scenery in the archipelago, which reaches a blinding white climax in the pumice and the turbulent obsidian flow structures of the Rocche Rosse. The remaining Aeolian Islands, Alicudi, Filicudi, Salina and Panarea are generally less populated and more isolated, and erosion has already smoothed their old and extinct volcanic features. Islets such as Basiluzzo, Lisca Blanca, Strombolicchio, and the various Faraglioni between Vulcano and Lìpari are extinct, and their original landforms have been eroded so much that they cannot be recognized. However, Vulcanello, between Vulcano and Lìpari, may still be considered an active vent. Several seamounts, which broadly prolong the arc formed by the upper part of the T-shape of the archipelago, could well be extinct, because rocks dredged from their surfaces seem to be among the oldest in the area.

Generally speaking, the first phase of volcanic activity built up the islands of Alicudi, Filicudi and Panarea, as well as the older parts of Salina and Lìpari. At the close of this phase, first Panarea became extinct, then Alicudi and Filicudi. After a long period of repose, the second phase began, when potassic materials erupted, which culminated in the present eruptions at Stromboli and Vulcano. The marine platforms around the older islands, at altitudes ranging from 105m above to 4m below the present shores, register both sea-level changes and volcanotectonic displacements. They are best preserved on Filicudi, Panarea, Salina, and in western Lìpari. Their absence from both Stromboli and Vulcano indicates the extreme youth of these islands.

The earliest historical eruptions in the Aeolian Islands are suggested when Homer describes the home of Aeolus, the god of the winds (*Odyssey* X: 1–5). Stromboli is usually considered to be the home of Aeolus, but the description is probably based on several islands. Aeolus and his family feast continually, and their home resounds to the noise of their festivities, which probably refers to the crater of Stromboli and its repeated explosions, vibrations, rumblings, roaring, smoke, hisses, and the pyrotechnic displays from incandescent ash and lapilli. Other Aeolian eruptions are mentioned or described with varying degrees of detail and accuracy by a succession of classical authors. In general, the eruptions on Vulcano are noted most often in antiquity, and the possible birth, and some subsequent eruptions of Vulcanello might also be inferred from these writings. Curiously enough, the first explicit reference to an eruption of Stromboli dates only from the second century BC (Polybius in Strabo VI: 2.10). Strabo also saw the difference between the eruptions of Stromboli and Vulcano. Stromboli, he said, was given over to the fire, but its flames were brighter than those of Vulcano, although they were less powerful. But references to what were remote islands are scarce and unreliable until the beginning of the nineteenth century, and most detailed scientific studies date from only the later parts of the twentieth century.

Stromboli

Stromboli is the northernmost and most famous of the Aeolian Islands, and it is the only volcano among them in continual activity. The island is the emerged summit of a large stratovolcano, 75km from the Sicilian coast, which has grown from a depth of 2000m on the floor of the Tyrrhenian Sea to reach an altitude of 924m. Stromboli is elongated along the regional tectonic trends, stretching almost 5km from northeast to southwest and about 3km across, with a total emerged area of 12km² (Fig. 2.20). The island forms a pyramid with overall slopes of 13–15°, with similar gradients continuing to the submarine base of the volcano. Apart from the active summit vents, its most notable feature is the Sciara del Fuoco ('street of fire'), the steep black swathe more than 1km broad that scars the northwestern flanks of the island. The main settlements, San Vincenzo and San Bartolo, lie on an esplanade on the northeast coast. They face the islet of Strombolicchio, an eroded remnant of an early cone that rises to 49m almost 2km offshore. The only other settlement, Ginostra, clings more precariously to the steeper downfaulted western coast dominated by other small satellite cones including the Timpone del Fuoco and the Vigna Vecchia. Stromboli has a double crest: Vancori at 924m is separated by a saddle from the Pizzo Sopra la Fossa at 918m, but neither marks an active vent. Together, they form a natural balcony overlooking the active craters, lying 750m above sea level, which erupt near the top of the Sciara del Fuoco. During the past 200 years, when eruptions have been described with reasonable accuracy, at least three, often five, and sometimes up to eight vents have functioned more or less regularly, and usually in unison, at intervals

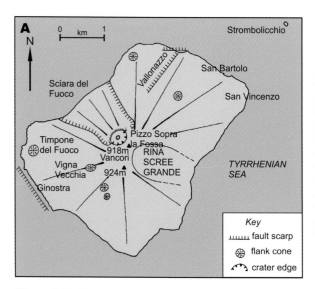
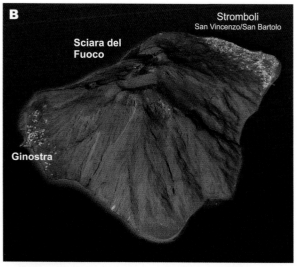

Figure 2.20 The Island of Stromboli. **A)** The main features of Stromboli. **B)** 3D topographic view of Stromboli constructed with elements from NASA (Shutterstock).

ranging from a few minutes to several hours. Although firm evidence is lacking, this kind of moderate, persistent and repeated activity is reputed to have operated for 2000 years or more, which would make Stromboli the most diligent, reliable and consistent volcanic performer in the world during that period, and most likely ever since the Sciara del Fuoco was formed about 5000 years ago. Thus, for centuries, the summit of Stromboli has apparently been the place to go in Europe to be sure of seeing a volcanic eruption.

The volcano can have periods of calm and periods of increased activity. Sudden outbursts still occasionally cause damage over the island. On 22 May 1919, blocks weighing several tonnes crashed onto houses, killing four people and injuring many more. On 11 September 1930, another vigorous explosion, accompanied by a small pyroclastic density current, caused six deaths. On average, one or two brief, but unusually vigorous, explosions have occurred more or less every year during the past 110 years. On the other hand, lava flows have been known to cascade down the Sciara del Fuoco for several weeks at a time, with or without any concomitant explosions from the craters above. Thus, close monitoring has revealed that Stromboli behaves less monotonously than has been supposed.

Nevertheless, most of the present Strombolian eruptions have modest vigour, modest dispersal and great frequency, but limited danger, forming cinder cones and, more rarely, lava flows. This kind of activity is, perhaps, the most common type of eruption on land. It has been associated with Stromboli ever since it was designated as a major eruptive type. But Stromboli is not as typical as a type should be. Elsewhere, such eruptions rarely last for more than a few years – enough to build up a cinder cone perhaps 100–200m in height and emit copious lava flows. These eruptions do not form stratovolcanoes, but Stromboli is composed of massive accumulations of lava flows, cinders and deposits of pyroclastic density currents that were emitted from a central cluster of vents that form a pile 3000m tall. Paradoxically, then, Stromboli is by no means a typical result of Strombolian eruptions; moreover, it forms a fantastical cone-shaped stratovolcano that rises from the waves in the Mediterranean (Fig. 2.21).

The growth of Stromboli

Stromboli lies on a basement of continental crust about 18km thick. The island grew up at the intersection of two faults, the most important of which runs from northeast to southwest. This is the predominant tectonic influence on Stromboli, dictating not only the elongation of the island but also marking the boundary between the older and younger volcanic eruptions. An older vent forming Strombolicchio rises at its northeastern end; the present active craters are aligned along it in the centre. Several spatter cones above Ginostra mark the southwestward continuation of the fault, and two submarine seamounts, representing satellite vents, reveal its extension some 4km offshore.

Stromboli is a young stratovolcano: it has few deep gullies and no marine terraces. The only deep valley, the Vallonazzo, clearly follows a fault down to the north

Figure 2.21 The classic view of Stromboli from a boat, as a typical cone-shaped volcanic island.

coast. The main features of degradation on Stromboli are the great screes, especially those flanking the southeastern summit area. The largest, the Rina Grande, marks the speediest path by which the visitor can descend from the crest. These morphological indications are corroborated by absolute dating. A potassium–argon age of 204,000 years has been obtained for Strombolicchio, the ancestor of modern Stromboli. The past 100,000 years of the history of Stromboli can be divided into seven phases.

In the first phase, which lasted until about 65,000 years ago, andesites and basaltic andesites erupted to form much of the southeastern coastal zone of Stromboli. The second phase, between 64,000 and 54,000 years ago, produced basalts and andesites, which are exposed on the lower southeastern flanks of the volcano. The third phase, between 54,000 and 35,000 years ago, expelled more andesites, transitional andesitic basalts and basalts. In the fourth phase, between 35,000 and 25,000 years ago, basalts and andesites lying on both the southwestern and northeastern flanks of the island were erupted. The fifth phase, marking the culmination of the old stratovolcano at Vancori, was one of the most eventful in the history of Stromboli. It began about 25,000 years ago with the collapse of the summit area. The subsequent eruptions formed the Vancori complex, which is composed mainly of thick piles of lava flows, ranging from basalts to latites and trachytes, which have armour-plated the southeastern two-thirds of Stromboli and also filled the first summit hollow. Fragments were rare at first, but they erupted more frequently later, so that they often occur on the Vancori summit itself.

The Vancori phase ended about 13,000 years ago with a collapse that probably included the partial sinking of the northwestern third of the island, and it may also have eliminated an upper reservoir where magmas had previously been able to evolve. The end of this phase concluded the emissions of the basic to intermediate lavas typical of the older periods in Stromboli's history.

The more recent history of Stromboli was marked by eruptions of basic, non-evolved lavas that clearly must have risen more rapidly from the mantle to the surface without a significant halt in a reservoir. This period begins with the sixth or neo-Strombolian phase, which began when the centre of activity shifted about 300m northwestwards and built up the Pizzo Sopra la Fossa. These new vents emitted the potassic basalts that now cover the surface of most of the lowered northwestern third of Stromboli. Towards the end of this phase, satellite eruptions formed the spatter cones of the Timpone del Fuoco and the Vigna Vecchia in the west of the island.

Then, about 5000 years ago, the most recent major tectonic event inaugurated the seventh and latest phase of activity on Stromboli, when a sector of the stratovolcano collapsed between the scarps now forming the Fili di Baraona and the Filo del Fuoco. Between them lies the smooth, straight slope of the Sciara del Fuoco, inclined at an angle of 40° like a large blue-grey scree, 1km wide at sea level and continuing to a depth of at least 700m. The terrace carrying the present active vents of Stromboli stands at its crest, about 750m above sea level (Fig. 2.22). Today's eruptions, mildly explosive in

Figure 2.22 A) The Sciara del Fuoco on the northwest of the island is the main thoroughfare for volcanic material from the vents down to the sea. **B)** Volcanic blocks and bombs bouncing down the Sciara del Fuoco and entering the sea.

Eruptions on Stromboli – the 'Lighthouse of the Mediterranean'

The famous moderate explosive eruptions are the glory of Stromboli and, especially at night, are among the most spectacular natural performances on Earth (Fig. 2.23). The first sign of another eruption is a low rumbling and a slight trembling of the ground. Then a red glow lights up first one and then another of the vents. The rumbling becomes a deep echoing roar of escaping gases. The red brightens inside the vents to pale vermilion, even yellow at times, as the molten rock is hurled to the surface. One small vent hisses out a blue flame like a huge Bunsen burner. The roar becomes high-pitched, raucous and hoarse.

Figure 2.23 Eruptions at Stromboli through day and night. **A** & **B**) Ash-rich eruptions; **B**) small Strombolian eruption at sunset. **D** & **E**) Classic Strombolian eruptions at night (photos **D** & **E** courtesy of Matteo Lupi).

Incandescent fragments rise 100m or more into the air in a constellation of sparks, hot ash and cinders that twist and twirl in the steamy air. Then they fall back gracefully and lie as glowing embers on the cones around the vents. Their vermilion deepens to red and the larger fading clots slide slowly down the pyre. The roaring calms to a subdued moan as the last cinders crash back down into the vent. Then silence returns. The whole sequence takes perhaps five minutes. Often, there is less than half an hour to wait for the next display.

These mild and almost rhythmic eruptive spasms depend on a constant supply of degassing magma, which is driven upwards by the gases that break out from the magma as the pressures upon it are released. The resulting explosions, which usually take place near the top of the vent, break up the lava into molten spatter, cinders, lapilli and ash, which are then thrown into the sky. In any one eruption it is mostly gases that are released, with minor amounts of molten material.

Occasionally eruptions are slightly different, and very occasionally, more dangerous. A lucky observer might see a lava flow erupt. In 2014 one such flow made its way down to the sea (Fig. 2.24). All the flows for at least the past 200 years have been confined to the Sciara del Fuoco. Some flows make their way downwards from the active craters, but many also rise from the higher reaches of the Sciara del Fuoco itself through short-lived satellite vents/fissures. Some solidified lava boulders break into angular rubble that becomes part of the scree slipping slowly down the 40° slope. But often, the molten stream cascades down the Sciara del Fuoco, and, as its surface solidifies, blocks break off and roll down into the sea, where they hiss like tempered steel in billowing clouds of steam.

Figure 2.24 The rare sight of a lava flow down the flanks of the Sciara del Fuoco. This one in 2014 reached the sea (photos courtesy of Angelo Cristaudo).

Dangerous eruptions, tsunamis and the monitoring of Stromboli

Although, for the most part, Stromboli offers people the chance to see one of the true wonders of the Earth, an erupting volcano, there are hazards. Some big eruptions occur, often called 'paroxysms' when they are at their largest. Landslides can cause tsunamis that threaten the local villages around the island, and occasionally there are injuries and, thankfully rarely, fatalities due to falling bombs. In 1986 (1 fatality) and 2001 (1 fatality), individuals were struck by bombs. In a more unusual incident in 2000, a man threw himself into the crater in an apparent act of suicide. These incidents are very rare, and visits to the summit area of the volcano are now supervised entirely by a strict mountain guiding system (Fig. 2.25A).

More violent eruptions do also occur sporadically on Stromboli. On 5 April 2003, around 9.15 local time (7.15 GMT), there was a strong explosion, which marked an unusual paroxysm eruption sending a big mushroom cloud of ash and debris several hundreds of metres above the volcano. Large bombs and blocks fell (e.g. Fig. 2.25B), and a couple of houses were damaged in Ginostra. Luckily, and probably due to the timing of the eruption, no people were killed or injured. Another paroxysm occurred on 15 March 2007, again with no

one injured. After this event the rules of tourist exploration on the mountain were somewhat tightened, and the authorities now work closely with the mountain guides and the volcano observatory to minimize the numbers of tourists allowed on the mountain at any one time and to maximize their safety.

A somewhat hidden threat from a volcano like Stromboli is the danger of tsunamis. One such event occurred on 30 December 2002, when two closely spaced landslides triggered tsunamis. A total estimate of 5,600,000m³ of rock detached from the Sciara del Fuoco and cascaded into the sea below. Two tsunamis hit Stromboli and Ginostra, causing damage to buildings and boats, but thankfully only causing minor injuries to people. Had this occurred at the height of the summer season it could have been worse. A tsunami warning system (with a buoy tiltmeter in the sea) has now been installed, and signs are in place around the town to direct people to higher ground in the event of an alarm. The volcano itself is extensively monitored with seismic stations, heat cameras, tilt meters, etc., all reporting to the observatory on the island and to those on the mainland of Sicily (see Fig. 2.25).

Figure 2.25 Hazard monitoring at Stromboli. **A)** A guide helps a group of volcano seekers down from the summit at night. **B)** Large volcanic bomb with the town of Stromboli in the background. **C)** Monitoring station with seismic and heat cameras; the solar panels keep the batteries charged for 24-hour monitoring. **D)** The main room at the observatory on Stromboli; linked with the mainland, they can report on any unusual volcanic activity and warn against tsunamis.

the main, are apparently a product of the formation of the Sciara del Fuoco and its consequent repercussions on the vent system of Stromboli.

Vulcano

Vulcano is the southernmost of the Aeolian Islands, forming an ellipse, 22km^2 in area, 3km broad and stretching 6km from north to south (Fig. 2.26). Apart from the ever-active Stromboli, Vulcano is the most recent of the Aeolian Islands to erupt (1888–90). It is apparently a young island, for it bears no trace of old marine terraces around the coast. Since Vulcano was built up above sea level, the migration of activity has created a varied but low-lying island that reaches only a modest 500m at Monte Aria, although its base lies at a depth of 2000m on the floor of the Tyrrhenian Sea. Indeed, the Fossa cone, which has been the focus of most of the eruptions during historical times, only reaches 391m. However, what Vulcano lacks in height, it makes up for in aesthetic quality. The volcanic landforms were often created side by side, instead of being superimposed upon each other, by a progressive and continuing northward shift of activity along a major fissure. Thus, historical eruptions in the straits between Lìpari and Vulcano

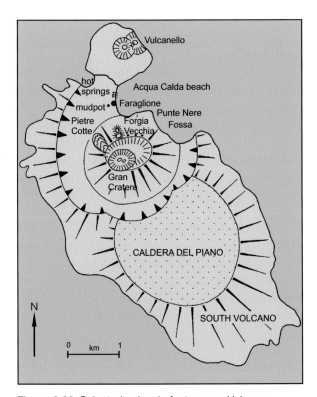

Figure 2.26 Selected volcanic features on Vulcano.

may have developed or given birth to Vulcanello, which became linked to the main island only after its latest eruption in about 1550. The remains of the stratovolcano; several calderas that formed, and then filled, to create the Caldera del Piano; the Fossa cone, with its deep crater, and two other satellite craters, as well as a recent obsidian lava flow on its flanks; and the separate cones of Vulcanello, not to mention a beach with hot springs and mudpots – all combined to give Vulcano its beautiful landscape (see Fig. 2.27A).

References to activity on Vulcano go back some 2500 years, although there are gaps in the historical record, especially in the early Middle Ages. Vulcano seems to have witnessed an eruption almost every century for which fairly reliable information is available, with about 30 active episodes altogether. The Vulcano of antiquity was intimately linked with myth, theology and geology, for it was reputed to be one of the major homes of the god of metalworking, fire and war, known to the Greeks as Héphaistos and to the Romans as Vulcanus. The ancient Greeks called it Hierá Hephaístou ('sacred to Héphaistos'); and, as Virgil described in the *Aeneid* (8: 416), the ancient Romans believed that Vulcanus worked his forges on the island. The geographer Strabo (VI: 2.10) described Vulcano at that time as 'deserted, wholly rocky, and entirely in the hands of the fire, with three erupting vents … the largest of which projects glowing blocks in the midst of the flames, that have already filled up much of the straits.' Strabo added that 'Polybius [who died about 120BC to 125BC] reported that one of the three craters had partly collapsed, but that the other two still remained intact. The largest of them, which was circular, measured five stadia (925m) around its rim, but it then narrowed down to no more than 50 feet (15m) … and it was one stadium (185m) from the sea.' This seems to be quite an accurate description of the Fossa cone, and all three craters could belong to it. On the other hand, one of the other craters could be Vulcanello; but the site of the third would still be a mystery.

South Vulcano

South Vulcano forms two-thirds of the island and is its largest and oldest volcanic accumulation. Its eruptions began at about 2000m deep on the floor of the sea. It grew above sea level and formed an elliptical stratovolcano that erupted innumerable lava flows, separated by relatively thin beds of cinders. One of these flows has been dated to 120,000 years ago. The whole mass is also

Figure 2.27 Views of Vulcano and the Fossa Cone. **A)** View of Vulcano Island from Southern Lìpari Island. **B)** View of the cone from Vulcanello. **C)** The Sulphur-stained Gran Cratere with fumaroles and Vulcanello in the distance. (Photo **A** by Brisk_g, photo **C** by Ghost-in-the-Shell).

riddled by radial lava dykes and some satellite vents. Fluvial erosion has eaten deep gullies into the flanks of South Volcano, and marine erosion has cut jagged cliffs, often 30m high, around its edges.

The second main episode in the history of Vulcano began when the summit of the stratovolcano collapsed about 100,000 years ago to form a caldera about 2.5km across. Subsequent eruptions then filled the caldera to the brim. However, even as the caldera was being filled, more collapses took place in its northern sectors. Successive calderas were formed, with each new rim biting into the previous caldera; but more eruptions compensated for each partial collapse. The infilling often included a few cinder cones, such as La Sommata, as well as the trachytic, latitic and, especially, rhyolitic domes and flows, about 15,500 years old, that form the Monte Lentia in the northwest. The infilled zone forms the Caldera del Piano, named after the piano ('the plain') about 300m above sea level, which supports most of the rural population of Vulcano.

Caldera della Fossa

When the Caldera del Piano had been filled, the northwestern flank of South Vulcano collapsed a fourth time and formed the Caldera della Fossa, which sank below sea level. This forms a slightly elliptical hollow, about 3km across, with a pronounced cliff, mostly 100m high, that marks its inner perimeter except on the northeast. This sector either suffered the greatest amount of sinking or was destroyed by a blast or landslide. The Caldera della Fossa has been the theatre of all the activity on Vulcano for the past 14,000 years. New eruptions, concentrated on a fissure running from north to south, have slowly carpeted the caldera floor, but more than a quarter of it still lies below sea level, and at least another quarter hardly rises more than 10m above it. The stage was then set for the activity that has given Vulcano its renown – the growth of the Fossa and Vulcanello. However, in the wider perspective of the history of the island these probably represent only initial contributions to the filling of its youngest caldera.

The Fossa and Vulcanello are tuff cones, and the proximity of the sea has had a distinct influence on their growth, for both have had hydrovolcanic components. The activity of the Fossa may indeed owe such distinctive features as it possesses chiefly to the interaction of sea water and an upsurging trachytic or rhyolitic magma.

The Fossa cone

The Fossa is a squat cone of trachytic tuffs that forms the backcloth to the Porto di Levante. Although it rises only 391m, its base is 1km across and its steep slopes range between 30° and 35° (Fig. 2.27A and B). Its northern flanks are scarred by the crater of the Forgia Vecchia, created by two successive satellite eruptions, and by the bristling grey lava flow of rhyolitic obsidian forming the Pietre Cotte. Two lava tongues, the trachytic Palizzi flow and the rhyolitic Commenda flow, emerge from the southern flanks of the Fossa, and the trachytic Punte Nere flow juts out seawards from its eastern base. However, these small lava flows seem only to mark late spurts of viscous emissions as the explosive phases closed. The Fossa is crowned by the Gran Cratere, which is 500m in diameter and no less than 175m deep (Fig. 2.27C). It has the typical dimensions of a tuffcone crater and a distinct funnel shape. In particular, its splayed upper northern walls are stained yellow by beautiful sulphurous crystals from many active fumaroles. The Fossa is a rather shy volcano, now erupting, on average, about once per century. It has probably been active for about 6000 years and it covers the layer of Lower Pilato rhyolitic pumice that exploded from Lìpari somewhere between 11,000 and 85,000 years ago. Its growth has been characterized by a westward migration of the vents by about 500m, by cyclical eruptive phases, and by a gradual decrease in the volume of materials erupted. The Fossa was constructed by four or more very similar eruptive cycles, in which a short eruptive outburst was followed a long period of repose. The initial cycle probably built up the bulk of the Fossa, which ended about 5,500 years ago, and eruptions during the past two centuries have expelled relatively little.

After a dormant interval, the second, or Palizzi, eruptive cycle began. The active vent shifted about 300m to the west, and the western sector of the cone was partly destroyed. Some of the earliest recorded observations of activity on the Fossa date from this time. The Palizzi cycle had two parts. First came hydrovolcanic eruptions of tuffs and then pumice, which have been dated to about 2200 years ago. Given the margin of dating error, one of these could have been the eruption noted by Thucydides, 'where great flames are seen rising up at night, and, in the daytime, the place is under a cloud of smoke' (*Peloponnesian War* III: 88). This eruption probably occurred in 425BC when the author was in Sicily. Explosive eruptions also took place between 370 and 350BC, and again in about 330BC, which might have been the activity noted by Aristotle and Callías. Aristotle (*Meteora* II: 8), described how the mountain swelled up and burst with a loud noise; fire escaped, and a violent wind hurled a great quantity of ash over Lìpari and as far as Italy. Callías (book III) noted that, when Agathocles (361–289BC) was tyrant of Syracuse, one of two craters on Vulcano could be seen shining very brightly from a long distance, and threw out glowing stones of monstrous size with a noise that could be heard 500 stadia away (92km). According to Pliny the Elder, 'an island', possibly Vulcanello, is reputed to have risen above the waves in 126BC.

Both Vulcano and Vulcanello were probably erupting between 29 and 19BC. One of these episodes could have inspired Virgil to compose the famous passage describing Vulcan's forge – the Gran Cratere of the Fossa – in the *Aeneid* (8: 416–453), where 'strong blows are heard resounding on anvils and, echoing their groans, … ingots hiss within the caverns and the fire pants within the furnaces'. The second part of the Palizzi cycle, dated about 1600 years ago, occurred when hydrovolcanic eruptions were followed by the emission of the Palizzi trachytic flow, which moulded the southern slopes of the Fossa cone. Activity was reported on Vulcano, for instance, between AD200 and 250.

The third, or Commenda, cycle used virtually the same crater as its predecessor. After resting for a few centuries, an unusually violent hydrovolcanic explosion cleared the plugged vent of the Fossa and expelled a breccia that gradually changed into a block and ash flow. These eruptions may be tentatively matched with reported activity perhaps in either AD526 or 580. This activity could correspond with a hydrovolcanic explosion on the northern flanks of the Fossa, dated to about 1400 years ago, that formed the first yellow tuff cone of the Forgia Vecchia. Those deposits were then covered by the Upper Pilato ash from Lìpari, which is believed to have erupted in AD729, but could be somewhat younger. This marker bed was, in turn, overlain first by dry surge layers and then by the Commenda lava flow that moulds the southwestern slopes of the Fossa cone.

Eruptions were also reported between AD900 and 950, and during the thirteenth century.

After a further interlude of repose, the fourth, or Pietre Cotte, cycle began when an eruption came from a new crater about 200m west of the Commenda crater and coated much of the Fossa cone with reddish tuff. This cycle probably started with the vigorous eruption on 5 February 1444, when 'Vulcano burned with perpetual fire… gave off thick fumes… sooty clouds, while pale flames rose up from the abyss and escaped from fissures and joints.' (Fazellus, 1558). The Fossa was active again in 1618, 1626, 1631, and in 1646, when Bartoli observed abundant fumes emerging from the incandescent crater, and again in 1651 and 1688. In 1727, d'Orville saw a small cone within the Gran Cratere of the Fossa, which erupted glowing stones and fumes with a noise like thunder; in the same year, a small tuff cone erupted on the northern edge of its larger predecessor, the Forgia Vecchia. In 1739, after about eight years of intermittent and unusually prolonged activity, the bristling rhyolitic lava flow forming the Pietre Cotte oozed down the steep northwestern slopes of the Fossa, but it solidified even before it reached the base of the cone. The Pietre Cotte cycle trailed on with the eruptions in 1771, 1786 (two weeks in March), 1873, 1876, and came to an end, for the time being, with the activity between 1888 and 1890. However, the roof of the magma reservoir still probably lies no more than 2km deep.

The Fossa has remained quiet since 1890, except for persistent fumaroles that occur, most notably on radial and concentric fissures, on the upper northern slopes of the Gran Cratere. They are associated with beautiful, if extremely fragile, sulphur crystals and little stalactites. A sulphur flow, 2m long, even issued from a small vent in late September 1989. The fumaroles have varied in intensity and temperature, with an initial maximum of 615°C in 1924, then decreasing to a minimum of 110°C in 1962. A period of unrest started in 1980, which was probably caused by injections of gas from the magma below that increased the pressures within the hydrothermal system near the surface. Temperatures rose again to nearly 700°C in 1993, but have subsequently lowered, and no eruption occurred. Such activity should be followed with care, because the development of tourism on the island implies that any new eruption could cause many casualties.

The Faraglione, mudpots and the Acquacalda beach

The fissure extending northwards from Vulcano to Lìpari did not remain inert while the Fossa was active. It gave rise to mudpots, hot springs, the Faraglione and Vulcanello, which together add much to the scientific and tourist interest of Vulcano. The oldest feature on this fissure is probably the Faraglione, which now forms a steep, bulky pinnacle rising 56m above the harbour of Porto di Levante. Near the Faraglione are the mudpots and hot springs that give their name to the Acquacalda beach, linking the Porto di Levante harbour with Vulcanello. The grey mudpots at the foot of the Faraglione have been made into a large bath for the benefit of

Figure 2.28 People enjoying the mudpools in Vulcano. The supposed healing powers of these volcanic soups have people travelling to Vulcano from afar (shutterstock_EugeniaSt).

The eruption of the Fossa cone on Vulcano (3 August 1888 to 22 March 1890)

For all its long list of recorded eruptions, only the latest outburst of Vulcano was described in any detail. After a period of complete calm, the initial vigorous explosion removed the old solidified material plugging the vent, deposited a thin, yellow breccia, and shifted the vent slightly southwards to the present Gran Cratere. But the main efforts of the eruptions were the powerful intermittent explosions, 'like the detonations of landmines', which sent billowing clouds of gas, steam, fine ash, and lapilli several kilometres into the atmosphere. These explosions varied in both intensity and frequency. Most expelled fine, pale grey or reddish trachytic ash, but the largest also ejected big blocks and especially 'breadcrust' bombs of rhyolitic obsidian, with a surface like the best French loaves, which were thrown as far as 1.5km from the vent.

The spasms were separated by irregular periods of rest, and there were abrupt changes from febrile outbursts to mere fumarole activity, or even total calm, that sometimes lasted several days. These eruptions expelled very thin alternating layers of fine trachytic ash and lapilli that apparently have no counterpart in the history of the Fossa. However, the eruption was a relatively mild affair, full of sound and fury, but signifying little change in the Gran Cratere, and producing only a layer of fragments around its rim, which over a century of intense Mediterranean showers have reduced from their original thickness of 5m to 1.7m. A few old photos actually exist of this activity (Fig. 2.29).

Figure 2.29 Photos of the last activity on Vulcano, 1888–1890 (photos Wikimedia Commons).

tourists. The muds vary in temperature from 20°C to 60°C and are reputed to cure rheumatism and ugliness if spread in the appropriate places. People flock to this island to cover themselves in the healing muds (Fig. 2.28), and the island has a laid-back, almost hippie feel, in part due to this natural spa. To the north, the Acquacalda beach and the adjacent shallows are riddled with small fissures, from which nearly-boiling water and carbon dioxide bubble out. The Acquacalda beach is itself part of the isthmus of black ash, about 500m wide, linking Vulcano to Vulcanello.

Vulcanello

Vulcanello is a buff-coloured cone, rising to 123m, that lies on the broad pedestal of two lava flows at the north end of Vulcano (Fig. 2.30). In fact, it is composed of three successively formed cones, with intersecting craters, which have joined together in a single elongated mass. During its growth, the location of the vent in relation to sea level was often crucial in determining what was expelled.

The eruptions of Vulcanello began below sea level, and the date of its emergence is not known exactly. Eruptions occurred on the new island perhaps in 126 and 91BC, and for some time between 29 and 19BC, and the latest eruption occurred in about 1550 (Fazellus, 1558), but it is uncertain when Vulcanello was active in between times.

The sea has eroded the eastern flanks of the easternmost cone of Vulcanello and provided a magnificent section of its alternating layers of reddish-brown leucite-tephrite lavas and cinders. The section also demonstrates

Figure 2.30 View to Vulcanello, with the port, Faraglione and Acquacalda beach in the foreground. The south coast of Lìpari can also be seen in the distance (photo – Shutterstock/silky).

that this cone erupted above sea level in Strombolian rather than Surtseyan eruptions. On the other hand, the second cone was formed by a Surtseyan eruption when sea water could penetrate the new vent, a little to the west of the first. It is composed of fine buff tuffs that are also scattered over the surface of the first cone. The ungullied surface of these cones confirms their recent origin. Then two eruptions of lavas ensued, which show that water could not penetrate the vents at this time. The first emissions have pahoehoe surfaces and form the broad plinth around the first two cones of Vulcanello. The much smaller second emission, of a more viscous trachyte with a more bristling surface, emerged from the northern flanks of the second cone and covers the edges of its predecessors. In places, the Upper Pilato ash layer from Lìpari lies on these flows.

The third, westernmost and latest cone was formed, in the Middle Ages, by two short Surtseyan eruptions that came from a vent situated some 50m west of the second crater. The third cone of yellowish trachytic tuffs rises only 100m high, but is remarkable for its fresh-looking, steep-walled crater. The first eruption probably occurred here in the thirteenth century. It was followed by a quiescent interval during which a soil formed on the cone. The latest eruption gave off yellowish tuffs in about 1550 (Fazellus, 1558). Marine currents are said to have redistributed the erupted fragments and formed the isthmus joining Vulcanello to Vulcano, but the ash making the isthmus is black, and is thus quite unlike any products expelled from Vulcanello. The bulk of the accumulation must therefore have come from Vulcano to the south. No subsequent eruptions have yet taken place on Vulcanello, and fumarole activity ceased in its craters at the end of the nineteenth century.

Lìpari

Lìpari is the largest of the Aeolian Islands, covering an area of 38km^2, rising more than 1500m from the floor of the Tyrrhenian Sea and 602m above sea level (Fig. 2.31). It represents only the emerged summit of a volcanic field that has seen marked shifts in the focus of its activity, and changes from andesitic to rhyolitic eruptions.

Volcanic products made Lìpari a major industrial centre from Neolithic times. Obsidian, with the sharp cutting edge given by its natural cleavage, was a valued tool throughout Neolithic and Graeco-Roman times (Fig. 2.32). The Pomiciazzo flow, which erupted

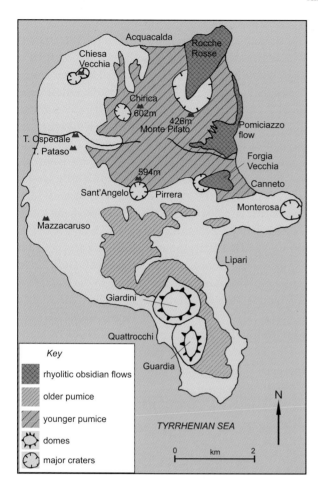

Figure 2.31 Volcanic features on Lìpari.

sometime between about 11,400 and 8600 years ago, was, for instance, the source of the most widely exported obsidian artefacts in the Mediterranean area during the Neolithic. Most of the abundant and easily accessible pumice deposits, such as those of Monte Pilato, erupted well after Lìpari was first colonized, and now cover 8km² in the northeast of the island. Pumice is a foundation material of toothpaste, soap, beauty creams, abrasive and polishing pastes (notably for computer screens), soundproofing and insulation – and a source of silicosis for those who extract it (Fig. 2.32).

The greater complexity of the development of Lìpari compared with its companions is at once evident from its much more varied skyline. There have been few re-peated eruptions from a central cluster of vents. Thus Monte Chirica, the highest point of Lìpari, rises a mere 602m above sea level. These vents emitted chiefly ash and pumice, short, thick and viscous lava flows, and domes of even more viscous material that are found notably in the south. Explosive eruptions have expelled much rhyolite in the more recent phases of activity. These rhyolites now tend to dominate the scenery, most remarkably in the Rocche Rosse, which forms a broad, twisted tongue of obsidian that juts out to sea at the northeastern corner of Lìpari.

Figure 2.32 The Volcanic treasures of Lìpari. **A)** Famous for its obsidian, the island formed a major trading post. **B)** Quarries of its pumice can be seen along the coast, like here in the north, with parts of the Rocche Rosse lava flow behind. (Photo **A** by Ji-Elle, photo **B** Shutterstock).

Lìpari first erupted from the sea floor, and an island emerged over 220,000 years ago. Lava flows, welded tuffs, and ash and cinders then formed at least a dozen low stratovolcanoes. These were largely composed of andesites, especially quartz latite-andesites and quartz andesites, which are now best exposed on the northwestern coast between Acquacalda and Quattropani. They also form the low, eroded stratovolcanoes of the centre and west: Timpone Carrubbo, Monte Mazzacaruso, Timpone Pataso, Timpone Ospedale and the Chiesa Vecchia.

As these stratovolcanoes were erupting, they were bevelled by a single interglacial marine terrace which, because of contemporaneous and subsequent dislocation, is now found at different altitudes above sea level, but mainly between 18m and 35m. Some eruptions continued in the east and formed the twin volcanoes that make up the Monterosa peninsula, north of Lìpari town. They were formed during a period of low sea level, and thus no raised marine terrace was developed around them.

There was then a long dormant interval before eruptions resumed about 60,000 years ago. This activity focused on Monte Sant'Angelo, where a long spell of explosive activity formed a stratovolcano in central Lìpari that was composed of quartz latite-andesite and quartz latite fragments and pyroclastic density currents. The palaeosols between the beds show that the eruptions were often separated by dormant periods. During the same episode, much more effusive eruptions formed the Costa d'Agosto, about 2km to the north, on the western flanks of the old Monte Chirica volcano. Thick rhyodacitic lava flows were then emitted from the southwestern flanks of Monte Sant'Angelo and came to rest on the west coast. Lastly, about 40,000 years ago, yet more explosive eruptions and the emission of a large quartz latite-andesitic flow were the prelude to the collapse of a caldera on the southern part of Monte Sant'Angelo, which seems to have extended, largely below sea level, from southern Lìpari to northern Vulcano.

The caldera became the main site of eruptions during the period between about 40,000 and 10,000 years

Figure 2.33 The Quattrocchi on the southwestern coast of Lìpari with Vulcano in the distance (photo by Ghost-in-the-shell).

ago. The change in the focus of activity in Lìpari was accompanied by a marked change in the materials expelled: rhyolitic pumice and lava domes and flows predominated during explosive eruptions that formed the southern peninsula, where the scenery is dominated by Monte Guardia and Monte Giardini. A major submarine hydrovolcanic eruption between Lìpari and Vulcano exploded the first layers of brown ashflow tuffs between 35,000 and 23,500 years ago. These fine-grained layers extended over Salina, Panarea, Filicudi and the northern coast of Sicily, and they covered at least 7km² on Lìpari itself. Their shoshonitic character suggests that they were derived from the Vulcano magma rather than that beneath Lìpari, and it is possible that they came from the still submerged Vulcanello vent.

Eruptions resumed on Lìpari itself with the formation of the Monte Guardia sequence. Soon after andesitic lapilli from Salina had blanketed the island about 22,480 years ago, several rhyolitic domes extruded, whose remains can be seen in the cliffs of Quattrocchi, which forms an often photographed vista (Fig. 2.33). They were then largely destroyed by a sub-Plinian explosion that covered most of Lìpari with an andesitic and rhyolitic breccia in a pumice matrix. Craters near sea level then exploded the bulk of the Monte Guardia sequence during a phase that occupied only short intervals in a period lasting no more than 2000 years. Eventually, the hydrovolcanic aspects decreased and explosive fragmentation gave way to viscous rhyolitic extrusions, to form the domes of Monte Guardia and Monte Giardini. Between about 20,300 and 16,800 years ago, Lìpari was blanketed by another layer of brown ashflow tuffs, which no doubt came from the submerged Vulcanello vent, like their predecessors.

Thereupon, the focus of activity changed to the northeastern quadrant of Lìpari, and it lasted from about 10,000 to 1300 years ago. Each eruptive episode was probably brief; each started with an explosion of rhyolitic pumice and ash, and each ended with extrusions of rhyolitic flows. The oldest eruption lies on a palaeosol on the uppermost brown tuffs and formed a small breccia cone, and then emitted an obsidian flow in the Canneto–Dentro area. This was soon followed by a larger hydrovolcanic eruption that formed a tuff ring and the Gabellotto–Fiume Bianco beds, which cover half the island with more than a hundred thin layers of pumice and ash. They probably also covered northern Vulcano, where they were identified as the Lower Monte

Pilato tephra. Between about 11,400 and 8600 years ago, activity concluded when rhyolitic lavas formed the Pomiciazzo dome and flow.

The period of repose that then ensued was marked by a palaeosol, 1.5m thick, which contains artefacts dated between 4800 and 1220 years ago. Activity resumed with the explosion of the Forgia Vecchia beds from two allied vents near Pirrera. They cover about 1km² and are composed of many thin layers of coarse, white ash and breccia. Further explosions at Monte Pilato formed a 150m high pumice cone, with a crater 1km wide, that covered most of the Pomiciazzo flow. They also expelled pumice as far as Vulcano, where they form the Upper Monte Pilato layers. Near the town of Lìpari, they also cover the Roman ruins of Contrada Diana, which date from the fourth and fifth centuries AD.

After the eruption of Monte Pilato reached its climax, the Pirrera vent, north of Lìpari town, emitted the rugged lobe of the rhyolitic Forgia Vecchia lava flow, which has been dated to about 1600 years ago. At Monte Pilato, rhyolite oozed up the vent, breached the northeastern sector of the cone, and spread in a bristling tongue of black obsidian 2km long, with a weathered red surface crust, which juts out into the Tyrrhenian Sea. This was the famous Rocche Rosse flow (see Figs 2.31 and 2.32), which marked the latest episode of volcanic activity on Lìpari. This could have been observed by St Willibald in AD729 or it could have taken place a century earlier, giving rise to the legend that San Calogero (AD524–62) expelled devils from the black stone of Lìpari to Vulcano. But recent archaeomagnetic research indicates two distinct dates for these eruptions of Monte Pilato – the first about AD700, and the second about AD1200. At all events, the most recent two volcanic formations of Lìpari occurred after the decline of the Roman Empire. Paradoxically, therefore, the obsidian that gave Lìpari its fame in antiquity is not that which can be seen so clearly in its present flows.

Alicudi

Alicudi covers an area of 5km² and forms the summit of a large conical stratovolcano rising from a depth of 1100m on the floor of the Tyrrhenian Sea to a height of 675m above sea level. Its eruptions seem to have come from a central vent devoid of satellites. The details of its growth are uncertain, but the oldest eruptions formed the Galera complex, which has been dated to about 120,000 years ago. It was a mainly effusive volcano, but its activity ended with the formation of a large caldera, which was eventually filled mainly by the Dirituso cone of ash and cinders. In turn, this cone developed a summit depression (perhaps a caldera) that was covered, between 27,000 and 41,000 years ago, by the Montagna complex of domes and andesitic flows. The last eruptions formed the short, thick lava flows of the Filo dell'Arpa on the southern flank of the stratovolcano. The rocks erupted range from basalts to potassium-rich andesites.

Filicudi

The island of Filicudi covers an area of 9.5km² and represents only the emerged portion of a complex volcanic structure that piled up from a depth of 1100m on the floor of the Tyrrhenian Sea. At least six centres formed stratovolcanoes and domes composed of basalts and potassium-rich andesite. At the Fili di Sciacca, a potassium-rich andesite has given a potassium–argon date of about 210,000 years. However, $^{40}Ar/^{39}Ar$ dating has produced ages of 1.02 million years for the Zucco Grande and 0.39 million years for the Filo del Banco. This suggests that Filicudi, and perhaps the Aeolian Islands in general, are older than previously thought. The steep-sided andesitic pile of Fossa Felci is the chief, and youngest, stratovolcano on Filicudi. It constitutes the summit, at 774m, and most of the rest of the island. In addition two lava domes, the Montagnola and Capo Graziano, erupted during the last phase of activity on the island about 40,000 years ago. Half the Capo Graziano dome has been downfaulted to reveal much of its internal structure. During the glacial period, several marine terraces were carved around the island after the eruptions had ceased.

Salina

Salina is the hub of the Aeolian Islands, at 27km² the second largest island after Lìpari, rising from a depth of 2000m on the Tyrrhenian sea floor. The Fossa delle Felci (962m), which is in fact the highest point in the whole archipelago, is separated by a saddle from the Monte dei Porri (860m) in the west. Together they form the distinctive double-hump profile that gave it its Greek name of Didýme – 'twin island'.

The relief reflects the volcanic development of Salina. Eruptions have been concentrated on two distinctive clusters of vents beneath the humps, both of which functioned about half a million years ago, as the older island grew up. The Capo and Rivi volcanoes in the east, and Corvo in the west, mark the first phase of basaltic

eruptions. Soon afterwards, the Fossa delle Felci grew up in the east – a more varied stratovolcano than its predecessors, with dacitic, andesitic and basaltic emissions. Glacial marine erosion has bevelled terraces into the flanks of these deeply eroded volcanoes. Potassium–argon dates of 430,000 years have been obtained from basalts in the Monti Rivi.

The western vents of Salina sprang into life again about 75,000 years ago when some of the most violent explosions in the history of the archipelago built up the andesitic cone of the Monte dei Porri (Fig. 2.34), and spread the grey Monte Porri tuffs all over Salina, Panarea and Lìpari. The latest volcanic event occurred in the far northwest of Salina about 13,000 years ago, when a great hydrovolcanic explosion of white pumice formed the crater that cradles the village of Pollara. It is one of the most impressive craters in the whole archipelago, 1.5km across and surrounded by a rim some 300m high. Salina has been quiet ever since.

Figure 2.34 Monte dei Porri, Salina (photo by Andrés Maneiro).

Panarea

A short ferry stop on the way from Lìpari or Salina to Stromboli, Panarea is an inactive volcano with a total surface area of only 3.4km². The highest point on the island, Punta del Corvo, is at 421m, and it is adorned with a number of quite up-market holiday villas, and some nice black volcanic beaches. The island's volcanic history started around 200,000 years ago, with calc-alkaline basaltic andesites to dacites forming domes and minor lava flows. The oldest parts of the island are in the northern part, and the rest of the island was made up of various domes, explosive products and flows up until around 130,000 years ago, including the dome that forms the Punta del Corvo. Volcanism then shifted offshore around the perimeter of the island, with the latest eruptions producing the island of Basiluzzo around 10,000 years ago. Thermal springs near the village of Punta di Peppe e Maria offer some evidence of the volcanic history of the island, and possibly point to a continued volcanic future.

Seamounts

Potassium–argon dates from lavas dredged from the Aeolian seamounts in the Tyrrhenian Sea show a range of ages. Sisifo seamount, in the west, yielded calc-alkaline basalts ranging from more than 1.5 million years old to 900,000 years old, as well as a rhyolite dating from 640,000 years ago. Eolo and Enarete seamounts, further east, seem to be rather younger. Their rocks have not, so far, produced potassium–argon dates exceeding 850,000 years old; several date from 770,000 to 790,000 years ago, and one sample from Enarete is 670,000 years old. Dates of 800,000 years were obtained from the submerged slopes of Panarea. A sample from Stromboli canyon is 530,000 years old, but another sample from its submerged northern slope is only 180,000 years old. These great ages imply perhaps that most of the Aeolian seamounts are extinct, and that, instead of being under construction, they might have been carried downwards as the floor of the Tyrrhenian Sea sank. However, a recent investigation has reportedly revealed fresh morphological features that may indicate renewed activity.

Figure 2.35 The grand volcano Etna on Sicily, the largest volcano in Europe. **A)** Etna rising majestically over Catania. **B)** Location of Etna on Sicily with eruption plume visible from space (October 2002). **C)** Plume from Mt. Etna taken from International Space Station at 4:25 pm GMT on 23rd June 2011 NASA/Astronaut Ron Garan. (Photo **A** by Ben Aveling – Creative Commons; **B** NASA/Goddard Space Flight Centre Scientific Visualization Studio; **C** NASA/Astronaut Ron Garan).

SICILY
Etna

Etna is the largest active volcano in Europe. It presides over the landscape of eastern Sicily and the Straits of Messina, and it forms the graceful backcloth to the Greek theatre at Taormina, as well as the constant threat on the northern skyline of Catania (Fig. 2.35). The Sicilian giant rises some 3320m above sea level, stretches 47km from north to south and 38km east to west, covering about 1,200km^2, with a volume of 350–500km^3. Its eruptions are plentiful, and the plumes from its larger eruptions can close airports, and are commonly visible from space (Fig. 2.35B, C).

Etna is a complex volcano, with four main large summit craters (Northeast Crater, the Voragine, the Bocca Nuova, and the Southeast Crater Complex) (Fig. 2.36), and more than 200 cinder cones, as well as many faults on its flanks. The people living on the mountain have always divided it into three parts. The piedmont or cultivated region is the lowest part, and it corresponds to the edge of the old shield that forms the base of Etna. Its broad and gentle slopes are scattered with white villages, set between citrus groves, vineyards and cereal plots that flourish on the fertile weathered lavas. Upslope, between 1000m and 2000m above sea level, lies the second, wooded, region, with many pine forests that eventually give way to Mount Etna broom, whose brilliant yellow blooms contrast with the black and reddish rugged basalts where it has gained a foothold. This is where many of the cinder cones have erupted, notably in the south, southeast and northeast. Some are already wooded, others still starkly bare. Elsewhere they would make imposing landmarks, but here they are dwarfed by the mass on which they stand. Eventually, at about 2000m, the third, bare and sterile region begins. The slopes steepen at last to 20° to form the upper cone of Mongibello, the Sicilian name given by Earth scientists to the most recent of the several stratovolcanoes that compose the mountain. Here, the snows melt in spring to reveal the naked black lavas erupted so recently that weathering has scarcely altered their surface. This wilderness, of a vastness seen on no other European volcano outside Iceland, extends up to an infilled caldera that forms the summit plateau, the Piano del Lago, at about 3000m. It is here that persistent activity most often changes the face of Etna. In its centre lies the summit cone, scarred by two large chasms, the Voragine and the Bocca Nuova, which joined together in 1999. Both chasms can be more than 200m deep, but lavas well up within them from time to time, as occurred, for instance, in July 1998. The Central Cone is flanked on the northeast and southeast by two sources of subterminal activity. Constant eruptions from the Northeast Crater

Figure 2.36 Aerial view of the Etna summit in 2003; SE crater cone in the right foreground, and behind it from left to right: the two vents of Bocca Nuova, the chasm or Voragine, and the fuming NE Crater. The village in the background is Randazzo. Recent eruptions around the SE crater have changed the morphology on this side, with a new crater building up and almost completely masking the original SE crater – signs of the ever-changing face of Etna.

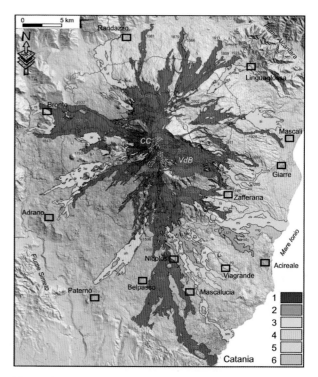

Figure 2.37 Map of Mount Etna and its historical lava flows: **1**, post-AD1600 flows and scoria cones; **2**, AD1300–1600; **3**, AD1000–1300; **4**, **AD**476–1000; **5**, 122BC–AD476; **6**, pre-122BC. The acronyms correspond to the names reported in the Italian geological map (Branca et al., 2011). CC = Central Crater, VdB = Valle del Bove caldera.

since 1911 built up cinder cones that grew even higher than the summit cone in 1978, whereupon their efforts began to wane, and the baton was taken up at once by new, persistent eruptions that quickly constructed the Southeast Crater. The flanks of the volcano contain many valleys where lava flows have been channelled. The most notable of these is the Valle del Bove (Fig. 2.37; Vdb), a great horseshoe trough, 5km wide and 1000m deep, which opens on its eastern flanks as if it had been scooped out by a giant bulldozer.

Etna provides a relatively safe and fascinating laboratory with the happy knack of always producing something new within its range of activity. Few volcanoes have been so obliging, and its bibliography is now probably the largest of any European volcano. Even as early as about 430BC, the philosopher Empedocles is reputed to have fallen into the crater during an investigative mission – an exploit that luckily has not been emulated since. But serious study reaches back more than a century, and Etna has been under intensive care by Italian, French and British research teams, especially since it took on a new lease of life in 1971. It has also been subject to practically every surveillance technique known to volcanology.

Etna has displayed almost continual activity throughout historical times, and it has the longest record of eruptions in the world, in part due to historical records

The eruption of the Monti Silvestri in 1892

The eruption of the Monti Silvestri (Fig. 2.38) was typical of the medium size flank activity on Etna. It lasted from 9 July until 28 December 1892 and emitted some 150 million cubic metres of magma. The eruption began on a flank fissure at a height of between 1800m and 2000m, where continuous explosions quickly built up three large cinder cones that have since been visited by millions of tourists. In the words of Riccò & Arcidiacono (1902–1904):

'We are stupefied by the prodigious scene before us, although we are 2km from the eruption. In the upper part, three craters are in flames and the one in the centre is giving off a double jet of fire. The cinders and glowing bombs are being violently thrown straight upwards and reach a height of about 400m, while a broad, fiery cloud forms a splendid purple veil above this cannonade. Lower down, an immense flood of fire stretches as far as the eye can see [and] an infinity of channels, streams and cascades of lava are running in all directions across [it]. The lava on the surface, which has already cooled and solidified, is being dragged along by the liquid current, and the fragments bump into each other and make a noise like a pile of falling tiles. There is a continuous rain of cinders and small hot lapilli, and hot suffocating air gusts out from time to time. Added to this inconvenience is the intense heat of the lava that is flowing at the foot of the hill where we are standing – but we are so absorbed by the terrible beauty of the scene before us that we never even notice it. The lava front is advancing so slowly that it is hardly perceptible. We estimate that its speed is about a metre a minute, but it varies according to the slope and the form of the land. We can feel only pity for all these fine trees that are condemned to be burned alive. The leaves shrivel up and lose their colour before they burst into flames. The branches twist. The lava arrives and surrounds and suffocates the unfortunate plant in its fiery grip. The victim gives out a kind of strident cry. At length, the rapid hydrocarbon distillation creates the naked flames that bring its torment to an end. We feel both sad and horrified as the enormous, sinister black mass, with its fiery base, advances into the wood.'

Figure 2.38 Tourists walking around the main crater at Monti Silvestri (photo by pjt56 – Creative Commons).

kept by many generations of Sicilians. It has registered eruptions from its flanks or summit, ranging from the formation of cinder cones and lava flows to emissions of gas, steam or ash, which have taken place in periods stretching from days to many months, in the course of about 250 of the 500 years that have elapsed since reports became more reliable in the early sixteenth century. Prolonged phases of persistent activity have marked the summit and, although flank eruptions have been more episodic, they have nevertheless occurred in more than 80 of the past 400 years. Etna is thus a treasure house of basaltic lava flows and has the finest collection of flank fissures and cinder cones in Europe.

The higher zones of Etna are, of course, most affected by eruptions, and it has been calculated that half the area above 2000m, mainly comprising the Mongibello cone, would be covered by new lavas every 250 years, whereas it would take 525 years to cover half the area between 1000m and 2000m, and almost 2000 years to cover half the area below 1000m with new lavas. These eruptions represent an appreciable long-term threat to property. Moreover, since 1971, Etna has enjoyed one of its most active phases in historical times, and it shows no signs of stopping, with practically every year bringing some form of activity (see lava map in Figure 2.37). Etna is thus one of the most active volcanoes in the world. Its average effusion rate of 0.8m^3 per second is exceeded only by Kilauea in Hawaii; in fact, Etna erupts more gas: in terms of tonnes per day some 200,000 of steam, 70,000 of carbon dioxide and 4,500 of sulphur dioxide.

In spite of all this agitation, violent outbursts on Etna have been both rare and brief. Etna mostly causes damage, not death; and it has not killed a hundred people in all its recorded history – even though a hydrovolcanic explosion killed nine tourists near the rim of the Bocca Nuova on 12 September 1979. The greatest damage has been caused by the inexorable progress of lava flows emanating from flank fissures. More than 35 eruptions during the past four centuries have damaged crops or property. Part of Fornazzo, for example, was buried in 1971, nearby Mascali was destroyed in 1928, and, most notorious of all, part of Catania was overwhelmed when the famous eruption of 1669 discharged one of the largest lava flows of historical times. In 1971, Etna was even impertinent enough to destroy the old volcano observatory and the upper parts of the Funivia cable track on the southern flank. A new observatory has now been built just below the Pizzi Deneri (2847m) on its northern flanks.

The geological setting of Etna

Most of Etna is young. In the first half of its existence, activity was hesitant and sporadic. But during the second half – the past 250,000 years – Etna has expelled nine-tenths of its volume, with about one third forming the upper cone in the past 40,000 years or so. The volcano rises in the midst of a zone of fearsome tectonic complexity. Sicily has long been the hub of repeated collisions between the African and Eurasian plates and their adjacent microplates. Associated with these collisions are the opening of the Tyrrhenian Sea, the opening and sinking of the Ionian Sea, and the subduction, bending and possible compression of the Aeolian–Calabrian arc. Then, as the subduction apparently waned, the edge of the Eurasian plate was thrust over the edge of the African plate, causing the displacement of the continental rocks of Sicily. Consequently, Sicily was shattered and transected by deep major faults. The Tyrrhenian side of Sicily is being crushed while the Ionian side is being extended. This extension keeps the faults open near the Ionian Sea. Etna rises close to, but curiously not exactly upon, the intersection of three of these major faults: the Mt. Kumeta–Alcantara fault, trending from east to west, that now probably marks the Afro/European boundary; the Messina–Giardini fault, running north-northeast to south-southwest and now delimits the coast north of Etna; and the conjugate Aeolian–Maltese fault, running from Vulcano to Malta. These faults probably transect the crust down to the mantle and thus direct magma to the surface with relative ease. It has been suggested recently that Etna could have originated by suction of asthenospheric material from beneath the African plate along the major Aeolian fault system.

Etna also lies on a zone that has undergone uneven uplift, and thus its volume is less than it seems at first sight, because, although the lava base lies at sea level in the south, it occurs at an altitude of over 1000m on its northwestern flanks. Uplift of 6m has taken place during the past 2000 years; and 30cm has been recorded within the past 50 years. The east-facing fault scarps, forming a staircase on its lower Ionian flanks, are among the main results of this displacement. The continued foundering towards the Ionian Sea has removed much support from the accumulating eastern flanks of the volcano, whereas the other flanks are bolstered by the Sicilian landmass. Thus a broadly arcuate area, 15km wide, including both Etna and its basement, has been gradually slipping seawards for some 100,000 years. Within this structure, the volcano was depleted further when gigantic landslips and debris avalanches formed the Valle del Bove. This zone of eastward displacement is now associated with a distinct rift zone curving from the northeast, through the Southeast Crater, and then extending southwards. Here, fissures have developed and have been responsible for many of the lateral eruptions of recent times. The most important result in human terms has been that the highly populated eastern and southeastern flanks have by far the most eruptions, and indeed earthquakes, whereas the western flank, in particular, has been relatively immune.

Ancient Etna

Eruptions began more than 500,000 years ago with emissions of tholeiitic lavas, which are still exposed on the southern fringes of the volcano. These eruptions are often considered to belong to a pre-Etna phase of activity. They took place from many vents aligned along fissures stretching inland from a silty Sicilian bay between the Peloritani Range and the Iblean Plateau. They expelled hot and fluid tholeiitic basalts that rose quickly to the surface. On land, they cover the upper-most terrace alongside the River Simeto near Adrano. Similar basalts erupted below sea level and formed pillow lavas and breccias that now provide the plinth for the castle at Aci Castello on the Riviera dei Ciclopi. The eruptions were separated by long intervals of repose, and they remained so scattered and infrequent that no marked single volcano ever developed.

Imperceptibly, ancient Etna began to grow up when the scattered and sporadic eruptions became more concentrated beneath the present summit. Alkaline basalts erupted from the deep reservoir situated just below the base of the crust, but differentiation within the reservoir soon caused the eruption of trachybasalt in both flows and fragments. It seems that by about 140,000 years ago, the volcanic accumulations had covered much the same area as that of present day Etna. Indeed, in all, these initial eruptions encompassed more than half the total history of Etna, although they produced only a fraction of its volume. These lavas had piled up on slippery beds of clay that dipped down eastwards to the Ionian Sea. Unlike the other flanks of Etna, which were buttressed by the bulk of Sicily, the eastern flank had little support; as a result, it began to slip eastwards, forming a horseshoe depression, about 15km across, which was of course open to the east. Thus began the long series of eastward slides towards the Ionian Sea that has persisted until the present day.

Several eruptive centres grew up within the vast caldera-shape depression of ancient Etna. One remains in the eroded piton of Monte Calanna (1325m); another is represented by many dyke and lava flow accumulations at Trifoglietto, about 4km from the present summit of Etna. At Trifoglietto, two large volcanoes apparently formed in succession about 70,000 years ago. The presence of amphibole crystals means that their trachyandesitic lavas can be readily identified; and it seems that they erupted from a small shallow reservoir that developed above the much larger reservoir at depth. After the Trifoglietto centres had been active for several tens of thousands of years, the focus of eruptions once more shifted further west. At the same time, the unsupported eastern flanks of the volcanic pile began to slip down towards the Ionian Sea. Each time that the magma arose to the west, the eastern flanks of Trifoglietto were destabilized a little further. As the pressure upon the magma was reduced, violent explosions and debris avalanches ensued. Thus, the enormous depression of the Valle del Bove began to form in successive stages on the eastern flanks of Etna.

The Ellittico caldera and the origin of the Valle del Bove

After Trifoglietto had lapsed into extinction, many more eruptions took place to the west, although the exact sequence of events is far from being well established. For instance, the Vavalaci, Cuvigghiuni and Leone centres are still largely the subject of debate. However, one major centre can be identified: the Ellittico volcano. It is the direct predecessor of modern Etna and was also called Ancient Mongibello by Earth scientists. At its apogee, some 15,000 years ago, the Ellittico volcano approached a height of 3700m. At about the same time, a shallow reservoir developed probably less than 6km below the crest of the volcano. It erupted the *lava cicirara* – the 'chick pea' trachybasaltic lava – which is so-called because of its large white plagioclase crystals.

About 14,000 years ago came the catastrophic eruption that gave off Plinian columns and pyroclastic density currents that cascaded down the flanks of the volcano, mainly to the northeast, northwest and southwest, where they form thick pink deposits near Biancavilla. This eruption eventually led to the collapse of the summit and the formation of the Ellittico caldera, 4km long and 3km wide. This marked the start of a new growth phase in the rebuilding of Etna (Fig. 2.39).

One of the major episodes in the history of Etna was the formation of the Valle del Bove. Its steep scalloped

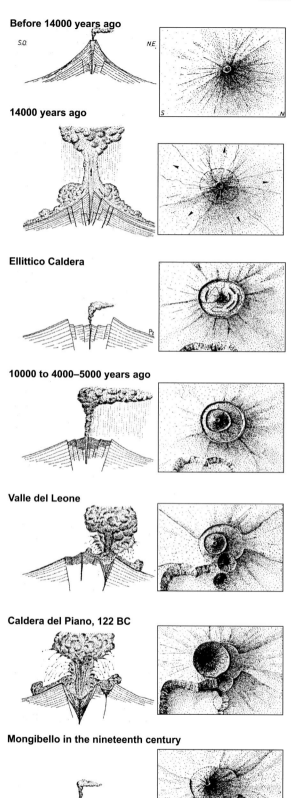

Before 14000 years ago

14000 years ago

Ellittico Caldera

10000 to 4000–5000 years ago

Valle del Leone

Caldera del Piano, 122 BC

Mongibello in the nineteenth century

Figure 2.39 The later stages of the growth of Etna.

walls rise 1000m above the floor to more than 2800m above sea level at its western end (see Fig. 2.40). Its floor slopes down towards the east and is coated by many recent lava flows, and the walls reveal the eruptive centres of ancient Etna. The Valle del Bove thus forms a huge trough that bites deep into the flanks of Etna and opens eastwards towards the Ionian Sea. It is 5km wide and more than 7km long, and perhaps as much as 10km³ of material have been removed from the trough and deposited on the lower flanks of the volcano and in the fan-shaped Chiancone conglomerate jutting out into the Ionian Sea.

The origin of the Valle del Bove is still debated. At present, it could be five or six coalescent landslides caused by largescale slope failure on the unsupported coastal flank of Etna, or a succession of debris avalanches generated by hydrovolcanic explosions. The landslides may have taken place in scallops – from the summit downwards – in rapid succession over a period of 2000 years, beginning between 5000 and 6000 years ago under wetter, warmer conditions. On the other hand, the Valle del Bove may have been formed by blasts and debris avalanches, which developed from the base upwards. They were apparently generated by hydrovolcanic eruptions that occurred at intervals over a much longer period, beginning about 50,000 years ago, and ending, for the moment, about 3000 or 4000 years ago.

Figure 2.40 Ancient Etna: the Valle del Bove (older than 9000 years **BP**) is a large horseshoe-shaped caldera filled with recent lava flows.

Mongibello

After a period of repose, eruptions resumed within the Ellittico caldera and gradually filled it to the brim, although its rim can still be detected as a slight shoulder at Punta Lucia on the upper western slopes, at Pizzi Deneri on the upper northern slopes of Etna, and near the Torre del Filosofo in the south. As the eruptions continued, they built up the large cone of flows and fragments that has been christened Mongibello. At first, these eruptions were silicic and explosive, and many pyroclastic density currents swept down the wooded slopes of the volcano. The carbon-14 analysis of the burnt wood indicates dates of eruptions at intervals of about a thousand years – 8460, 7100, 6100, 5000 and 4280 years ago. Although most of these eruptions came from the central cone of Mongibello, others sprang from the last phases in the growth of the Valle del Bove.

The last explosive episode of the series caused the formation of the Caldera del Piano when the upper reaches of Mongibello collapsed. This occurred just over 2000 years ago, probably in 122BC. Eruptions soon began to fill the Caldera del Piano; but Etna reverted to the style that had prevailed before the explosive interlude that had marked the growth of Mongibello. Basaltic eruptions returned and violent phases became relatively few. Lavas filled the caldera to the brim and formed the plain (piano) that gave it its unusual name. Then they built the present central cone. However, at the same time the eastern flanks of Etna continued to slip down towards the Ionian Sea along a whole

Empedocles and the Torre del Filosofo

The southern flanks of the volcano, rising from the city of Catania, are the most accessible sector of Etna. Most tourists take the cable car and the bus up these slopes (Fig. 2.41), and reach the foot of the central cone near the abandoned refuge of the Torre del Filosofo at an altitude of 2918m. This name comes from the ruins of a narrow tower, which, legend has it, used to be the observatory of the Sicilian Greek philosopher, Empedocles, who lived at Agrigento. He claimed to be the equal of the gods and, therefore, wished to give the impression that he had rejoined the immortals on his death. Thus, he is said to have thrown himself into the crater of Etna, so as to leave no trace of his remains on Earth. Unfortunately, one of his sandals was recovered and gave his game away. In fact, there is no other evidence that he ever came to Etna. Some assert that he died in Messina after breaking a leg in a road accident, others that he was strangled. Aristotle, perhaps jealous of such a glamorous demise, said that Empedocles had died a natural death in Greece at the ripe old age of 60. The tower itself was only 1m across and was much too narrow to have been anything more than a temporary refuge. It might, for instance, have been built for the visit of the Emperor Hadrian, who, of course, really was a god. Today the Torre del Filosofo is disappearing beneath the many lava flows emerging from the Southeast Crater (see Fig. 2.41B).

Figure 2.41 Tourist access along the southern flanks. **A)** The new SE Crater cone and its lava flows in June 2015. The old Central Cone is barely visible in the left background. Mountain hut and buses of Etna guides for scale. **B)** Last remnants of the Torre del Filosofo (photo **B** by Ji-Elle – Creative Commons).

series of rift faults, whose displacements are helped by Hyblean–Maltese tectonic movements. They now form clear fault scarps, which the Sicilians call timpa, the best-known of which are the Timpa d'Acireale, and the Timpa di Santa Tecla, further north.

The Central Cone

Sir William Hamilton and Alexandre Dumas figure among those who have written graphic accounts of their ascent of the Central Cone. But although the thrill still remains, the morphological details change with almost every passing year. At the start of the twentieth century, the Central Cone rose 300m above its base and had a deep crater some 500m across. Persistent eruptions completely filled this hollow and formed an undulating summit plateau. Thereupon, in 1945, its northeastern part again collapsed in a great abyss, the Voragine, which widened as its walls crumbled and gas, fumes, steam and occasional lavas welled up within it. Several smaller vents also erupted little cones of spatter and fragments on the southwestern part of the summit plateau, notably in 1955 and in 1964.

On the night of 9–10 June 1968, a new narrow abyss collapsed in the western part of the summit plateau. This new vent, called Bocca Nuova, widened eventually until it joined up with its predecessor. Both are the source of occasionally upwelling lavas, almost continuous emissions of fumes and steam, and of episodic, often violent, hydrovolcanic explosions, as well as of spectacular lava fountains, as in July 1998. In September and October 1999, strong Strombolian activity almost entirely filled the Bocca Nuova with fragments, so that it once again formed a rugged platform. From it, copious lava flows fanned out for 5km over the western flank of Etna and descended to 1700m.

The Northeast and Southeast Craters

Some persistently spectacular eruptions have come from two sub-terminal vents lying at the foot of the Central Cone, which have been blessed with the banal names of Northeast and Southeast Crater. The Northeast Crater started as a collapse pit on 27 May 1911 at 3100m at the base of the summit cone. After emitting gas and lava flows for decades, it began to construct a typical cinder cone in 1955. Between 1966 and 1971, rather viscous hawaiite lavas also accumulated in a fan 4km wide and 200m thick. The cone is now horseshoe-shaped because upwelling lavas removed a sector in October 1974 and again in January 1978. Nevertheless, by January 1978 the

cinder cone was 250m high and had become the highest point of Etna at 3345m, and even 3350m in 1981. Upon this achievement, activity waned remarkably, although there was a vigorous eruption in September 1986. However, this crater was again active in 1995 and 1996, and it gave off lava fountains on 27 March 1998. During 1999, the vent stayed open and produced continuous emissions of gas and frequent explosions of cinders.

The Southeast Crater had first exploded as a pit in 1971, but it burst into magmatic activity only on 29 April 1978. Its initial eruptions were punctuated by periods of repose, but persistent activity began in 1980. It has expelled lava flows and built up a cinder cone faster than its predecessors, especially since 1998, when it rose almost as high as its neighbour, the Central Cone.

Hydrovolcanic eruptions

The summit craters are prone to hydrovolcanic eruptions, which are characterized by sudden brief explosions that usually last from a few minutes to a few hours. They form plumes of ash and steam rising more than 10km into the air. They often disperse ash for 40km around, damage crops, disturb travel and sometimes cause the closure of Catania airport. This type of eruption is particularly dangerous when it occurs at the Voragine. For example, the eruption on 17 July 1960 threw glowing lava clots halfway down the northeastern flanks of the volcano. The eruption on 4 September 1999 not only showered bombs of up to 5m in diameter all over the Central Cone, but ejected cinders of up to10cm across as far as the Ionian Sea beyond Fornazzo and Giarre. Moreover, these explosions can also occur from the Bocca Nuova, the Northeast Crater (on 24 September 1986) and the Southeast Crater, from which several metres of bombs and cinders were expelled on 4 January 1990.

These violent outbursts are caused when magma intrudes laterally into a wet zone of the volcano and suddenly forms masses of steam. They also occur when damp materials collapse from the walls of the vent and form a thick blockage deep in the conduit, where temperatures may approach 1000°C. Pressure builds up for several days until a vigorous explosion shatters and expels the blockage, and often also widens the vent. These eruptions only expel older, reheated volcanic materials. In other instances, the damp blocking material may enter into contact with molten magma already standing relatively high in the vent. The resulting explosion comes more rapidly and more violently. These hydrovolcanic eruptions are more frequent on

Etna than on similar volcanoes, probably because of the permeability of the upper walls of the vents and the presence of a basement layer of impermeable clays.

Explosions can also occur when a tongue of lava flows over snow or a damp land surface. This is how the eruption of 1843 produced the largest number of deaths recorded on Etna. Near the summit of Etna, hydrovolcanic eruptions constitute a major element of danger, not only because they multiply the violence of its usually relatively mild activity, but also because they develop very rapidly. In addition to the nine people killed and the 23 injured on 12 September 1979, two people were killed at the Northeast Crater on 2 August 1929, and another two were killed at the Southeast Crater on 17 April 1987. These explosions are difficult to predict because the preliminary conditions for their development – the lateral injection of magma, water infiltration, or vent blockage – cannot often be witnessed.

Lateral eruptions and flows

The lateral eruptions of Etna are almost invariably those that cause damage to crops and property, and have sometimes destroyed villages. They are heralded by localized earthquakes as magma rises up vents branching at depth from the main conduits that generate fissures radiating from the summit zones of Etna. Most of the fissures concentrate together on the rift zone, forming a band 25km long and 2–3km wide. Most of the lateral vents occur between 1500m and 3000m, but some of the most voluminous and dangerous eruptions, such as that in 1669, have occurred below 1000m. Curiously, lateral eruptions often start in November, March or May, possibly in relation to high rainfall or snowmelt. Since 1600, there has been an average of a dozen lateral eruptions per century, but their frequency has increased fivefold since 1971. They emit fluid hawaiite lavas that form long flows, but also construct many cinder cones and some spatter cones, because they have a relatively high gas content considering their composition. Almost two-thirds of the lateral eruptions have built cinder cones, varying in height from 10m to over 200m, which tend to form on the upper reaches of the fissures, whereas spatter cones, hornitos or quiet lava effusions characterize their lower reaches.

Half the lateral eruptions continue for less than 25 days, but they can last from as little as a few hours, as on 29 April 1908 in the Valle del Bove, to a few days, as when the Monte Leone was formed from 22 to 24 March 1883, to a few weeks, such as between 2 and 20 October 1928, when Mascali was destroyed, or to several months, as from 14 December 1991 to 31 March 1993 in the Valle del Bove. The eruption lasting from 1614 to 1624 was almost as exceptional in duration as the eruption of 1669 was in intensity.

The effusion rate of lava has varied with time. Output was apparently low from 1500 to 1610 but unusually high from 1610 until 1669. However, the great eruption of 1669 seems to have impoverished the system for a while, because output remained low for almost a century thereafter until 1763. Output was moderate from 1763 to 1971, but since then eruptive activity has never been so brisk. The lateral eruptions with slow effusions are also the most prolonged. They may start with an explosive phase that forms a cone, but soon give way to fluid lava effusions that build up a thick cover of thin flows close to the vent, chiefly because they can often last for more than a year, as in 1991–3, or even more, as in 1614–1624, when 1km^3 was erupted, or in 1651–4, when 0.5km^3 was erupted.

The total length attained by lava flows on Etna depends critically on both the effusion rate and the volume of erupted lava. Thus, the flows in 1669 had effusion rates of 50–100m^3 per second and travelled more than 17km in less than a month, whereas the flows in 1991–3 had average effusion rates of 6m^3 per second and did not exceed 8km in length, although the eruption lasted a long time, as did those in 1614–24. This has direct implications for volcanic risk, because careful calculation of the effusion rate could indicate the probable maximum length of the flow and, therefore, the possible threat to the areas down slope. Although such measurements cannot always be taken, simply watching the advancing lava fronts during the early stages of an eruption would offer a useful guide to what might occur.

Most flank lava flows develop aa surfaces, but pahoehoe surfaces have been formed on flows produced by more prolonged eruptions with lower rates of effusion. The lavas of 1614–24 are composed of piles of thin pahoehoe flows featuring tumuli, pressure ridges and lava tubes, as well as the well-known large hornitos forming the Due Pizzi. Many flows have pahoehoe surfaces where they first emerge, but develop aa surfaces further down slope. Effusion rates are high enough to maintain the forward impetus of the snout while the sides congeal rapidly and restrain lateral flow. Thus, most of the lava flows retain their tongue-like outlines, even where they could easily spread sideways over gentle slopes.

Lateral eruptions are most common and most voluminous in the southeastern and southern sectors, and least common in the northern, and especially the western, sectors. Half the lateral eruptions of Etna occur on the rift zone that curves through the summit area, and most of the remainder occur in the Valle del Bove. The area most vulnerable to eruptions in the south and southeast is where the population is most concentrated, on the richest agricultural land. Unfortunately, the eruptions also tend to occur lower down in this sector and to produce the greatest volumes of the most fluid lava, discharged at the fastest rates. They reached their climax during historical times with the destruction of part of the city of Catania in 1669 (Fig. 2.42).

Figure 2.42 Contemporary picture in the Catania Cathedral of the 1669 eruption on Etna.

The eruption of 1669

The greatest eruption on Etna in modern times began on 11 March 1669 and ended 122 days later on 11 July. It was heralded from 25 February 1669 by earthquakes that severely damaged Nicolosi, a village 700m high on the southern slopes. A gaping, glowing 2m wide fissure opened on 11 March. It stretched 12km from an unusually low altitude of 850m, just north of Nicolosi, to 2800m at the foot of the summit cone. During the night, the vent that was to form the cone of the Monti Rossi burst into activity as the fissure spewed out lava that overwhelmed Malpasso. The hot, fluid hawaiites surged forth at an exceptionally fast rate that approached 100m3 per second, almost ten times the average discharge of lavas on Etna. Within a few days, village after village was threatened and swamped.

On 16 March, the lavas reached San Giovanni di Galermo, 6km from the Monti Rossi. On 25 March, there was a violent hydrovolcanic explosion at the summit crater, and the summit cone collapsed in a shudder of tremors. Meanwhile, cinders and lapilli were quickly building the twin cones of the Monti Rossi and, at the same time, the lava flow divided into three arms, with its main trunk directed towards Catania. Nine villages had been destroyed by the time the lava flow reached Catania, 17km from the vent, on 12 April 1669. During the next three days the lavas wrapped around the city

walls, rose above one stretch and pushed it down. The flow reached the sea south of Catania on 23 April, and by 30 April it had started to invade the western part of the city. On 6 May, with courage and scientific acumen, Diego Pappalardo and at least 50 Catanians seem to have succeeded in diverting the flow by opening its solidified sidewalls so that the molten lava in the core would turn sideways and stop its forward surge. Unfortunately, this newly created branch advanced towards Paternò, whose alarmed inhabitants drove off the Catanians before they could cause them any more trouble.

This first known attempt at volcanic hazard control was thus thwarted, because the new breach solidified and the lava flow resumed its original course. No lives were lost in this eruption, because, after all, the lavas had travelled an average of only 500m a day; but they had advanced inexorably, sustained by the rapid supply of fluid magma. It is very doubtful whether Pappalardo and his men could have saved west Catania, unless they had managed to divert the whole flow onto Paternò. When the eruption ceased on 11 July 1669, lavas between 12m and 15m thick had covered an area of nearly 40km2, and the twin Monti Rossi had reached a height of 250m – among the highest cinder cones on the flanks of Etna. The lava and fragments had a combined volume of about 600–800 million cubic metres.

Historical eruptions on Etna

Reports of eruptions on Etna go back more than 2500 years. Eruptions on Etna were perhaps first described obliquely in the Polyphemus story of Homer (*Odyssey*, X), which was written about 750BC. However, contrary to an oft-held opinion, Hesiod makes no reference to Sicily in his

Theogony. Pindar's *Pythian Odes* (I: 19–28) and then Aeschylus' *Prometheus Bound* (365–74BC) are the oldest works in which Etna is specifically mentioned. They describe the eruption that took place at the same time as the Battle of Plataea in August 479BC, as was recorded in the Greek text known as the Arundel Tables. Another eruption, in the spring of 425BC, was recorded by Thucydides (*Peloponnesian War*, III: 116). Diodorus Siculus (*The library of history*, XIV.59.3) described how, in 396BC, a lava flow reached the sea and prevented the Carthaginian army from marching along the Ionian coast. The great Plinian eruption of 122BC, which rained ash on Catania and probably formed the Caldera del Piano, was described by Lucretius (*On the nature of things*, VI: 639–46), St Augustine (*The city of God*, III.31) and Orosius (*Against the pagans*, V: 13). The large eruption of 44BC features prominently in Virgil's *Aeneid* (III: 571–82), and its cloud of ash probably darkened the Roman sky when Julius Caesar was assassinated. Ash erupted in AD38 frightened Caligula, even though he was viewing Etna from the safety of Messina (Suetonius, *Caligula* 5.1). A lava flow that erupted in AD252 was apparently arrested at the gates of Catania, when the veil of the newly martyred St Agatha was waved at it. In 1169, a widespread earthquake caused 15,000 deaths in Catania and the collapse of the eastern flank of the summit cone, but it is doubtful if an eruption ensued. In 1329, the Monte Rosso was built at an altitude of only 550m on the southeastern slopes of Etna, and its flows threatened Acireale on the Ionian coast. The eruption in 1408 destroyed part of the village of Pedara and another perhaps happened in 1444, but no major activity then seems to have occurred for about a century.

A large fissure eruption on the southern flanks in 1536 was followed in 1537 by an effusion of lava that initiated the collapse of the summit cone. From 1607 to 1610, repeated eruptions occurred on both the northwestern and the southwestern flanks. The ten-year period of repeated eruptions between 1614 and 1624 marked the longest episode of lava flow emissions during historical times. The lavas covered 21km², and formed many lava-tube systems, the notable hornitos called the Due Pizzi, as well as megatumuli such as the Monti Collabasso, in a lava field composed largely of pahoehoe surfaces. Prolonged flank eruptions also occurred intermittently between 1634 and 1638, and again between 1651 and 1653. This seventeenth century activity was merely preparing the terrain for the eruption of 1669, which was the largest and most famous eruption recorded during historical times on Etna.

After the eruption of 1669, the flanks of Etna saw very little activity for almost a century, although weak emissions of gas and occasional lavas occurred intermittently from the summit zone. The chief event on Etna during this period was the great tectonic earthquakes of 9 and 11 January 1693, which severely damaged eastern Sicily and caused over 60,000 deaths, including 16,000 in Catania. The present summit cone was perhaps initiated in 1723, whereas a flank eruption in 1763 formed the Montagnola, which is still a prominent landmark on the upper southern flanks of Etna. It was soon followed in 1764–5 by the prolonged emission of the vast pahoehoe lavas coating its upper northwestern flanks. In July 1787, the crater on the Central Cone gave a spectacular display of lava fountaining rising up to 3km high, and an explosion at the beginning of 1792 created the Cisternazza pit near the western brink of the Valle del Bove. Intermittent lateral eruptions and frequent persistent activity at the summit marked the first half of the nineteenth century. Two episodes in 1832 and 1843 affected the western flanks and threatened Bronte. During the eruption of 17 November 1843, the snout of the lava flow invaded a wet area near Bronte, and the resulting explosion killed 59 unwary observers.

The second half of the nineteenth century saw more vigorous, rather more frequent, lateral eruptions. After those of 1852 in the Valle del Bove and 1865 on the northeastern slope, an unusually vigorous eruption to the north on 26 May 1879 created Monte Umberto–Margherita and a lava flow 9km long. On the southern slopes, Monte Leone formed at 1100m during a three day eruption in March 1883, Monte Gemmellaro grew up in about a week in late May 1886, whereas the Monti Silvestri accumulated over five months in 1892, and both these eruptions were accompanied by extensive lava effusions. However, Etna was calm at the time of the Messina earthquake (28 December 1908), although a small eruption had occurred on 29 April in the same year.

In March 1910, very fluid lavas burst out at about 2000m on the southern flanks of Etna and quickly reached the area around Nicolosi and Belpasso. On 27 May 1911, a pit collapse formed the Northeast Crater, which was to become the major active focus on Etna for the next 60 years, although it also experienced hydro-volcanic explosions. Meanwhile, lateral fissure eruptions took place on the northeast slopes in September 1911 and in June 1923. On 2 November 1928, a fissure opened on the northeastern flank and descended to the unusually low altitude of 1200m within three days. Torrents of lava surged forth, reached and overwhelmed Mascali in less than 36 hours, and stopped near the coast on 20 November. During the next 15 years, effusions gradually and intermittently built up the Central Crater, but the Voragine reappeared in its northeastern parts in October 1945. In 1955 the Northeast Crater began to build up what was to become a considerable cinder cone surrounded by a thick lava apron during the next 20 years. A week of eruptions alongside the Voragine in May 1964 formed a small cone and gave off lavas that escaped by a notch from the rim of the summit cone and flowed down its flanks. On 10 June 1968, the Bocca Nuova appeared on the western side of the old crater of the Central Cone as a hole, about 4m wide, that exhaled hot gases. Its walls collapsed again in the winter of 1970 to form a chasm 100m across, which widened intermittently until it joined up with the Voragine.

In 1971, Etna began its most active period of modern times. The lateral eruptions beginning on 5 April 1971 occurred from unusually high fissures at the southern base of the summit cone, and their lavas buried the Volcano Observatory and the upper parts of the cable-car line. The flows often ran like rivers down the volcanoes flanks (Fig. 2.43). Lavas from other fissures on the outer crest of the Valle del Bove also travelled 7km down to the outskirts of Fornazzo. From 30 January to 29 March 1974, what was, for modern Etna, an unusually explosive eruption formed the two Monti De Fiore cinder cones on its western flanks. It was followed by vigorous activity from the Northeast Crater, which

lasted until just after its cone had become the highest point of Etna in January 1978. On 29 April 1978 began the eruptions of the Southeast Crater, which had built a cinder cone by late summer and, on 3 August 1979, another fissure sent more flows towards Fornazzo. Part of the walls of the Bocca Nuova collapsed on 2 September 1979 and the blockage was removed by the lethal hydrovolcanic explosion on 12 September.

The lateral fissure eruption on 17 March 1981 occurred in an area of the northern slopes that had not been active for at least 400 years. Effusion was very rapid and fluid lava flows soon covered 6km², entered the River Alcantara and threatened, but spared, Randazzo, before they halted on 23 March. On 27 March 1983, another lateral fissure opened on the southern flanks of Etna. These thick lavas also covered 6km², but with a much slower rate of effusion, for the eruption lasted until 6 August 1983. For the first time since 1669, serious attempts were made to divert these flows. Another prolonged low-discharge eruption from the Southeast Crater lasted from 28 April to 16 October 1984. It was followed by eruptions from both the Southeast Crater and the adjacent southern flanks, which lasted from 8 March 1985 until 13 July 1985, with a smaller eastern flank eruption at Christmas 1985. Yet another eruption started in the same area on 30 October 1986.

Eruptions resumed on 29 August 1989 when ash was ejected as far as 19km from the summit vent. Two weeks of spectacular eruptions from the Southeast Crater began on 11 September 1989, which included lava fountains rising 600m. Two fissures opened in the east: one on 25 September produced a 6km-long lava flow towards Milo; the other formed an open rift, reaching 1m wide, from which no lava emerged. More cinders and lava fountains

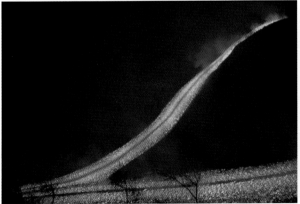

Figure 2.43 The 1971 flank eruption on the NE slopes. The main lava flow is descending with an initial speed of 20km/h at vent.

were expelled from the Southeast Crater in January and February 1990, before a brief respite ensued until the end of 1991.

After the eruption of 1991–3, Etna was relatively quiet for two years. There were several phases of lava fountaining from the Northeast Crater in November and December 1995. Then, a phase of explosions and effusions between 21 July and 19 August 1996 sent lava flows cascading into the nearby Voragine in the Central Crater. All four summit craters were active throughout 1997, during which the Southeast Crater filled with lava flows. Activity increased during the winter of 1997–8, when the Southeast crater exploded vigorously and the Bocca Nuova glowed continuously, emitting a plume of sulphur dioxide. On 27 March 1998, seismic tremors accompanied explosions from the Northeast Crater for two hours, but the Southeast Crater soon resumed its more moderate behaviour, which lasted for several months. In June 1998, the Voragine started an episode of Strombolian activity, and it was soon joined by the adjacent Bocca Nuova (Fig. 2.44). These vigorous eruptions came from cones that had formed on the floors of the chasms, and they often expelled bombs and large clots of lava, notably between 12 and 13 July. On the evening of 13 July, the easternmost cone in the Voragine emitted three lava flows that carpeted the floor of the chasm. On 22 July, lava fountaining and vigorous explosions from the Voragine sent a column of ash and gas 10km into the air, scattered ash all over the eastern flanks of Etna, and closed Catania airport. The Voragine then ejected two lava flows, one reaching the road between the north and south flanks of Etna, and the other cascading into the Bocca Nuova. On 15 September, the Southeast Crater exploded much ash, followed on 16–19 September by

Strombolian eruptions, ending with lava fountains and small flows. They were repeated at intervals of five or six days, and separated by periods of almost total calm, until the end of January 1999. A vigorous eruption from the Southeast Crater on 4 February was followed by the opening of a fissure that almost reached the base of the cone and by the emission of quantities of lava until November 1999. A strong lava fountain shot from the Voragine on 4 September and large eruptions from the Bocca Nuova in September and October 1999 entirely changed its morphology. Then from 26 January 2000, the Southeast Crater started some of the most violent and frequent eruptions in its recent history (Fig. 2.45), which continued into 2001, producing its first flank eruptions since 1991–1993 in July–August 2001.

Eruptions over 2002–2003 are notable in that they damaged the tourist stations at Piano Provenzana and Rifugio Sapienza on the volcano's flanks, and they were imaged by the ISS station in space (Fig. 2.46). Lava flows

Figure 2.45 Eruption column from 16 April 2000 from the Southeast Crater. The Northeast Crater can also be seen with white smoke (right).

Figure 2.44 View into the Voragine with some strombolian activity, 1998.

Figure 2.46 Photo of Etna erupting on 30 October 2002, taken from the International Space Station (NASA).

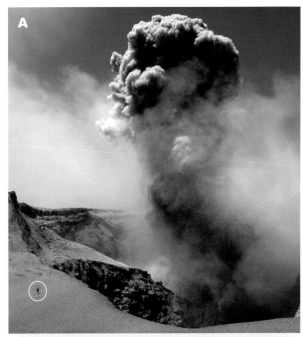

Figure 2.47 Recent spectacular activity on Etna. **A)** Ash explosion in 2011 from the Central Crater as seen from the NE cone (circled man for scale). **B)** 2014 Strombolian activity at night with the moon in the background (Shutterstock/silroby80). **C)** The spectacular paroxysm eruption, December 2015 (Shutterstock).

in 2004–2005 were followed by more intense eruptions between 2006 and 2007. The longest flank eruption to have occurred since the 1991–3 episode lasted some 417 days between 13 May 2008 and 6 July 2009.

The most recent activity has been watched even more readily through the digital age, and in many cases, even the most modest of eruptions can get snapped and placed into the public area, sometimes overstating its relevance. That said, through most of 2011 and into 2012 Etna was particularly active (e.g. Fig. 2.47A), with many instances of the airport at Catania being closed due to the threat of ash, and with eruptions also threatening the refuges (e.g. the Sapienza Refuge). A new flank eruption started in 2014 with lava flows and strombolian activity, with examples that even challenge Stromboli as the type locality for these (Fig. 2.47B). Another large eruption started on 3 December 2015, with a massive ash cloud visible from a great distance and some spectacular eruptions around the vent areas (see Fig. 2.47C), and further eruptions continued on into early 2016. By the point of writing (May 2016) there seemed no end to the activity at Europe's most famous volcano, and Etna promises to be in the limelight for many centuries to come.

To some extent, by stressing the unusual at the expense of the more mundane elements, or of more recent, well-documented activity, these summaries of eruptions give an unbalanced picture of Etna's behaviour. For example, throughout the period from 1971, persistent activity – ranging from gas jets, ash and cinder explosions, to lava that formed lakes, flows and fountains – was often registered from the terminal and subterminal vents, and rapid changes have taken place on the volcano. But lateral eruptions are emphasized

The eruptions of 1983, 1991–3 and 2001 and the diversion of lava flows

The eruption on 28 March 1983 occurred, in the midst of tourist installations, on a fissure between 2200m and 2400m on the southern flanks of the volcano. The effusion rate was only about 10m³ per second and, in spite of the claims of certain journalists, no settlements were in danger. Indeed, the flow took over a month to extend beyond Monte San Leo, which the flow that erupted in 1910 had reached in less than a day. But this was the first eruption in this sector for 73 years, and there was strong political pressure to halt this terrifying scourge.

The diversion started at the beginning of May. The aim was to use bulldozers to thin the solidified walls of the flow and then explode a hole in them in order to divert it at source, and thereby expose more molten lava to the open air and make it solidify more rapidly. Enormous earth barriers were also built to stop the lava from invading land alongside it. When the explosion occurred on 14 May, molten lava flowed from the new hole, but stopped after two days when its surface solidified. A mere 50,000m³ had been diverted out of the total of 79 million cubic metres that the eruption eventually gave out. On the other hand, the barriers did play a useful role in preventing the new flows from spreading sideways. The lava continued to flow until 6 August, without further human interference, but it never extended beyond the position that the front had already reached the previous April.

On 14 December 1991, lava fountains burst forth from a fissure at the foot of the Southeast Crater. During the following night, activity extended down slope and several vents opened at about 2200m in the Valle del Bove. Lava soon spread down and invaded the rich orchards of the Valle di Calanna. In January 1992, the Italian army built a barrier of ash and cinders, 21m high, across the exit of the Valle di Calanna to try and stop the lavas from reaching the lower valley leading directly to the town of Zafferana. Just before they reached the barrier, the lavas halted for three months. Then suddenly, at the end of March, they resumed their progress and swept the barrier away on 8 April 1992. The citizens of Zafferana now feared the worst and, by the end of May, the snout was within 1km of the town. But it was a false alarm: the snout of the degassed flow was now 8km from its source, and it proved too viscous to reach the town, although molten lavas broke out near the destroyed barrier several times during the following month. Meanwhile, during April and May, concrete blocks were dropped onto the flow near its source, and explosives were used five times to stem the supply to the advancing snout. This tactic seemed to work eventually, although the most successful attempt took place on 29 May, which coincided with a marked reduction in the rate of lava discharge on 1 June. Thereafter, the flows were confined to the Valle del Bove and the eruption weakened considerably, although emissions lingered on until March 1993.

In 2001 a new eruption at Etna, from a fracture system running down from the southeast summit cone, sent lava flows down, covering roads and isolating some country houses. Again intervention plans were put in place to divert lava from key areas such as tourist facilities of the Sapienza and Silvestri. Thirteen earthen barriers were constructed, with a maximum length of 370m, height of 10–12m, base width of 15m and volume of 25,000m³ (see Fig. 2.48). These were made of loose material excavated from the side of the mountain at an angle to the flow, and were able to divert lava movements successfully. This type of intervention is costly and may not always work, but with increasing experience and understanding of the volcano's slopes and behaviour, they can help to stem the effects of Etna on Man.

Figure 2.48 2001 activity and the need for barriers. **A)** Fountain and flow in 2001 (inset with commemorative stamp). **B)** Construction of an earth barrier in 2001, in order to protect the cable-car and several buildings near 1900m elevation. (stamp image sourced from UPU – Universal Postal Union).

because of their potential danger to the people and their property. For instance, there were 14 flank eruptions between 1971 and 1996, or five times the average for the previous three centuries, and the average output of lava was almost three times that of the previous century. The apparently unprecedented scale, variety and frequency of activity have been matched by discharge rates higher than average and, since 1981, by longer eruptions. This vigour has, in itself, made Etna the best European volcanological laboratory during the past few decades. However, in spite of the scientific interest in all the present spectacular activity on Etna, the real danger to the area around Catania lies in the threat of an earthquake that could kill tens of thousands of people. Hence, Italian Earth scientists have concentrated their research on developing both seismological and volcanological observatories throughout the region.

Pantelleria

The Italian island of Pantelleria rises in the straits between Sicily and Tunisia. In spite of its almost African climate, it is green and attractive, for its weathered volcanic soils flourish with capers and vineyards. These and the fishing boats provide the main livelihood of its few thousand inhabitants. Pantelleria is about 13km long and 8km broad, and it reaches a height of 836m at Montagna Grande (Fig. 2.49). It is wholly volcanic in origin and began to form over 300,000 years ago in a submerged rift, running from northwest to southeast. About half the total accumulation now rises above sea level. Many different vents have taken part in the growth of Pantelleria, and its complicated history is reflected in its mountainous and varied relief. The emerged parts of the island are largely composed of unusual hyperalkaline rhyolites known as pantellerites. They form layers of ash, pumice, welded and unwelded ignimbrites, cinder cones, domes, flows, and even little shields of lava. In fact, although these pantellerites had a silica content of about 70 per cent, they were emitted at quite high temperatures of between 900°C and 1020°C. Thus, they were less viscous than rhyolites and they behaved more like andesites, and the island has also erupted trachytes and alkaline basalts.

It seems that, when Pantelleria first emerged from the sea, it formed one or more shields, which outcrop in part on both the southeast and southwest coasts, where a lava flow at Scauri has been dated to about 324,000 years ago. Similar shields apparently formed the Cala

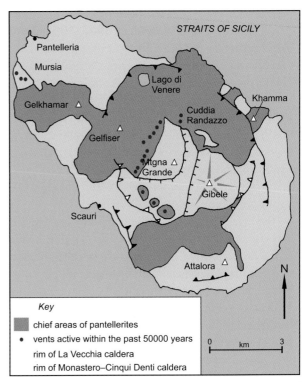

Figure 2.49 Structure and main volcanic features of Pantelleria.

dell'Altura and the Cuddia Khamma. The shields also contained thick layers of welded tuffs (ignimbrites). About 114,000 years ago, a powerful eruption led to the collapse of the broad La Vecchia caldera, which covers an area of some 40km². Further eruptions, notably from the Cuddia Attalora and the Cuddioli di Dietro Isola, filled the hollow with pumice, voluminous lava flows, and at least three layers of welded tuffs.

The second major eruption on Pantelleria took place over 50,000 years ago, when the expulsion of the green tuffs covered the whole island in a layer that reaches up to 35m thick and effectively hides much of the earlier history of the island from geological scrutiny. The eruption, from a stratified reservoir, first expelled pantelleritic pumice, followed by thick ignimbrites, and lastly trachytic pyroclastic density currents. The collapse of the Cinque Denti, or Monastero caldera, brought the episode to an end. It lies within the La Vecchia Caldera, but nevertheless has a diameter of 6km.

The eruption of the green tuffs started the first of six cycles of activity that have produced 80 eruptions during the past 50,000 years. Each began with rather moderate explosions of rhyolitic fragments and ended with effusions of trachytic lavas from a

Figure 2.50 Panoramic view of the thermal waters in the Lago di Venere in Pantelleria, Sicily (Shutterstock).

shallow reservoir of stratified magma. The dispersal of the fragments was often restricted to the formation of cones of grey or yellowish pumice, ranging from 25m to 100m in height, the craters of which were then often filled by lava emissions. The second cycle has been dated to between about 32,000 and 35,000 years ago, and it produced mainly trachytes, with some pantellerites, and built the cone of Monte Gibele. The third cycle occurred about 22,000 years ago. Eruptions of fragments, domes and flows took place chiefly outside the caldera, ejecting, for example, the pantellerites forming Monte Gelkhamar. The fourth cycle occurred about 20,000–15,000 years ago with eruptions around the edge of the caldera. Monte Gelfiser shows their typical development from initial explosions to concluding lava effusions. The fifth cycle, between 14,000 and 10,000 years ago, saw eruptions that were limited to the north of the island and came from fissures trending either north–south or northwest–southeast.

The sixth and latest cycle of activity began about 8000 years ago and ended about 3000 years ago. Its products outcrop in the centre of the island, especially around the Montagna Grande, and include pyroclastic density current deposits, cones, domes and lava flows. These eruptions built the Cuddia Randazzo pumice cone, whereas the pumices of the Sibà Montagna and the Roncone erupted at different times from vents aligned on the same fissure. The basaltic Mursia cones and flows on the northwest coast also belong to this sixth cycle. Although submarine activity occurred in 1891 about 4km to the northwest of the island, no historical eruptions have apparently been recorded on Pantelleria, but active fumaroles, hot springs and thermal waters show that the island is not extinct (Fig. 2.50).

Graham Bank and Foerstner Bank – Ferdinandea and the birth and death of a volcano

In the Mediterranean channel of the Straits of Sicily between Sicily and Tunisia, submarine banks have displayed volcanic activity during historical times, mainly in relation to the widening of the rift that gave rise to the Italian island of Pantelleria. Eruptions, for example, were recorded with varying degrees of reliability in 1632, 1707, 1831, 1845, 1846, 1863 and 1891, and possibly also in 1801 and 1832. The eruption of Graham Bank in 1831 was the largest and best documented of these events. It began midway between Sciacca, in Sicily, and Pantelleria, at a depth of about 250m, and it now forms a circular hump rising 100m from the sea floor. Its summit platform carries three tuff cones, one of which rises to within 8m of sea level.

Graham Bank was first noticed in 1632, and it is possible that an eruption in 1707 was bulky enough to form a temporary island. But its largest recorded eruption announced its debut when tremors were felt in Sicily and a ship reported 'unusual agitation' when it sailed over the spot on 28 June 1831. In early July, there was a foetid odour at Sciacca, and fishermen declared that the sea was 'boiling and full of dead fish' about 50km offshore. During the following days, columns of water spurted 25m skywards at 15–30 minute intervals. On 12 July, pale trachytic pumice washed up on the shore at Sciacca, and an island 3m high emerged on 16 July. By early August, vigorous explosions had built up a cone of hawaiitic fragments that rose 65m above the waves and was 3700m in circumference.

The news soon attracted the attention of the commander of the British Mediterranean Fleet, Vice Admiral Sir Henry Hotham, who sent out a boat to take charge of this strategically placed island. On 2

August 1831, Captain Humphrey Senhouse landed on the islet, raised the Union Jack, claimed it for King William IV, and named it Graham Island, in honour of the First Lord of the Admiralty. This enterprising seizure excited international alarm. King Ferdinand of Naples sent the *Etna*, whose captain also claimed the island and naturally called it Ferdinandea. Some time later, the French geologist, Constant Prévost, also landed and called it Julia, because it had first been seen in July.

But nature avoided a crisis. The eruption did not expel enough tuffs to prevent sea water from invading the vent, and the Surtseyan explosions continued unabated until the magma stopped rising. Therefore, no lava flows could be emitted that might have protected the fragile tuffs from wave attack. Fortunately for international peace, the eruption stopped and the winter Mediterranean storms destroyed the island on 28 December. The eroded crest of the tuff cone now lies some 8m below sea level. A brief additional eruption was also reported from the Graham Bank on 12 August 1863, but no cone was built above sea level. However, new international trouble may be brewing. In the early weeks of 2000, the waters above the shoal began bubbling and spouting up, and there were frequent tremors in the area. It is therefore possible that a further eruption could form another island.

Other submarine eruptions have also been reported from this area of the Straits of Sicily. In 1890, the people of the port of Pantelleria noticed an increase in fumarole activity, several tremors, and an uplift of 80cm on the northeast coast of the island. They felt stronger earthquakes from 14 October 1891, which, however, decreased markedly on the morning of 17 October. That very morning, a submarine eruption began on what was later called the Foerstner Bank, about 4km west of the little port of Pantelleria. Amid repeated rumblings, columns of smoke and steam rose into the air for about 1km on a line trending from northeast to southwest. Close observers saw black cinders and large glowing bombs being thrown skywards in the steam. It is possible that an island about 1km long was built about 10m above sea level for a day or two, but it may also have been no more than a mass of floating pumice. In any case, the eruption stopped on 25 October before a cone could be built above sea level.

Chapter 3

Greece

Introduction

Featuring one of Europe's, possibly the world's, most impressive sea-flooded calderas (Santoríni), the Hellenic volcanic arc is the best-developed island arc in Europe. It swings for 500km across the southern Aegean Sea, through the Cyclades, from the Isthmus of Corinth to Bodrum in Turkey (Figs 3.1; 3.2). It comprises the smaller volcanic centres of Aegina, Póros and Méthana, and the larger accumulations of Mílos, Antíparos, Santoríni, Nísyros, Yalí and southwestern Kós. But only Méthana, Nísyros and Santoríni have recorded eruptions of more than fumaroles during historical times. However, the Minoan eruption of Santoríni in the Bronze Age was not only one of the greatest volcanic events in Europe during the past 10,000 years but it also made an impressive contribution to one of its most spectacular landscapes.

The volcanic arc was generated by subduction of an oceanic part of the African plate beneath the small Aegean plate on the southern edge of the Eurasian plate. The volcanic arc rises some 220km north of the Hellenic

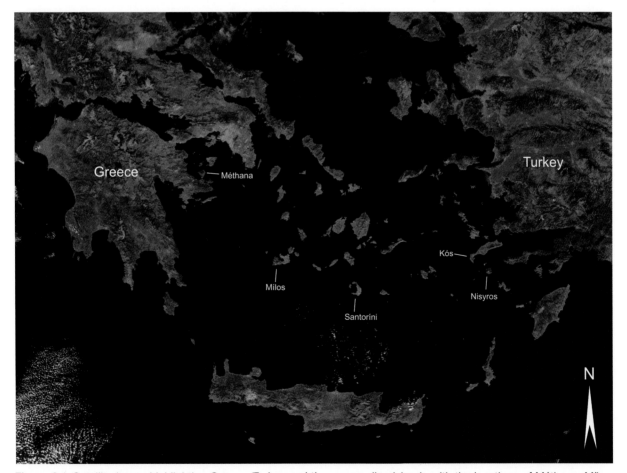

Figure 3.1 Satellite image highlighting Greece, Turkey and the surrounding islands with the locations of Méthana, Mílos, Santoríni, Nísyros and Kós indicated (image – Jacques Descloitres, MODIS Rapid Response Team, NASA/GSFC).

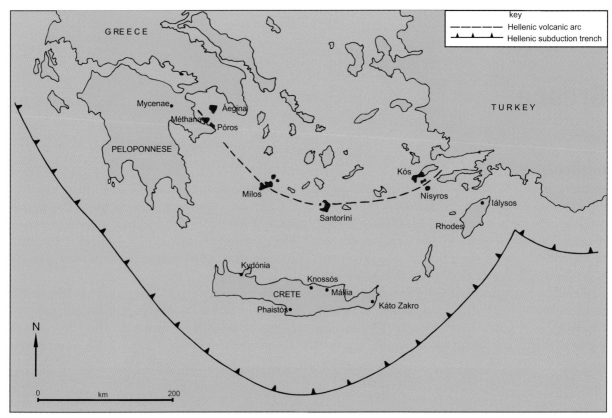

Figure 3.2 Generalized volcano–tectonic map of Greece, Turkey and surrounding islands. The subduction zone offshore to the south results in the Hellenic volcanic arc running from Greece, through the Cyclades, to Turkey.

trench, where the upper surface of subducted slab has reached a depth of 130–150km. The oldest emerged rocks, on Mílos, are 3.5 million years old. During historical times, the Kaméni Islands in the Santoríni caldera have marked the most voluminous output from any centre in Greece. The Hellenic volcanoes have generally erupted calc-alkaline, andesitic, dacitic and rhyolitic rocks, and, among them, andesites and dacites each constitute about a third of the volcanic products ejected. Only Santoríni has preserved significant quantities of basalt from some of its eruptive episodes.

In the west, the eruptions have been smaller and the lavas have usually been viscous. Thus, domes and thick flows of basaltic andesite and dacite, erupted about 1–2 million years ago, make up the hills in the central and southern two-thirds of Aegina and the volcanic southern peninsula of Póros; and viscous flows also accumulated as low shields in western Mílos. However, greater magmatic differentiation, in relatively shallow reservoirs, in the centre and east of the arc occasionally generated more explosive eruptions of rhyodacitic and rhyolitic ash and pumice, and the collapse of calderas notably off southwestern Kós, in Nísyros and, of course, in

Santoríni. Since the latest eruption in the arc took place in the Kaméni Islands in 1950, activity has been limited to mild fumaroles and hot springs that are manifest in all the volcanic centres, especially in Méthana, the Kaméni Islands and Nísyros. Further explorations of the waters around these volcanic centres are also revealing more volcanic material, and broadening the understanding of the volcanic activity in Greece.

Santoríni

Steep rock walls, classic white- and blue-adorned buildings and a sunset view to die for – Santoríni is the jewel in the catalogue of Greece's volcanoes. From a distance, Santoríni appears as a round island that rises to a broad central focus, with scattered white villages surrounded by vines and tomatoes in terraced fields of buff pumice. In the south, the rugged Cycladic limestones of Mount Profítis Elías form its highest point at 565m. In fact it is an island chain that is sculptured by, and hosts a record of, episodes of sometimes violent volcanic eruptions, that are recorded in the geological deposits preserved on the older island segments and in the newly emerging Kaméni islands in its centre (Fig. 3.3). Santoríni is

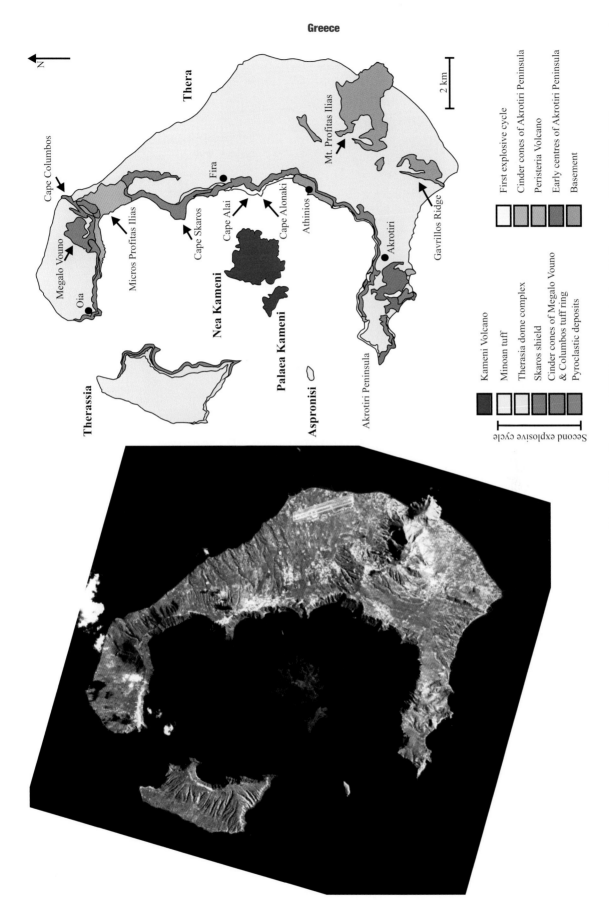

Figure 3.3 Satellite location and simplified geological map of Santorini (satellite map ASTER image acquired on 21 November 2000, NASA JPL).

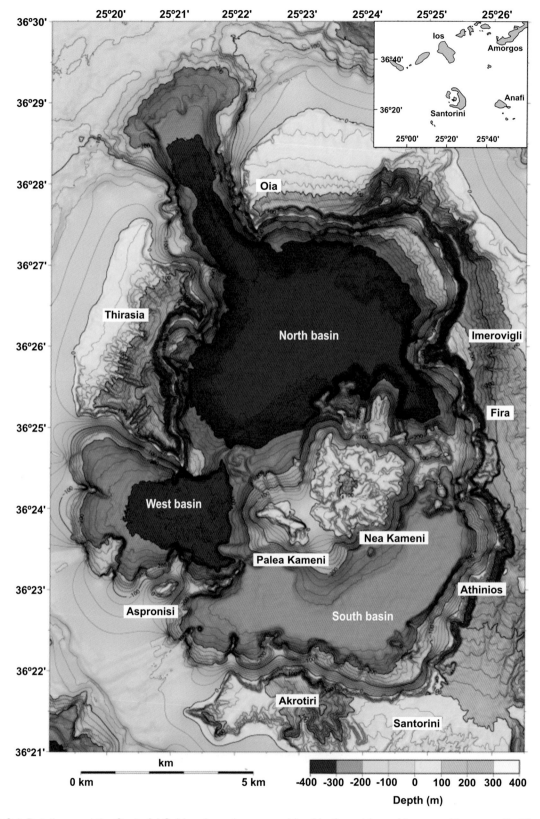

Figure 3.4 Detail around the Santoríni Caldera based on a combined bathymetric and topographic map with 15-m grid resolution (image courtesy of Paraskevi Nomikou).

a group of islands that acquired this name during the Venetian occupation, but in classical times it was called Théra, which is also often used today. However, it seems clearer to use Théra for the principal island and retain Santoríni for the whole group.

The glory of Santoríni is its sea-flooded caldera, about 85km² in area, whose rim marks the formidable inner edge of three islands. Recent bathymetric data reveals the extent of this impressive structure and paints further detail on the central emergent volcano that is starting to refill the caldera (Fig. 3.4). On the main island, Théra,

the northern part of the rim makes a rampart of red, brown and black layers of lava and cinders 400m high, but its southern part forms a less forbidding wall of buff pumice rising no more than 200m. The town of Théra, where spectacular views of the caldera can be seen, is accessed by a zig-zag trail up the side of the caldera wall, and sills and dykes can be seen to cut through the older volcanic deposits at many parts of the excavated volcano (Fig. 3.5). A similar wall faces it on the white pumice islet of Aspronísi (white island) to the west. Therasía, which has a rampart like northern

Figure 3.5 The spectacular Santoríni caldera. **A)** View looking north along the caldera walls. **B)** The zig-zag route up the caldera wall from the port to the old town of Thera. **C, D)** Dykes and sills cutting through older volcanic units exposed in the caldera wall.

Théra, then completes the circle. In the midst of the caldera, intermittent activity since 197BC has formed the Kaméni (Burnt) Islands. They have been the site of all but one of the eruptions of historical times in Santoríni, and they make a low, stark agglomeration of recent lava flows, domes and cones that offer a splendid view of the whole caldera.

The event that gave Santoríni its notoriety – and much of its spectacular landscape – was the Minoan eruption in the Bronze Age. The ejected pumice buried a city, excavated at Akrotíri, which had formed part of the Minoan civilization named after the legendary King Mínos, who reigned at the Palace of Knossós in Crete. The Minoan eruption has also been blamed for the sudden end of this civilization.

The growth of Santoríni

Before the present Hellenic volcanic arc ever existed, Mount Profítis Elías marked the summit of an island of limestones and schists that formed the most southerly of the Cycládes. Eruptions began at a depth of about 1000m on the adjacent Aegean sea floor, perhaps as much as 3 million years ago. Santoríni apparently never formed a single stratovolcano, but separate vents erupted a complex volcanic field over a long period. The oldest volcanic rocks are exposed in the Akrotíri Peninsula on the southern arm of Théra where, between 550,000 and 650,000 years ago or earlier, a dozen or so vents erupted dacitic domes, andesitic lava flows, spatter cones and cinder cones. Volcanic activity then shifted northeastwards and constructed the Peristeria volcano over 430,000 years ago, while emissions of andesitic and dacitic lavas built the bases of four shield volcanoes to a height of some 200m: Megálo Vounó, Micró Profítis Elías, Therasía and Skáros. Events then occurred in two long cycles. The first cycle began 360,000 years ago and culminated 180,000 years ago in the powerful eruptions that expelled the rhyodacitic Lower Pumices. The second cycle lasted from 180,000 years ago until the rhyodacitic Plinian Minoan eruption in the Bronze Age. Large silicic magma reservoirs developed during the last 50,000 years of each cycle.

Santoríni had dozens of violent eruptions of rhyodacitic pumice during the second cycle. Many of these eruptions took place along two parallel fissure or fault systems, trending from northeast to southwest, one passing through Cape Kolómbos in the north, the other passing through the Kaméni Islands in the centre.

The caldera of Santoríni was formed not all at once, but seems to have resulted from four main episodes of collapse. After the Lower Pumices erupted about 180,000 years ago, a caldera collapsed and was probably flooded by the sea. Mainly andesitic eruptions then supervened and probably filled the caldera with both lava shields and layers of exploded fragments, such as the distinctive Middle Pumice and the Upper Scoria. A second caldera then collapsed, chiefly in the north, about 70,000 years ago. Again, eruptions of basalts, andesites and rhyodacites filled the hollow and extended well beyond its walls. These developed the Skáros shield and the Therasía shield or dome complex.

About 21,000 years ago, both these lava piles were severely damaged when a rhyodacitic Plinian eruption brought about the collapse of the third caldera, centred on Cape Riva in northern Therasía. Subsequent eruptions then seem to have formed a volcanic island in the centre of this caldera, similar to the present Kaméni Islands. Many pyroclastic deposits can be found around the present-day island that depict some of this activity (e.g. Fig. 3.6 A & B).

Several millennia of quiescence allowed at least two (and perhaps many more) Minoan settlements to grow up on the island. One, near the southern tip of Therasía, was exposed when pumice was being quarried to make the dykes alongside the Suez Canal in 1869, but it has not yet been systematically excavated. The main Minoan settlement yet discovered is being excavated at Akrotíri on the southern arm of Théra. It is possible that the fine naval flotilla fresco unearthed there could depict the Minoan landscape of Santoríni, although its lack of perspective, not to mention its artistic licence, makes it hard to fit the details into a coherent picture. But, if the fresco is at all accurate, then a large body of water occupied central Santoríni, and the chief Minoan settlement on the island must have stood near the present southern tip of Therasía, and not at Akrotíri itself.

The Minoan eruption of Santoríni

The most famous eruption of Santoríni was not the one that sculptured its caldera as described above, but rather the one that ended its occupation by the Minoans. The aptly termed 'Minoan eruption' was the largest on any European volcano in recent millennia, and it destroyed the Minoan Bronze Age settlements on the island. As the magma pressure rose, the inhabitants were warned by earthquakes that were powerful enough to damage

Figure 3.6 Santoríni's pyroclastic rocks. **A**) Examining layers formed from pyroclastic density current and fallout from Plinian clouds. **B**) A volcanic bomb deforming the layers of volcanic sediment it fell on. **C**) The remains of the Minoan eruption exposed in quarry walls on the side of the caldera.

sturdy two- and three-storey buildings in Akrotíri. These were being repaired when a hydrovolcanic eruption, perhaps lasting for several weeks, deposited a thin layer of pale lithic ash over southern Théra. The Akrotírians abandoned their town, and no human bones have been found there. But it must be very doubtful whether any inhabitants of Santoríni survived the ensuing cataclysm.

The Minoan eruption took place in a shallow, flooded caldera during four phases that occurred without respite and lasted for a total of about four days. The first phase was marked by a continuous Plinian eruption, forming a column 36km high, that spread up to 6m of coarse, roseate rhyodacitic pumice. It was followed by a thin layer of fine white ash and then by a layer of pumice ejected by further Plinian eruptions. The parent vent was probably situated near the present position of the Kaméni Islands. The pumice was strongly concentrated

in southeastern Théra, but it formed only a thin layer on Therasía, which indicates that the Plinian column was winnowed by northwesterly winds, possibly when the meltemi was at its height in the summer. The first phase might have lasted up to about eight hours. The second phase was marked by a gradual change to distinctive hydrovolcanic characteristics as erupting vents opened in the sea. Its fine ash and pumice attain a maximum thickness of 12m. Sometimes, pyroclastic density currents knocked down any walls in Akrotíri still emerging from the previous blanket of pumice. This second phase probably lasted about a day.

The third phase of the Minoan eruption was also hydrovolcanic and came from a vent filled with a slurry of ash, pumice, steam and water. It produced more than half the Minoan materials on Santoríni. They comprise up to 55m of chaotic layers of white pumice and ash (e.g. Fig. 3.6C), with many lithic fragments, including breadcrust bombs. They are thickest in the quarries south of Phíra, near the probable vent. The Minoan caldera may have started to collapse at this time. As more water reached the vent, ashflows merging into mudflows were generated, which radiated over most of Théra and Therasía. Most of the lithic fragments that erupted during the third phase seem to be derived from the dacitic lavas of the Therasía complex, as if they came from new vents situated to the west of those in operation during the first two phases. The third phase might have lasted a day.

The fourth and final phase might have lasted for several hours. It expelled ashflows, mudflows and ignimbrites, which mantle the outer flanks of Théra and Therasía with up to 40m of fine ochre pumice. The close of the fourth phase saw the elaboration of the Minoan caldera, whose floor now lies some 380m below sea level in the north. It enlarged the previous calderas considerably, and submarine slumping also extended two submerged arms on either side of Therasía. Estimates of the volume of magma ejected have varied from 19km³ to 39km³, depending on views about the nature of the landscape before the eruption, and on different interpretations of the extent and thickness of Minoan deposits; but every estimate indicates that this was a massive outburst.

The effects of the Minoan eruption extended physically beyond the limits of Santoríni and scientifically well beyond the confines of the Earth sciences. The eruption presented two interrelated problems: when did the eruption take place, and what were its effects on Minoan civilization in Crete, 125km to the south? Excavations at Akrotíri showed that the eruption had buried the town in the period named the Late Minoan IA, traditionally assigned by archaeologists to about 1550–1500BC. Further research cast doubt on this date, but generated as many problems as solutions. Radiocarbon dates obtained from shortlived shrubs killed by the eruption ranged between 1675 and 1525BC, with a preference towards 1645BC, although the data pointed to about 1629–1622BC when tested statistically. Thus, if anything, the radiocarbon dates indicated that Akrotíri was buried a century before the dates suggested by the traditional archaeological record.

Cores taken from the Greenland ice cap showed aerosol acidity peaks about 1645BC. They were linked to Santoríni because it was believed to be the only large eruption that could have caused such a concentration at about that time. Attempts to prove or disprove this

link have generated much dispute, and this problem is far from being solved. For example, analysis of the same fragments in the ice cores led different authors to conclude that they did, or probably did not come from Santoríni.

Dendrochronological studies (whereby absolute dates could be established by the analysis of tree rings) were also brought into controversy. Bristlecone pines in the southwestern USA registered severe frost damage in about 1627BC, which was attributed to an unusually cold spell. Ancient oaks preserved in bogs in Northern Ireland also showed poor growth of tree rings, which was linked to a cold spell in about 1628BC. Exceptional growth in a tree ring in a 1503-year chronology established in Anatolia was also assigned, perhaps rather arbitrarily, to 1628BC. The eruption of Santoríni was invoked as the source of a volcanic aerosol or dust veil that would account for these biological anomalies. But, in each case, the link to Santoríni was only circumstantial. Many other explosive volcanoes, including dozens scarcely studied, would have been just as likely as Santoríni to supply the requisite stratospheric veil and cause a cold spell. Moreover, not all large eruptions cause frost damage to vegetation, and not all severe cold spells are caused by volcanic eruptions. Indeed, during the better-documented centuries after AD550, acidity peaks in the ice cores and restricted tree-ring growth coincided only seven times.

If the Minoan eruption caused acidity peaks in the Greenland ice cap and the anomalous tree-ring growths, then it probably occurred in about 1628BC. But this is over a century older than the dates traditionally adopted by archaeologists, and its acceptance would apparently require the addition of 130 years to the history of Egypt, and a revision of the Bronze Age chronology throughout the eastern Mediterranean area. Thus, these still-conflicting strands of evidence mean that the exact date of the Minoan eruption remains to be established. The research involves the confrontation between new and traditional techniques and attitudes in several academic disciplines. This is truly the stuff of a prolonged academic polemic, and a consensus should not be awaited with bated breath, as the outcome may be long in coming.

Eruptions of Santoríni during historical times

Eruptions of Santoríni have been recorded for more than 2000 years, and they seem to have happened with increasing frequency in recent centuries. All but one of the historical eruptions have taken place from vents in the centre of the present caldera, where they have

The collapse of the Minoan civilization

The Minoan civilization was the most sophisticated in the European Bronze Age. It was named after the legendary King Minos, whose palace at Knossós was the largest of several that have been excavated in Crete. Akrotíri was one of at least two Minoan settlements on Santoríni. Minoan civilization collapsed when these Cretan palaces seem to have been suddenly destroyed in the Late Minoan IB period, which has been traditionally assigned to between about 1500BC and 1450BC. Only Knossós survived, under the occupation of Greek Mycenaeans. They were certainly in charge at Knossós just after the catastrophe, because their clay inventory tablets were written in the Mycenaean Greek Linear B script, whereas earlier documents had been written in the different language of the Linear A script.

The collapse of Minoan civilization has long puzzled the experts. It has been attributed, *inter alia*, to a military takeover by Mycenaeans, to extensive earthquakes, to volcanic ash from Santoríni, and to civil war in Crete. However, the most controversial solution emerged when Marinátos proposed that Minoan civilization had been annihilated when the eruption of Santoríni had generated a tsunami that had destroyed the coastal settlements of northern Crete, while associated earthquakes had severely damaged those inland. This view proved to be a red herring, but it stimulated much research and academic dispute about the role of the eruption of Santoríni, which are reviewed in more detail elsewhere.

Marinátos's own excavations soon showed that the eruption had buried Akrotíri (Figs 3.7, 3.8) in the Late Minoan IA period, assigned traditionally to between about 1550and 1500BC. Thus, this eruption could not have destroyed the Cretan palaces at the end of the Late Minoan IB. To overcome this crucial point, it was suggested the eruption had taken place in two episodes separated by an interlude of about 50 years. But the geological consensus now indicates that this outburst took place within days, not decades, during the Late Minoan IA.

Neither could the eruption have caused tectonic earthquakes over 125km away in Crete. Volcanotectonic earthquakes would have affected only Santoríni and its immediate vicinity. And even at Akrotíri, the damage caused by these precursory earthquakes was being repaired when the great eruption was unleashed. Moreover, previous Cretan earthquakes had not annihilated Minoan civilization; any palaces that were destroyed were simply rebuilt to levels of even greater splendour.

Deluges of pumice and ash over Crete might have killed crops and herds and caused civil disruption, famine and death; but they could not have destroyed the substantially built Cretan palaces. And, for instance, a pumice layer 10cm thick at the Minoan colony of Iálysos in Rhodes does not even seem to have broken the continuity of settlement. It also lies beneath the Late Minoan IB pottery layer, which again indicates that the eruption occurred during the Late Minoan IA. In addition, most of the ash and pumice was blown over Turkey, and indeed over the Black Sea, thus leaving Crete relatively unaffected on the very fringes of the area of deposition.

The tsunami from Santoríni would certainly have been strong enough to demolish the palaces near the north coast of Crete; and Knossós, 5km inland, could also have been badly damaged. But this tsunami could not have touched Phaestós and Hághia Triáda, standing 100m high near the south coast. The tsunami might not even have travelled mainly towards Crete, for the Santoríni caldera is breached in the west and northwest, which would direct it primarily towards Mycénae and the Peloponnese. The Mycenaeans would, therefore, scarcely have been in a position to profit from any weakness in Crete. And, of course, the tsunami would have developed when the caldera collapsed: it would have reached Crete in 25 minutes, in the Late Minoan IA and not about 50 years later, at the end of Late Minoan IB.

Figure 3.7 Fresco of a ship procession from the excavation at Akrotíri, Santoríni (photo by Smial, Wikimedia Commons).

Santoríni seems irrelevant to a purely political Mycénaean takeover or to internecine Cretan struggles. But new evidence indicates that the destruction of the palaces was selective, and that administrative buildings seem to have been special targets. Moreover, recent research has also indicated that, in fact, the Cretan palaces might not have all been destroyed at the same time during the Late Minoan IB. If these interpretations of events are substantiated, war or sporadic earthquakes would seem to be the chief suspects, but the eruption of Santoríni could be entirely eliminated from the enquiry. Thus, it now seems impossible to blame Santoríni for the fall of Minoan Crete. But the Minoans were, no doubt, greatly impressed by the Plinian column rising 36km above Santoríni, the shock waves, the sea waves, and the loudest noise in the Mediterranean Bronze Age, which would have reverberated all over Europe.

Figure 3.8 Excavations at Akrotíri reveal the sophisticated and detailed lifestyle of the Minoan civilization (Shutterstock/NikD90).

formed the Kaméni Islands, a strikingly barren landscape of black and grey aa lava flows, and several domes and explosion pits that still give off fumaroles. The Kaméni Islands rise a total of 520m from a depth of 380m on the caldera floor, and they represent an accumulation of 2.5km³ during the past 2100 years (Fig. 3.9).

In 197BC, a new island called Hierá erupted in the centre of the caldera between Théra and Therasía. 'Flames came bursting up from the sea for four days, causing the water to seethe and flare up. Gradually an island emerged and was built up, as though it had been forged by implements out of a red-hot mass' (Strabo I: 57). This island formed a tuff cone that was quickly destroyed by the waves, and it probably now remains as the Bánkos shoal, to the northeast of the present islands. The eruptions that followed, from AD46 until 1950, have provided a timeline that can be mapped in the exposed islands and their underwater constructions (Fig. 3.10).

In AD46, lava eruptions formed the islet of Thía, which marked the inception of Palaeá Kaméni. In AD726, an emission of lavas enlarged Palaeá Kaméni, and the pumice that exploded from a vent on its east coast covered the surrounding sea and floated off as far away as Turkey. The next active episode, which took place in 1570 and seems to have lasted for three years, formed the islet of Mikrá ('small') Kaméni. It was 500m

Figure 3.9 The Kaméni Islands. **A**) View of the Kaméni Islands from Théra (Shutterstock), inset image of commemorative stamp depicting the last eruption at Santoríni in 1950 (stamp image courtesy of Josef Schalch mountainstamp.com). **B**) The boat dock on the Neá Kaméni. **C**) Sampling the lava flows on the island in search of clues to their eruptive history. **D**) View of Palaeá Kaméni from Neá Kaméni (Shutterstock/Philip Lange).

long, 320m wide and 71m high, but it foundered a little to 66m in 1707. Effusive and explosive eruptions between Palaeá Kaméni and Mikrá Kaméni took place from 23 May 1707 to 11 September 1711 and led to the formation of Neá (New) Kaméni.

The next eruptions lasted from 26 January 1866 until 15 October 1870. Each started with effusions and ended with explosions. Submarine eruptions off the southern coast of the original Neá Kaméni led to the emergence of the dacitic Geórgios dome, which eventually rose to 131m above sea level and was flanked by several lava flows. This still constitutes the highest point of the Kaméni Islands. Two adjacent vents also erupted, but most of their lavas were eventually covered by those expelled from Geórgios. A small ash ring also formed around this dome, which then gave off only fumes from 15 October 1870. By the end of this episode, Neá Kaméni had trebled in size.

The period between 1925 and 1950 was the most active in the history of the Kaméni Islands. On 11 August 1925, submarine eruptions marked the beginning of a new episode, and new lavas finally joined Mikrá Kaméni to Neá Kaméni on 12 August. Then, the Dáfni dome arose, a 96m-high tuff ring exploded around it, and three lava flows formed a rugged apron almost surrounding the old islet of Mikrá Kaméni. This eruption ceased in January 1926, but another, lasting from 23 January to 17 March 1928, formed the small 15m dome of Naftílos, southeast of Dáfni.

Longer eruptions began on 20 August 1939, when submarine effusions west of Neá Kaméni formed the Tríton dome, which rose to 13m above sea level during the first week. On 26 August 1939, the pit exploded that was to become the main centre of activity for the rest of the episode. On 26 September 1939, the Ktenás dome was extruded to a height of 12m near the crest of Geórgios. Ktenás was soon surrounded by a horseshoe-shaped cinder rampart and also gave off lava flows that spread into the sea. On 13 November 1939, the large Fouqué dome was extruded from the explosion pit. This eventually rose 122m above sea level; its lava flows covered most of the Ktenás lavas, and parts of the flows erupted in 1866–70. This dome was also linked to the violent explosions that took place between 12 July 1940 and early July 1941. Many explosive phases were soon followed by dome extrusions and lava-flow eruptions. Thus, the small Smith A and Smith B domes formed from 15 July to 8 September 1940; Reck dome nearby was extruded between 17 July and 17 October 1940,

and it was surrounded by a cinder rampart. But an explosion shattered the dome, and thus only the rampart survives. On 24 November 1940, another explosion formed the Níki dome, which eventually rose 125m above sea level. When these eruptions ended on 5 July 1941, Neá Kaméni had been consolidated into a round mass 2km across.

The latest eruptions, of explosions and lava effusions, lasted from 10 January to 3 February 1950, and formed only the small 9m-high dome of Liátsikas and two lava flows. Since then, the emanations have been limited to solfataras and fumes at about 70–86°C.

All the historic activity of the Kaméni Islands has been concentrated along a band trending northeast, 600m wide and 4.5km long, following the dominant trends of fissures throughout Santoríni. The dacitic lavas came from a reservoir about 3km deep at a temperature of 950–1000°C. Each episode began with extrusions of viscous dacites and was followed by the formation of cinder cones and lava flows. The triggering of the eruptive events is thought to be caused by the intrusion of new magma that acts as a trigger for the eruptions. These magmas contain olivine crystals, which have recorded the time-line of eruption from the injection of the triggering magma until the overturn of the chamber and ensuing eruption (Fig. 3.11). These 'crystal clocks' show that triggering of eruptions in Santoríni occurs on a monthly scale, which means that the monitoring of shallow intrusive events could provide good precursors to eruptions with enough time for action. Interestingly, there is anecdotal evidence of earthquake unrest in Santoríni occurring a month or so prior to known recorded eruptions.

In view of the marine location of the Kaméni vents, the general absence of Surtseyan eruptions is remarkable. This may be due to high effusion rates, which prevented water from entering the vents. The main explosive episodes, in AD726, 1650 and 1925–8, were probably related to the highest rates of discharge, which enable the magma to retain its gases more easily. The kind of activity witnessed over the past 500 years will probably continue in the Kaméni Islands. With earthquakes occurring as pre-cursors to possible activity, and with the enhanced networks of ground deformation, seismic and other modern monitoring, we are sure to be aware when the next eruption strikes.

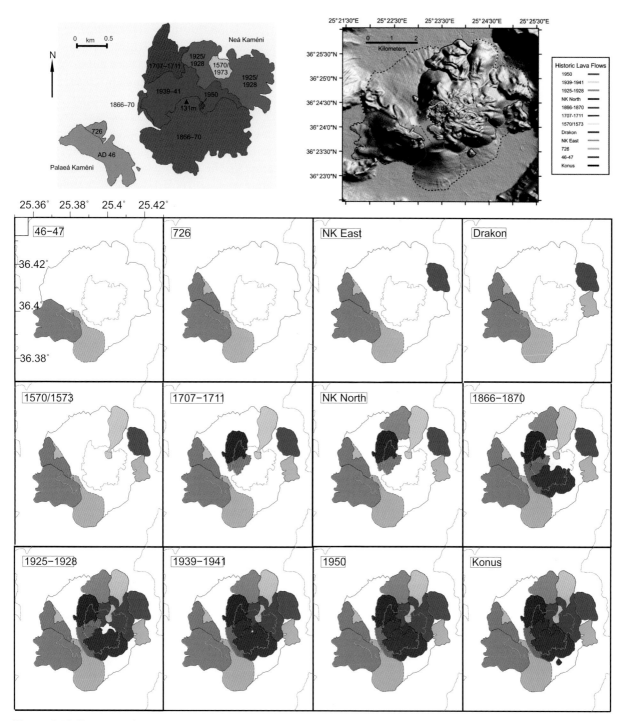

Figure 3.10 Evolution of the Kaméni Islands. **A**) Onshore outline of the historic Kaméni lava flows. **B**) Lava flow outlines for historic lava flows drapped over the topographic and bathymetric data, outlining extent both on and offshore. **C**) Evolution of the historic flows both onshore and offshore (Images for **B** & **C** courtesy of Paraskevi Nomikou).

Mílos

Mílos is the south-westernmost island of the Cyclades, where it forms a small group with Kímolos, Polýegos and Antímilos. It covers an area of 160km² and is composed of ochre plains and bare, tawny hills mantled with pumice and ash. Mílos and its neighbours are almost entirely composed of volcanic rocks, among which are those dated at about 3.5 million years as the

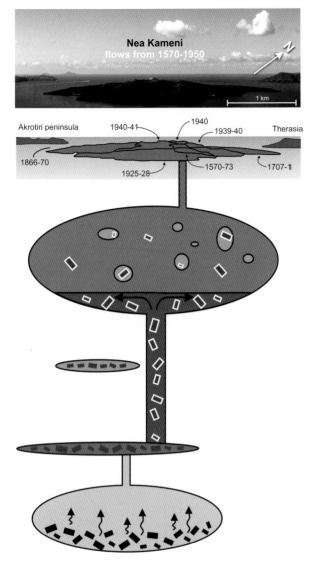

Figure 3.11 The complex plumbing system beneath Santoríni involves several magma reservoirs that are sometimes separated and sometimes in communication, as evidenced by a complex crystal population and different magma batches on eruption. Many of the lava flows contain enclaves of more mafic magma, which is thought to act as the trigger for the eruptions (photo courtesy of Vikki Martin).

Is Santoríni re-awakening?

With an active record of eruptions from recent historical accounts that goes back hundreds, even thousands, of years, the question arises, 'When will Santoríni erupt next?' In January 2011 a series of shallow earthquakes was recorded around Santoríni, fuelling discussions about the possibility of shallow magma movement in the volcano. As the earthquake swarms continued and more preliminary data came in about ground deformation around the island, scientists mobilized more ground-based equipment, used airborne LiDAR surveys and available satellite data to monitor these new activities around Santoríni.

The main period of activity was between January 2011 and April 2012. The more detailed networks of GPS equipment set up from the NERC Geophysical Equipment Facility were supported by the crew from NERC's Airborne Research and Survey Facility (ARSF) with high resolution LiDAR laser surveys, producing state of the art 3D mapping of Santoríni. A multinational team including the Greek National Technical University, University of Athens, and Oxford University, re-surveyed the main trig points of the island (Fig. 3.12A), recording daily changes in the island's movement. This, along with the LiDAR and available Satellite data, was combined to produce a 3D deformation map highlighting the main areas of movement (Figure 3.12B). Although this movement subsided shortly after with no new eruption, it provided a valuable test of the monitoring available for Santoríni and was the first major event since 1950, and one that produced both great excitement and also a fantastic collaborative effort by scientists to monitor it closely.

Figure 3.12 Monitoring the activity at Santoríni. **A)** Differential GPS on a trig point. **B)** Map of the ground deformation in 2011 using Satellite Radar Interferometry (InSAR) and detailed GPS measurements (images courtesy of David Pyle, ESA).

Meet the Scientist – Professor David Pyle

David is a volcanologist at the University of Oxford, and works on a number of projects with collaborations in other universities in the UK, Europe and beyond. He is interested in the impacts of volcanism and volcanic degassing, with current projects including: Strengthening Resilience in Volcanic Areas (STREVA), Rift Volcanism, Mantle Volatiles Consortium, Tempo of Volcanism in Southern Chile. He has worked and published extensively on Santoríni and the Greek volcanoes and he expands on his passion for Santoríni and volcanoes below.

What is so special about Santoríni volcano?
'It took me a long time to realize this, but Santoríni is quite simply one of the best-preserved active volcanoes in the world. In terms of its life cycle, Santoríni is now quite a mature volcano. It has had a dozen or so spectacular explosive eruptions in the past quarter of a million years; and the deposits left behind after every one of those eruptions are laid out in the most wonderful "layer cake" fashion in the cliffs that trace out the edge of the huge crater, or caldera, left behind after its last major eruption about 3600 years ago. Santoríni is also a volcano that is very much alive – even though it hasn't erupted since 1950. The dusty, rocky "Kameni" islands that lie in the centre of Santoríni's sea-filled crater might look ancient – but most of the rocks on these islands are less than 300 years old; and the volcanic landscape here is one of the most dramatic anywhere in the world.'

What has been your most exciting discovery about it?
'I first went to Santoríni to try and piece together the stories of the first half of its eruption life cycle, which was a great way to learn about how volcanoes behave; and good fun. At that time, Santoríni was quiet and hadn't erupted for about 40 years. The youngest part of the volcano – the Kameni islands – was, as far as I could see, a tourist trail over some dusty grey lavas with some half-hearted patches of steamy ground, and some rusty mudpools. Over many return visits – many to run field trips – my patchwork understanding of the volcano matured, but still the Kameni islands remained quiet. Then, in 2010 or so, we were looking for a dormant volcano to study using a relatively new satellite radar technique, to find out a bit more about background "inactivity" at volcanoes. Just a few months later, in early 2011, Santoríni started to show signs of restlessness, with swarms of tiny earthquakes, and about 15 centimetres of bulging that we could see developing very clearly from the satellites, but which was barely noticeable on the ground (Fig. 3.12). After about 18 months the earthquakes stopped and the bulging began to relax, and a minor "volcano seismic crisis" was over. This was a fascinating time, though – the first such event since the last eruption in 1950 – an episode where no one quite knew what was going to happen next, and yet one that we could watch from a satellite orbiting about 800 kilometres above the Earth's surface.'

What is the most important thing we still need to find out?
'We still don't really know how eruptions start, particularly at volcanoes which have been quiet for many decades; and before they start erupting, we really don't how they are going to behave when they do erupt. We also don't really know how to tell when eruptions have finished!'

How did you first start getting into research on volcanoes?
'Volcanoes were a childhood fascination, which eventually turned into the best day job in the world. The formative volcanic events while I was growing up – like the May 1980 eruption of Mount St Helens, USA, and the tragic consequences of the November 1985 eruption of the Nevado del Ruiz, Colombia – left an indelible mark on me; and I have been hugely privileged to travel to some amazing places all around the world during my research into volcanoes. As a seven-year-old, I lived in Chile for a while. It was here that I got my first taste of a real live volcano, when our summer holiday travels took me to some recently-formed volcanic mudflow deposits (lahars) on the slopes of Volcan Villarrica.'

How important do you think the volcanoes of Greece have been in our understanding of volcanology?
'Santoríni has a very important place in the recent history of volcanology. It had a major eruption in the mid-1800s that led to a lot of interest from geologists at the time, who visited to make new observations of the lavas, gases and other phenomena. The 1866 eruption of Santoríni led to the first study of the medical consequences of volcanic gases for humans; and helped to establish some ideas – that are still current – about how large volcanic craters form. Santoríni has also become very important in archaeology, in particular in the human and political consequences of the huge "Minoan" eruption of Santoríni 3600 years ago. This eruption buried and preserved an amazing Bronze Age city at a place now called Akrotiri, in the south of the island, giving a unqiue glimpse of what life had been like in this important trading centre, before the island was buried under metres of pumice.'

Any other comments about the volcanoes of Europe?
'From our vantage point in the UK we might think of Europe as being geologically quiet once we get away from the active volcanoes of southern Italy and southern Greece. But in fact there are traces of geologically-recent volcanic activity extending from northern Spain (Garrotxa), to central France (Auvergne) and western Germany (Eifel and the Rhine Valley). And, as we all found out in 2010, northern Europe lies within range of volcanic ash from Icelandic eruptions.'

oldest yet determined in the arc. The geology of the island reflects this long volcanic history, with a diverse set of deposits within such a small area (Fig. 3.13).

Eruptions that would eventually build the volcanic island of Mílos began, like many of the volcanoes of Greece, under the sea in deep water. The preserved activity occurred first in the southwest and then in the northeast, before finally switching to the centre. Many eruptions sprang from fissures and created a volcanic field that never developed a central focus. Domes, lava flows and small pyroclastic density currents built up the western peninsula of Mílos and the islands of Kímolos and Polýegos. The eruptions that built the undulating northeastern arm of Mílos began about 1.8 million years ago when obsidian-rich rhyolite domes were expelled, as well as layers of ash and pumice. It was at this stage that deposits accumulated sufficient thickness to breach the sea surface, and Mílos Island as we see it today began to emerge. This phase ended about 900,000 years ago, when two dacitic domes extruded: Halepá dome forms a *malpaís* of rough lava, whereas the Plákes dome forms the plinth of the village of Pláka (Fig. 3.14A).

After a dormant interval, hydrovolcanic eruptions began some 380,000 years ago and ended with effusions of rhyolitic lavas that now compose the promontory jutting northwards from the Plákes dome. At about the same time, andesitic, dacitic and rhyolitic eruptions formed Antímilos Island. In the south, hydrovolcanic eruptions began 140,000 years ago, followed by effusions

of rhyolitic lavas, and they ended 90,000 years ago when more hydrovolcanic eruptions formed a tuff ring 1.5km across. These eruptions joined the eastern and western islands together, leaving the Gulf of Mílos between them. Less than 20,000 years ago, more hydrovolcanic eruptions formed a dozen small, often intersecting, maars on the eastern peninsula, which brought activity to a close. However, fumaroles still persist, such as those on the coast near Adámas. The island has a wealth of volcanic deposits, which are exposed as the domes that make up some of the higher ground, as well as beautiful white tuff beaches, cliffs and sections through its rich volcanic geology (Fig. 3.14).

Mining on Mílos

The rich volcanic rocks of Mílos were of use as more than just picturesque beaches and hidden coves. The islands have a rich mining history that can be seen with old mine workings as well as recent excavations and activity. Sulphur, bentonite, manganese ore and even obsidian have played a part in the mining history of Mílos, as man has sought to use the land to his advantage. Mining activities date back several thousands of years to when obsidian was traded around the Mediterranean. The sulpur was used by the ancient Greeks, with the oldest records from the fifth century BC, as well as rock materials for building stone and milestones. Mining was undertaken at a more organized scale after the 1860s,

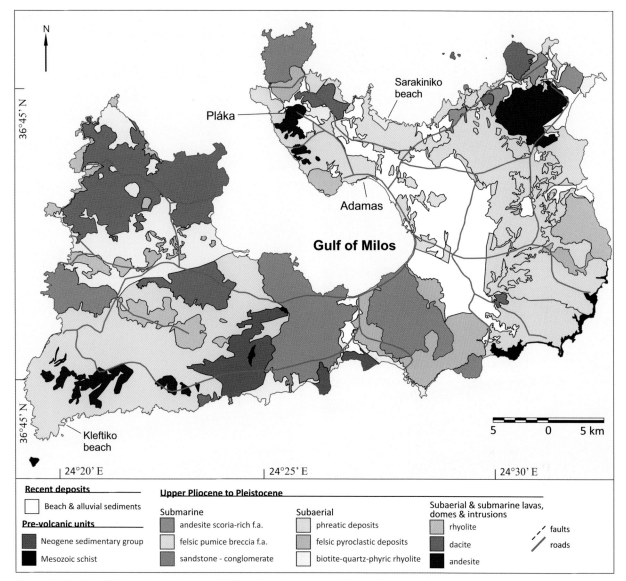

Figure 3.13 Simplified geological map of Mílos showing geographic localities and main geological units (map courtesy of Sarah Gordee).

when mining concessions were first issued, first with sulphur (e.g. the mine at Paliorema) and later with manganese ore, with big operations starting in the 1890s (e.g. Cape Vani mine). The mining of manganese was relatively shortlived with the closure of the Cape Vani in 1928, but sulphur mining continued until the late 1950s, and now the remnants of these mining activities can be seen around the island (e.g. Fig. 3.15). Recent mining activity has concentrated on large open pit mines for the likes of bentonite (altered volcanic ash, which can be highly absorbent and sometimes used in cat litter), perlite (a volcanic glassy rock, used in building), and ashes used in cement production.

Méthana

The isthmus linking the Méthana Peninsula to the Peloponnese is so low that it looks like an island from most directions. It forms a triangle 10km across, made up of an array of rugged domes and thick, viscous lava flows, rising to a hilly and inaccessible central area, where several peaks exceed 500m. Méthana is younger than its Sarónic neighbours, the oldest rocks being only 900,000 years old.

The first three eruptive episodes all began with small effusions of basalts and basaltic andesites that were succeeded by larger and more violent eruptions of dacites and rhyodacites. However, the latest phase began with

Figure 3.14 Views of Mílos. **A**) The Plákes dome forming the plinth of the village of Pláka. **B**) Concentric columnar joints in porphyritic dacite pumice block, northeast Milos, central coast. **C**) Graded, waterlain tuff beds in the Filakopi Pumice Breccia – central coast of northeast Milos. **D**) Medium beds of waterlain ash and pumice lapilli of the Sarakiniko formation, central coastline, NE Milos. **E**) Sea stack of volcanic tuffs, Kleftiko beach. **F**) White tuffs sculptured by the sea, Sarakiniko beach (photos **A**, **E** & **F** Shutterstock/Lefteris Papaulakis/Josef Skacel/s_kaisu; photos **B**, **C** & **D** courtesy of Sarah Gordee).

Figure 3.15 Mining on Mílos. Main photo of abandoned mine workings at the Paliorema Sulphur Mines. Inset commemorative stamp showing large bentonite mine at Aggeria (main photo ©Shutterstock/Denizo71; stamp image courtesy of Josef Schalch mountainstamp.com).

extrusions of dacitic domes that now form the thickest accumulations in the centre of the island. The latest eruption extruded the Kaméni Vounó (Burnt Hill) dome in northwestern Méthana. As it extruded, the western sector collapsed and a rugged brown latite–andesite flow, up to 50m thick, stretched 1.5km northwards and extended out to sea. A smaller flow, with a similar aa surface, turned southwestwards, where it now dominates the village of Kaméni Xorió. These eruptions were most probably those noted by Strabo (I: 3–18) in about 250BC: 'In Méthana a mountain, seven stadia [1300m] in height, was raised up as a result of a fiery eruption. It was unapproachable in daytime because of the heat and smell of sulphur, and at night it shone for a great distance. It was so hot that the sea boiled for five stadia [925m] and was turbid for as much as twenty stadia away [3.7km]'. The peninsula still has several active hydrothermal areas, especially those forming large hot springs, and the baths at Loutrá Méthana in the southeast.

Nísyros

Nísyros is the largest volcanic island in the Dodecanese. It constitutes the emerged parts of a stratovolcano, 8km across and 42km^2 in area, which rises to a height of 698m (Fig. 3.16). Its smooth truncated cone, often clothed with almond and olive groves, seems all the more luxuriant beside the bare, rugged limestones of its neighbours. But the main eruptive features of Nísyros are hidden away in the 300m-deep summit caldera. It is here that the fumaroles in the circular pit of Stéfanos provide the most evident source of contemporary activity on the island, although reddish-grey domes have recently filled the western half of the caldera, and several yellowish cones have erupted at their base.

Nísyros is a young stratovolcano, for some of the oldest exposed rocks erupted only 66,000 years ago. Its landforms are often clearly much younger: cliffs formed by marine erosion are rare; marine terraces are completely absent, soils are still thin, the domes have retained their bristling surfaces, weathering has scarcely

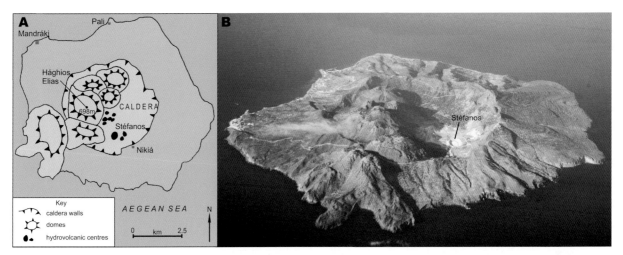

Figure 3.16 Nísyros island. **A)** Main volcanic features of Nísyros. **B)** View of Nísyros from the air highlighting location of the crater of Stéfanos (photo courtesy of Brian McMorrow).

affected many lava flows, and even the friable pumice layers have developed only shallow gullies. However, records of activity on Nísyros throughout historical times are sparse and uncertain, and the volcano could possibly have been quiescent for much of that period.

The eruptions on Nísyros ranged from basic andesites to dacites, rhyodacites and rhyolites, and they occur both in thick lava flows and as layers of ash and pumice. Many came from a reservoir less than 8km deep, although some of the more recent activity, which formed rhyodacitic domes within the caldera, probably originated in a small reservoir that was less than 2km deep.

Submarine basaltic andesite eruptions piled up fragments and pillow lavas from a depth of 1000m on the Aegean sea floor. As the volcano emerged, mainly effusive eruptions of similar composition formed much of the emergent stratovolcano, where one of the oldest flows has been dated to 66,000 years old. Subsequently, however, more andesitic or dacitic materials erupted, and explosions of fragments became more common. For instance, they form a blanket about 250m thick in both northeastern and southern Nísyros. Plinian outbursts expelled dacitic and rhyodacitic 'lower pumices' that may have been linked to an early caldera collapse. At about the same time, Mount Hághios Ioánnis, near Nikiá, expelled thick, viscous rhyodacitic lava flows that were also voluminous enough to reach the southeast coast. Slightly later, several Plinian eruptions expelled about 10km³ of ash and pumice that now form a yellow blanket, 100m thick, on the northern slopes.

These eruptions probably led to the collapse of the caldera less than 24,000 years ago – which is the age of

the latest dated rocks of the stratovolcano proper. But the caldera must have been in existence about 5000 years ago, because Neolithic artefacts have been found on its floor. The original caldera was circular and 3km in diameter; its floor stood only 100m above sea level, and it was enclosed by walls ranging from 150m to 400m high. When the eruptions resumed, the chemical nature of the magmas hardly changed, but fewer fragments were expelled. Instead, several domes of reddish-grey dacites extruded from fissures extending from inside the caldera southwestwards onto the flanks of the stratovolcano. Five domes extruded within the caldera and almost completely filled its western half. The domes of Nífios and Hághios Geórgios are unusually large, and Hághios Elías, which is almost 600m high, is probably one of the largest domes in Europe. Their rocky surfaces, still bristling with their original pinnacles, suggest that they were formed only recently. At about the same time, dacitic domes also extruded on the southwestern flanks of the stratovolcano. The two southern domes distended, flowed down slope and solidified as two rocky promontories projecting out to sea. Neither weathering nor marine erosion have yet been able to alter them significantly, and they seem to be no more than a few thousand years old.

Kós

On Kós the volcanic rocks are generally quite old and limited in outcrop. Initially dacitic and then rhyolitic domes extruded in the southwestern Kéfalos peninsula between about 3.4 and 1.6 million years ago. Then an

The eruptions on Nísyros in 1871 and 1873

The only well-documented eruptions on Nísyros occurred in 1871 and 1873. After an earthquake had shaken the island, explosions like thunderclaps followed in late November 1871; red and yellow flames rose into the air, and rock fragments whistled over the highest peaks and rained down on the coast. Two craters opened at the foot of the Hághios Elías dome, spread white ash over the floor of the caldera, and covered the whole island in a curious mist. But calm returned and soon the vents were giving off no more than fumes.

On 2 or 3 June 1873, several earthquakes rocked Nísyros, a fissure opened on the floor of the caldera, and another crater exploded near those that had formed in 1871. Salty water gushed out first and then combined with the mud, ash and rock fragments that erupted during the next three days to form a mudflow that was 500m long and 3m thick. When the water evaporated, it left behind a salty crust, like frost, on the ground and trees. During the summer, tremors were felt, and hydrogen sulphide and steam emerged regularly from the vents. On 11 September, an earthquake opened a submarine fissure between the island of Yalí and the capital, Mandráki, where buildings were damaged and the sea turned milky. On 26 September, the new craters in the caldera once again threw out salty water, mud and rocks, and continued to gush out steam and hot water from time to time during the ensuing months. At one stage, even the prospect of abandoning the island was mooted, because several people had been injured

and almost everyone had been forced to leave home and camp out in the open air. But the eruption stopped before the decisive steps had to be taken. There was a postscript to these eruptions in September 1887, when a small pit nearby briefly ejected mud, rocks and steam.

These hydrovolcanic and hydrothermal eruptions expelled buff fragments around steep craters, or explosion pits, the largest of which reaches 180m across. These fragile features have already suffered much erosion, and six other older craters, probably formed several centuries ago, have been so degraded that they can now be identified only from air photographs. The crater of Stéfanos, which had developed 'a long time before', was certainly active in 1873, when a small vent also opened on its southwestern edge. Stéfanos ('the crown') is a circular hollow, 240m across and 25m deep, with nearly vertical sides. It is also the main source of the contemporary hydrothermal activity. It may have resulted from a hydrovolcanic explosion, but its wide, flat floor also suggests that it may have originated partly by piston-like collapse. The floor of Stéfanos has to be crossed with care, for it contains many fumaroles (at temperatures approaching 100°C), mudpots, and solfataras with fine sulphur aureoles (Fig. 3.17). Other fumaroles rise from a fissure stretching from the caldera walls below Nikiá to the lower flanks of Hághios Elías dome, where they have stained the dacites yellow, and separate vents on the north coast have also produced the hot springs at the little spa of Páli.

Figure 3.17 The Crater of Stéfanos, a popular spot to look at the hydrothermal system on Nísyros. Inset photo highlights the position of the larger Stéfanos crater amongst other vents. Inset commemorative stamps of Stéfanos also shown (main image ©Shutterstock/Jiri Vavricka; stamp images courtesy of Josef Schalch mountainstamp.com).

eruption about 500,000 years ago expelled vast quantities of pale yellow pumice and ash from a centre in the nearby Bay of Kamári. Another major eruption, about 160,000 years ago, ejected the Kós Plateau tuff, a blanket of rhyolitic pumice and ash 30m thick that covers much of southern Kós. Its total volume probably reached 100km³, and it spread across the sea as far as Kálymnos to the north and Tílos 40km away to the south. Its source was a vent near the island of Yalí, and the eruption was accompanied by the collapse of a caldera, between 5km and 10km across, which now lies below sea level. Other eruptions occurred on the island of Yalí about 31,000 and 24,000 years ago.

Offshore volcanism – the hidden extent of the arc

Recent work has enabled extensive surveys of the sea floor around the Hellenic volcanic arc. This in turn has unearthed some wonderful images and examples of submarine volcanism and helped extend our knowledge of the extent of volcanic deposits. New maps are being drawn up which extend the known boundaries of recent volcanic deposits (Fig. 3.18), concentrating on the main volcanic areas (Méthana, Mílos, Santoríni and Nísyros).

Offshore volcanism was, of course, known about historically due to the eruptions inside the Santoríni caldera, but it also occurred in other surroundings.

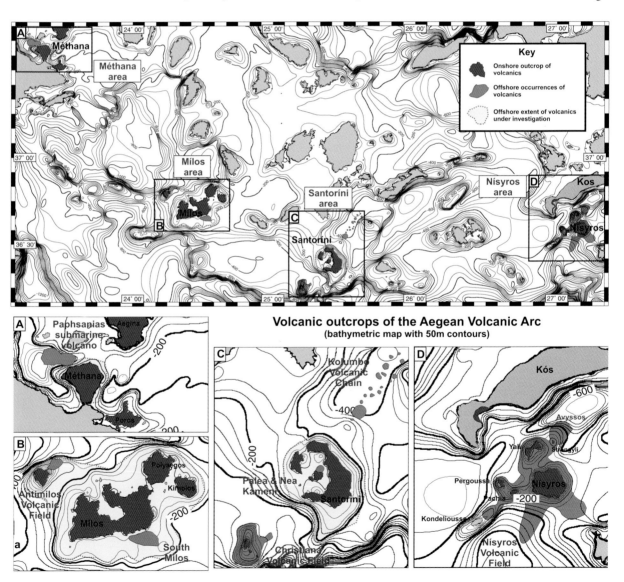

Figure 3.18 General bathymetric map with 50m contours covering the locations of Méthana, Mílos, Nísyros and Kós, highlighting onshore and offshore volcanics. **A**) Detail of Méthana area. **B**) Detail of Mílos area. **C**) Detail of Santoríni area. **D**) Detail of Nísyros area (images courtesy of Paraskevi Nomikou).

In 1650, the only recorded activity in historical times outside the Santoríni caldera occurred on a fissure on the sea floor, 6.5km northeast of Cape Kolómbos. Violent earthquakes shook Santoríni for several days in early March 1650 and again in mid-September. Many homes were damaged, the ground roared, the sea turned green, and rocks crashed from the cliffs around the caldera. On 27 September the most severe earthquake rocked the houses as if they were cradles. That day, a Surtseyan eruption began offshore, when lava emissions were followed by explosions of snow-white pumice and ash. In the midst of deafening explosions that were heard 400km away in the Dardanelles, a small tuff cone formed an islet. This was the most damaging of all the historical eruptions, because it generated a small tsunami on 29 September that caused much destruction on the nearby coasts, knocking down orchards and five churches in Santoríni, and throwing a Turkish naval vessel onto the shore on the island of Kéos. Pumice floated all over the Aegean. The vent also gave off noxious fumes that blew over Théra, where many people were blinded for several days, and about 50 inhabitants and more than 1000 farm animals were asphyxiated. On 2 October, a boat coming from the island of Amorgós sailed too close to the vent and was enveloped in gas. The next day, the people on the island of Íos saw it wandering aimlessly about in full sail, and went to investigate. They discovered the corpses of all nine members of the crew, with their eyes inflamed, their heads grossly swollen, and their tongues hanging from their mouths. After this macabre climax, the vigour of the eruption diminished. However, on 4 November an explosion covered the northern part of Théra in black smoke, and about 20 agricultural workers lost consciousness. They did not recover their senses until that evening, just as they were about to be buried. The eruption then waned and stopped altogether on 6 December. A short time afterwards, the sea destroyed the tuff cone and left behind a shoal, the Kolómbos Bank, which is about 18m deep. The detail of this Kolumbo volcano in its present form shows columns, ash layers and submerged fumaroles (Fig. 3.19)

The story of the volcanoes of Greece is clearly an unfinished book that promises to provide some cracking and explosive chapters in the future. There are the obvious candidates in Santoríni and Nísyros, where the recent historical activity would focus our attention on the possibility of future eruptions. However, the offshore realm is equally as exciting as new volcanic stories are unravelled, and the prospect of new underwater volcanic eruptions in the Mediterranean might be realized.

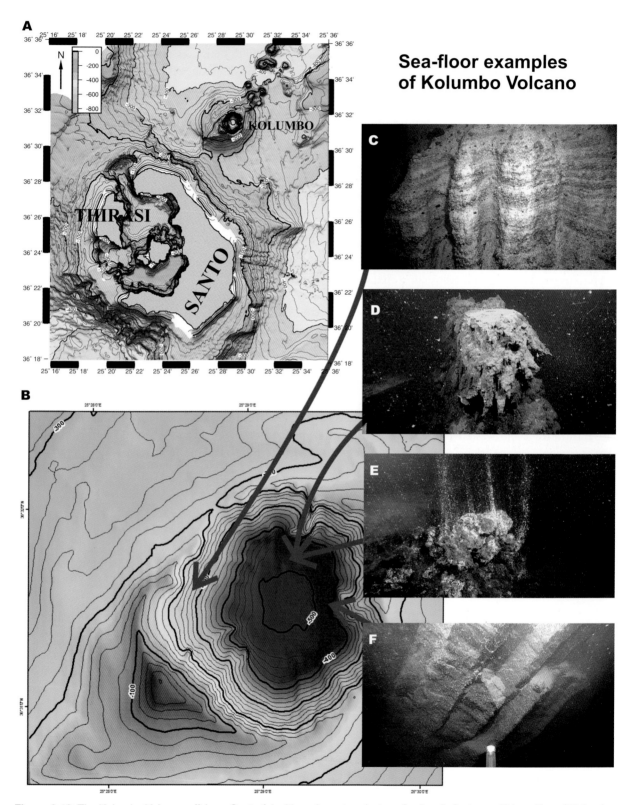

Figure 3.19 The Kolumbo Volcano offshore Santoríni with underwater photos of volcanic features. **A**) Location of Kolumbo NE of Santoríni. **B**) Bathymetry around the Kolumbo crater. **C**) Volcanic tephra layers. **D**) Hydrothermal vent in the northern crater floor of Kolumbo Volcano. **E**) High-temperature hydrothermal vent still discharging gases. **F**) Columnar joints on lava dyke on the western slope (Images courtesy of Paraskevi Nomikou).

Chapter 4

Spain

Introduction

The Canary Islands form Spanish provinces, situated off the Atlantic coast of Africa, between 1200 and 1750km southwest of Cádiz. The archipelago consists of seven main islands. Tenerife, covering 2058km², is the largest; El Hierro, only 275km² in area, is the smallest; and Fuerteventura lies only 115km from Africa (Fig. 4.1). All the islands belong to the African plate. They fall into two groups. Fuerteventura and Lanzarote form the eastern group and belong to the same volcano-tectonic unit that trends parallel to the African coast. Here the climate is arid, the vegetation is steppe-like, and fissures have dominated the distribution of eruptions to such an extent that neither island has a marked scenic focus. In contrast, the central and western group of islands – El Hierro, La Palma, La Gomera, Tenerife, and Gran Canaria – are more mountainous. Thus, even little El Hierro reaches 1051m and Tenerife rises to a majestic climax of 3715m in the Pico de Teide. They lie in the path of the northeast trade winds, which bring cloud, humidity and a luxuriant vegetation to their northern windward shores, although their leeward southern slopes are often arid. Their lower parts enjoy an equable and mild climate throughout the year, and several have become some of Europe's major tourist attractions. Thus, in any season, the volcanologist is unlikely to be alone on the summit of Teide, and may even encounter tourist-laden camels in Lanzarote. On the other hand, few areas beyond the shadow of Vesuvius have done more for popular appreciation of the impact of volcanism on mankind and the environment.

The Canary Islands contain a great variety of volcanic forms ranging from plateau basalts, which are the remains of large basal shields, to recent cinder cones and rugged lava flows that have often joined into *malpaís*; and their stratovolcanoes have sometimes been decapitated by large calderas. The islands have been considered as the type-locality of the caldera ever since Von Buch published his controversial work on their landforms in 1825. Indeed, the Caldera de las Cañadas in Tenerife is, with Santoríni, among the most spectacular in all Europe. Fortunately, this well-populated archipelago has not undergone a caldera-forming eruption in historical times.

The Canary Islands used to be inhabited by the Guancho peoples, who passed on some references to eruptions before they were exterminated by Spanish settlers. Lanzarote, Fuerteventura, La Gomera and El Hierro were settled from 1402, Gran Canaria from 1483, Tenerife from 1491, and La Palma finally from 1493. Thus, historical times in the archipelago amount to less than 600 years, during which activity has been dominated by basaltic eruptions along fissures, which occurred in Lanzarote, La Palma and Tenerife. In addition, parts of Fuerteventura, Gran Canaria and El Hierro all have cones and flows of such remarkable freshness that they scarcely seem to be more than a thousand years old. Indeed, all the islands except La Gomera have had activity within the past 10,000 years. The recorded eruptions in the Canary Islands have occurred at average intervals of 30–35 years. The latest occurred onshore in La Palma in 1971, offshore El Hierro in 2011–2012, and by far the most prolonged and extensive historical eruption took place in Lanzarote from 1730 to 1736.

The volcanic activity in the archipelago has perhaps historically been the most difficult to explain in all Europe. The islands lie on the passive continental margin of northwestern Africa, on one of the oldest parts of the Atlantic Ocean floor, where the basaltic oceanic crust ranges from 180 to 155 million years old from east to west. A basal complex outcrops on Fuerteventura, La Gomera and La Palma, and probably lies hidden beneath the remaining islands. It contains some sediments, some plutonic intrusions, and some submarine basalts, but chiefly assemblies of dykes. However, most of the Canary Islands are much younger, and many parts of the surface are probably no more than a few thousand years old. The islands were born separately from different sources of magma and did not all develop in the same way. Thus, for instance, the oldest dated volcanic rocks occur in

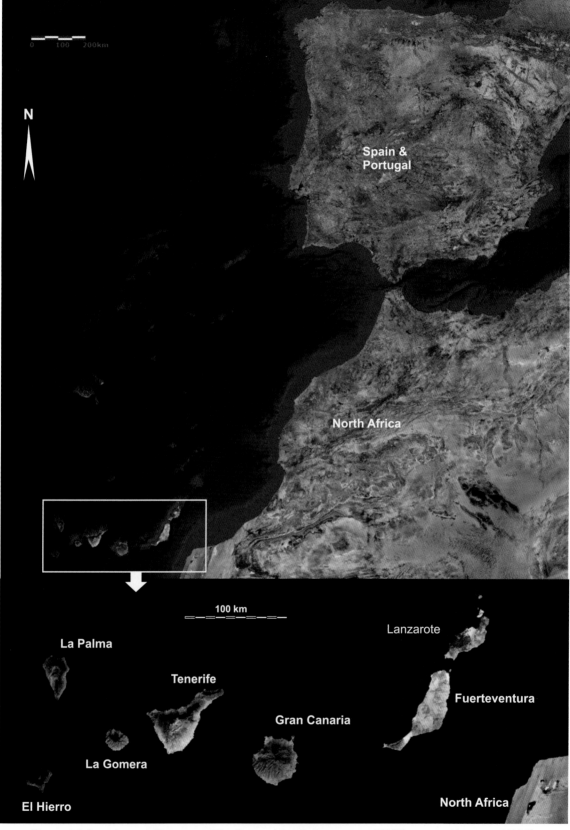

Figure 4.1 Location satellite maps of the Canary Islands (large map – USGS, zoomed in map – NASA).

Fuerteventura, where they are about 20.6 million years old, but they are no more than 2.0 million years old in La Palma and about 1.12 million years old in El Hierro. The rocks themselves cover a wide range from basalts to rhyolites. Basalts, chiefly emitted from fissures, account for the longest and most prolific eruptive episodes; and the more evolved rocks developed largely beneath the stratovolcanoes, probably in relatively shallow magma reservoirs.

Tectonic movements might have played a role in the growth of the Canary Islands by opening up the oceanward prolongation of the South Atlas Fault of North Africa. Nevertheless, the archipelago probably owes most of its growth to a hotspot, which has given rise to a broad westward development of activity in the islands (see Fig. 4.2). It seems probable that masses from the rising magma have formed individual basal shields, and, at times, domed up their surfaces until major three-arm rift systems have developed, separated by angles of 120°. The fissure systems on these rifts then enabled yet more magma to reach the surface; and they have, for example, become the zones of the most marked concentrations of recent emission centres in the islands. Thus, the rifts have been built up into large, high ridges, which reach spectacular proportions in Tenerife, where they form a Y-shaped pattern centred on Teide. These rift ridges grew up so quickly that they sometimes became unstable. Consequently, in El Hierro, La Palma and Tenerife, parts of these ridges have collapsed into the Atlantic Ocean in major landslides. Most of the eruptions in the Canary Islands during historical times have occurred on the fissures developed along these ridges. During these eruptions, fragments explode and form cones on the upper parts of the fissures, while fluid basalts, and perhaps spatter, emerge lower down. However, they have produced only tiny volumes of broadly alkaline basalts, and formed but small cones and thin lava flows.

Tenerife

Tenerife is the largest of the Canary Islands, covering an area of 2058km² and stretching 84km from northeast to southwest and 50km from north to south (Fig. 4.3). Tenerife is the central island, with the greatest volume of erupted materials in the archipelago: more than 15,000km³, of which some 2000km³ lie above sea level. It rises from a depth of more than 3000m on the Atlantic Ocean floor, where eruptions began perhaps as much as 11.6 million years ago. The island is Y-shaped, with three main ridges, extending northwestwards, northeastwards and southwards, which have

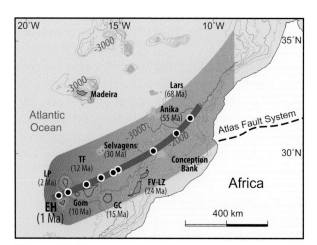

Figure 4.2 Canary Islands hotspot evolution map. The ages imply an east to west age progression in the Canaries, forming a classic hotspot alignment (plume trace) (map courtesy of Val Troll).

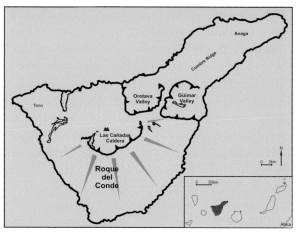

Figure 4.3 Location map, features and outline of recent activity on Tenerife (Satellite Image – USGS).

been formed primarily by basaltic fissure eruptions along rifts. The crowning glory of Tenerife, the Pico de Teide, rises where these ridges intersect and soars from the spectacular Caldera de las Cañadas to a height of 3715m, the highest peak in Spain and in the whole of the Atlantic Ocean, and the highest volcano in Europe outside the Caucasus. It dwarfs its companion, the Pico Viejo, which, at 3134m, is itself almost as high as Etna. But they rise so high because they formed on the floor of the Caldera de las Cañadas, which itself stands 2100m above sea level. Teide has been an unmistakable landmark for six centuries of mariners, ever since the northeast trade winds led the first sailing ships to the Canary Islands. The trade winds also brought their equable climate and orographic rainfall to the northern slopes of Tenerife, which gave rise to rich vegetation and agriculture, in contrast with the aridity of the rain shadow in the south, which is fit only for tourist resorts. Teide itself rises well above the trade winds and enjoys a dry regime of clear skies and westerly breezes.

Teide and Pico Viejo grew up less than 200,000 years ago inside the Caldera de las Cañadas, which has been famous ever since it was described by Humboldt and Von Buch in the early nineteenth century. It formed at the crest of a huge volcanic pile, now called the Cañadas volcano, which had, itself, grown up on top of basal shields built up by basaltic eruptions from the floor of the Atlantic Ocean. Teide is crowned by a small cone, El Pitón, with a white crater emitting fumes. It is unlikely that the Pico de Teide itself has seen any activity since Columbus, en route for the Americas, witnessed an eruption in August 1492. However, several eruptions have occurred in Tenerife since the Spanish settlement, and the latest formed the cinder cone of Chinyero in 1909. These are but the most recent of a whole series of eruptions that have formed dozens of cones and lava flows in the past few thousand years on the long, high ridges that form the Y-shaped backbone of the island.

As in the rest of the Canary Islands, the lavas of Tenerife are predominantly basaltic, and they emerged chiefly from fissures in fluid flows. But occasional evolution in shallow reservoirs produced eruptions of more evolved compositions (e.g. trachybasalts, trachytes and phonolites – see chapter 1, Fig. 1.3 for classification nomenclature). These more evolved eruptions expelled extensive ignimbrites and pumice or ash layers, with a total volume of some 70km³, but also produced lava flows such as those decorating the southern slopes of Teide, as well as some domes in the Caldera de las

Cañadas. Several recent lava flows of both phonolite and basalt in the caldera also have conspicuous blocky obsidian surfaces. Teide itself has a very shallow magma chamber and its eruptions have been basaltic, trachytic and phonolitic.

The basal shields

Eruptions concentrated at first in three separate zones of the ocean floor, where basaltic flows issuing from fissures had built three island shields by about 3.28 million years ago. They now form massifs on the three extremities of Tenerife: the Anaga peninsula in the northeast, the Teno peninsula in the northwest, and the Roque del Conde in the south, near Adeje. The basal complex probably occurs below these basalts, because fragments have been thrown up as xenoliths during historical eruptions.

The Anaga Massif is an accumulation of some 1000m of tabular flows of alkali basalt and basaltic trachyandesite, interspersed by cinder or ash layers and transected by innumerable dykes, and even occasional domes. It was formed 6.5–3.28 million years ago by three similar cycles of activity separated by intervals of repose. The Teno Massif is also composed of tabular basaltic flows, with some explosive breccias, mudflows and cinders in the lower levels and some trachytes in the upper, again transected by innumerable dykes, which can be seen along the Teno (Fig. 4.4). The eruptions began about 6.7 million years ago and ended when the Roque Blanco phonolitic dome erupted about 4.5 million years ago. Near Adeje, in southwestern Tenerife, about 1000m of basalts are exposed in and around the Roque del Conde. The youngest eruptions formed the trachytic dome of the Roque Vento about 3.8 million years ago.

A largely dormant period followed 3.28–1.9 million years ago. Barrancos cut deep into the shields, and the basaltic plateaux were reduced to ridges, knife-edge *cuchillos*, or pinnacles, such as the Roque Imoque near Adeje. At the same time, marine erosion trimmed the faulted seaward flanks of the shields, often revealing spectacular cliff cross-sections of the geology (e.g. Fig. 4.4).

The Cañadas volcano

When activity resumed on Tenerife, it changed in both location and eruptive style. The new concentrated vents built two successive stratovolcanoes where the regional rift fissures of Tenerife intersect. The first stratovolcano, now called Cañadas I, erupted for nearly a million years.

Figure 4.4 Examples of the basal shields; **A)** view along the Teno coastal peninsular looking north with dykes cutting through the volcanics; **B)** students mapping dykes on the Teno Massif; **C)** close-up of dykes cutting through lavas and volcaniclastic rocks at the base of cliffs along Teno peninsular (photo **C** from Shutterstock/ricok).

It grew up from vents on the west of the central area, and its deposits now lie in western Tenerife. Its initial basaltic eruptions, 1.89–1.82 million years ago, were followed about 1.5 million years ago by mixed basaltic and more evolved eruptions. Explosions of pyroclastic density currents about 1.24–1.05 million years ago were probably related to caldera collapse, which brought the main eruptions of Cañadas I to an end.

Eruptions from close-knit vents to the northeast immediately started to build Cañadas II stratovolcano. Its deposits now lie in eastern Tenerife and are well exposed, for instance, in the eastern wall of the Caldera de las Cañadas. They form many layers of ash and lava flows, ranging from basalts to trachytes and phonolites, which are about 650,000 years old. However, effusive phases were interspersed with both explosive episodes and dormant periods, when soils formed. About 600,000

years ago, Cañadas II underwent a phase of collapse, revealed by many layers of fragments exposed in the Cañada de Diego Hernández section of the caldera rim. Nevertheless, the eruptions of Cañadas II continued unabated until they had built up a large and rather unstable stratovolcano, which almost certainly rose well above 3000m, for its highest remaining flank still culminates at 2717m at the Montaña Guajara.

The Cañadas volcanoes together represented a considerable eruptive output. A million years of eruptions probably gave the Cañadas I stratovolcano a volume of 350–400km³. The Cañadas II stratovolcano probably reached between 150km³ and 200km³ in volume during the ensuing 800,000 years. In comparison, however, the Teide–Pico Viejo volcanoes have erupted twice as fast, for together they are already approaching a volume of 150km³ after less than 170,000 years of activity.

The Caldera de las Cañadas

Arguably the most spectacular feature of the Island of Tenerife is the Caldera de las Cañadas. The formation of the Caldera de las Cañadas was the last and most spectacular event in the evolution of the Cañadas II stratovolcano. This majestic caldera is elliptical, double and asymmetrical, and its main axis trends some 17km from northeast to southwest (see Fig. 4.5). The Caldera de las Cañadas harbours some of the most striking volcanic landforms in Europe, where the crystal-clear atmosphere, the brilliant sunshine and the absence of any continuous vegetation combine to emphasize the varied colours of the rocks, displayed in almost pristine splendour like a painted desert. Here steel-blue, brown, or black and glassy lava flows, and grey or yellow pumice piles and dark red cinder cones decorate the base of the grey Pico de Teide and its black companion, the Pico Viejo, which

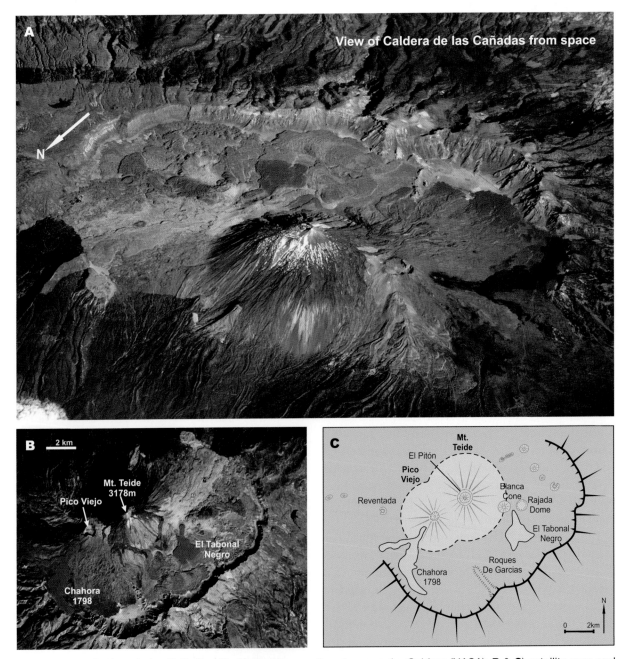

Figure 4.5 The Caldera de las Cañadas: **A)** view looking southeast across the Caldera (NASA); **B** & **C)** satellite map and outline of features and recent activity in the Caldera de las Cañadas (satellite map, USGS).

produced the latest eruption within the caldera in 1798. And, in the south, the rim of the Caldera de las Cañadas protects them all with a cliff that reaches 700m high and curves for more than 20km from El Portillo in the northeast to beyond the Boca de Tauce in the southwest. The Roques de García jut out from this cliff and divide the caldera itself into a larger eastern hollow and a smaller, lower, western hollow. But there is no sign of the northern rim of the caldera, and opinions differ about what might have happened to it. Fine views of the caldera can be found from the caldera rim or during the cable car ascent of Teide and from its summit (see Fig. 4.6).

The Caldera de las Cañadas is a classic among European calderas, and the many opinions about its formation illustrate the changing evaluations of the tectonic, erosional, explosive, collapse and mass-movement processes that have been proposed to explain calderas during the past 200 years. It was first considered a crater of elevation; then an erosional amphitheatre similar to its supposed counterpart, the Caldera de Taburiente in La Palma; and, even later, as the headwaters of two river systems, whose exit northwards had been later covered by Teide and Pico Viejo. Later still, it was thought to have resulted from a massive explosion, like that which was then believed to have destroyed Krakatau in 1883. But it became accepted that it had formed when the summit of the volcano had collapsed after massive eruptions of magma, either in one single or several repeated catastrophes. The intricate network of galleries built to extract underground water show that the northern rim of the caldera is entirely missing. Thus, if collapse formed the northern rim, landslides might have removed it later. In a simpler view, the Caldera de las Cañadas has been seen as entirely the result of a gigantic landslide, whose uppermost arcuate scar forms the majestic southern rim of the caldera. Thus, the upper 100km³ of the unstable

Figure 4.6 Views of the Caldera de las Cañadas. **A**) Panoramic view into caldera with Teide and Pico Viejo in view (photo courtesy of Davie Brown). **B**) View of caldera wall during cable car ascent (inset showing the cable car).

Cañadas II stratovolcano apparently suddenly slipped north-westwards into the Atlantic Ocean, where it seems to be related to vast debris-avalanche deposits on the submarine flank of northern Tenerife. A seaward-sloping layer of chaotic breccia, 100m thick, could represent the sliding plane. Although the landslide was not caused by direct volcanic action, it did, however, unleash pyroclastic density currents. Their deposits still crown the summit of the south-eastern rim of the caldera at the Cañada de Diego Hernández, and have been dated to 130,000–170,000 years ago. They indicate the oldest possible date for the birth of the Caldera de las Cañadas.

The Roques de García, separating the two parts of the caldera, form a ridge that has been dissected into a number of individual pinnacles. The most famous of these is the much photographed Roque Cinchado ('tightened-belt rock'), which is also often referred to as the Arból de Piedra ('tree of stone'). The ridge is about 200m high and 2.5km long, and is composed of an array of brightly coloured, altered volcanic rocks, notably at Los Azulejos. They form an intrusive complex that is apparently older than its surroundings. The Roques de García jut out from a distinct saddle in the southern rim of the caldera and may represent some of the original vents of the Cañadas stratovolcanoes, which have survived in the landscape because they were plugged with agglomerates that offered greater mechanical resistance as the caldera developed.

Soon after the Caldera de las Cañadas formed, the Pico Viejo and the Pico de Teide grew up as a pair of large stratovolcanoes in their own right, along a major fissure trending from northeast to southwest from the Cumbre Ridge to the Santiago area. They erupted from separate vents that issued from a common shallow reservoir, where the three rift-fissure zones of Tenerife intersect. The magma evolved towards a trachytic and phonolitic composition, but explosive eruptions of ash and pumice were punctuated by frequent emissions of viscous lava flows that form the armour-plating on both volcanoes. Pico Viejo and Teide were active during the same period and their products are often interbedded, but they are dissimilar twins, and their different eruptive styles created very different crests. Pico Viejo has a wide crater, whereas the crater of Teide, La Rambleta, has been completely filled by the summit cone of El Pitón. Both volcanoes have remained active during the past millennium.

Pico Viejo

Pico Viejo is broader, lower and more gently sloping than Teide, but still forms a considerable volcano, rising 1034m above the floor of the caldera to a height of 3134m. It erupted many aa lava flows, but many recent flows, like those emitted in 1798, have pahoehoe surfaces. However, some of the youngest phonolitic flows were so viscous that they solidified in thick elongated masses within 300m of their parent vents, even on slopes of 20°.

Pico Viejo is crowned by a circular crater 750m across and 150m deep, which is rimmed by a ragged scarp (see Fig. 4.7). It could have been formed by a large hydrovolcanic explosion, but it seems more likely to be a small caldera formed by recent collapse along ring fractures, because few exploded fragments occur on its flanks. The latest events on the summit occurred when hydrovolcanic explosions blasted out two pits, one 140m and the other 75m deep, into the floor of the crater.

Figure 4.7 Views of the crater at Pico Viejo looking down from Tiede. The island is well signposted about the geology with many information boards, including ones at the top of the volcano (inset photo of Dougal Jerram at one of the information boards about Pico Viejo).

The flanks of Pico Viejo have also witnessed eruptions recently. The oldest occurred on a fissure radiating northwestwards from the stratovolcano and formed the phonolitic dome of the Roques Blancos, which sagged down on its lower, northwestern side and formed a stubby lava flow. The fissure later extended further down slope and expelled a black obsidian flow that eventually reached 3km in length. The second eruption formed Los Gemelos ('the twins') along the saddle linking Pico Viejo to Teide. The twin craters were produced by a brief hydrovolcanic explosion that expelled pumice, ash and breadcrust bombs. Their sharply defined outlines indicate their youth. Most recently, in 1798, the only eruption recorded within the Caldera de las Cañadas took place 1km further down the western flanks of Pico Viejo and formed the Chahorra vents, or the Narices ('nostrils') de Teide (Fig. 4.8).

Figure 4.8 View of the Chahorra or Narices de Teide (nostrils), which erupted in 1798 on the flanks of Pico Viejo. Lavas can be seen in the foreground, and as dark flows down from the side vent on Pico Viejo (photo courtesy of Francis Abbott).

Pico de Teide

The Pico de Teide is a majestic volcano in a splendid setting (see Fig. 4.9). It is a large symmetrical cone, streaked with dark grey lava flows spilling like paint from its summit. Eruptions of trachybasalt and especially of phonolite have built Teide within the past

Figure 4.9 Views of Mount Tiede **A**) view of Teide and Pico Viejo with snow covering; **B**) El Pitón on Teide with Montaña Blanca; **C**) Mount Teide summit crater: spectacular views of the Las Cañadas caldera from the summit of Mount Teide stratovolcano at 3718 metres altitude. The small 80-metre wide crater has hundreds of active fumaroles emitting sulphurous water vapour (photos **A-C** courtesy of Francis Abbott). **D**) Classic photo of Teide with the Roque Cinchado in the foreground.

Meet the Scientist — Valentin Troll

Val is an award-winning volcanologist with a wide range of interests. He is currently the Chair of Petrology at Uppsala University, where he works with a team of exciting scientists doing research on volcanoes around the world. He has been working on the Canaries for many years, and here he provides some insight into his passion for these volcanic islands.

How long have you been working on this volcano/volcanic system?
My first trip to the Canaries was on a research vessel over Christmas 1998 and it rained. I have worked on the volcanic landscapes of the Canary Islands ever since and earned my PhD with a thesis on the *Evolution of large peralkaline silicic magma bodies and associated caldera systems: a case study from Gran Canaria, Canary Islands.*
To date, I have published over 50 scientific articles on the Canary Islands in international scientific books and journals and led about one dozen fieldtrips for students and/or scientists from international organizations to the islands. The recent publication of our Springer book on Teide Volcano (Carracedo and Troll, 2013) and the new *Geology of the Canary Islands* (Elsevier) represent the pinnacle of my efforts over the last 15 years.

What has been your most exciting discovery about it?
The increasing knowledge on the interior of the islands. Volcanologists can actually look inside a volcano, by reading the signals the volcano directly or indirectly emits (such as erupted rocks, emitted gasses, deformation pattern and stratigraphy). Volcanoes grow like living beings. There are young volcanoes, like on La Palma, there are mature volcanoes, like Teide on Tenerife, and there are 'emeritus' volcanoes like Gran Canaria. Piecing together the life cycle of a typical island, seeing the links between places and similarities in processes, has likely been the biggest revelation to me so far.

What is the most important thing we still need to find out?
We still need to find out 'exactly what does what inside a volcano?' We really have not been there (apart from Jules Verne) and can only speculate on 'cause and effect' relationships for many aspects. Scaled experiments will be vital in the next decade, as will be comparative studies between eroded plutons and active volcanic systems (e.g. Fuerteventura vs. Teide). This includes the peculiar floating stones erupted during the first week of the 2011 El Hierro events and their significance on a regional scale in the Canaries, an exciting new aspect that is still to be explored.

What inspired you to study volcanoes?
I am inspired by their explosive force and internal dynamics. Explosions occur by rapid, unchecked volume expansion of materials. Dynamite, for example, undergoes a virtually instantaneous volume expansion of factor 60 during an explosion. This phenomenon occurs at volcanoes too, and leaves its mark on the eruptive products. The textures and chemical signals preserved in the rocks can then be interpreted to better understand the inner workings of volcanoes, though, admittedly, much of the 'vocabulary' and 'grammar' of this language still remains to be fully understood.

170,000–130,000 years. Its lower northern flanks are clothed with distinctive reddish-brown or bluish-grey aa and blocky flows of fairly viscous phonolites, which form tongues, about 12m to 15m thick, with lobate ridges, lava moraines and steep rubble-covered margins. On the steepest slopes in the east, lava masses have sometimes detached themselves from the snouts of the flows and have rolled forwards, like snowballs, down slope.

Examples of these can be seen as the contrast of black lava block 'eggs' on the light-coloured surface of Montaña Blanca (see Fig. 4.10). In contrast, the flows reaching the floor of the Caldera de las Cañadas sometimes spread out in wide lobes with crescent-shaped pressure ridges that run parallel to their margins. Most of these flows have suffered such limited atmospheric alteration that they cannot be many thousands of years old.

Figure 4.10 **A**) Val Troll on Montaña Blanca, Tenerife, with one of Teide's 'eggs' in the background. These are large fallen blocks from the nose of the last summit eruption at Teide, the lavas Negras (summit of Teide is further to the left); **B**) Close-up view of a fallen block with slopes of Teide in the background (photo **A** courtesy of Val Troll, photo **B** courtesy of Francis Abbott).

Like Pico Viejo, Teide probably had an open summit crater for a long time, the rim of which can still be discerned as the shoulder at La Rambleta. The cone of El Pitón now fills this crater and crowns Teide like a giant sandcastle. It is itself a small stratovolcano 150m high, which encloses a crater, La Caldereta, which is 70m in diameter and 40m deep. Apart from the contemporary fumaroles that have changed its crater walls to yellow and brilliant white, El Pitón has been as quiet as its parent, Teide, for the past 500 years. The latest magmatic eruption gave off fresh black, glassy, phonolitic aa lavas that spilt from El Pitón and flowed in ridged tongues down the flanks of Teide as far as the Caldera de las Cañadas. This emission was then followed by the throat-clearing explosion that gave El Pitón its present funnel-shaped crater.

Seven satellite vents surround Teide. Five of these vents lie on a circle, 3km from Teide, which probably represents a ring fracture in the Caldera de las Cañadas. The eruptions probably occurred where this ring fracture intersected fissures radiating from Teide.

The Pico Cabras, the Montañas de las Lajes and the Montaña Abejera, in the northeastern sector of Teide, have much in common. Each eruption began with the extrusion of a brown phonolitic dome, which distended and breached on its northward downslope side, and delivered a more fluid phonolitic flow extending towards the northern shore of Tenerife. In contrast, the Montaña Majua and the Montaña Mareta, which erupted on the gentle slopes beyond the southern base of Teide, produced only lobes of unusually fluid black phonolitic lava that are 10m thick and over 2km long and wide. The Montaña Majua lavas eventually accumulated thickly above the vent, but the Montaña Mareta remained as only a low mound. It is very similar to El Tabonal Negro ('the black table'), which issued from a vent low on the southeastern flanks of Teide and emitted thick, blocky black phonolitic flows, decorated by arcuate ridges, that spread over 3km across the caldera floor.

The two remaining flank eruptions of Teide took place about the same time. They formed the Montaña Blanca and the Montaña Rajada, which both had more complicated histories that are reflected in their more varied relief. The first eruptions of Montaña Blanca came from five vents along a fissure radiating from Teide and gave off flows of brown phonolite. They were followed by a subPlinian explosion of frothy phonolites, which was the most violent eruption on Tenerife since the formation of the Caldera de las Cañadas. It formed the Montaña Blanca, a mound about 300m high and 1km across, and composed of loose yellow pumice, with lumps attaining up to 30cm in diameter. The eruption ended with a return to calmer conditions and the emission of small domes and flows. All the flows from the Montaña Blanca are notably rugged, with rough angular blocks on their aa surfaces, and their margins are invariably very steep and rubble-strewn. These eruptions took place about 2000 years ago.

The Montaña Rajada, 1km east of Montaña Blanca, erupted bulky phonolitic lava flows. These travelled 5km, wrapped around the older Montaña Mostaza cinder cone, and formed the remarkable chaos of blocky and shining phonolites of the Valles de las Piedras Arrancadas ('valley of the uprooted stones'). Succeeding emissions were even more viscous and travelled scarcely more than 200m from the vent. The very latest lava emissions from this vent formed the Montaña Rajada itself ('the split mountain'), which is about 250m high and 1km in diameter. It forms a rugged dome of reddish-brown phonolites, whose summit was burst open by an explosion that formed a ragged central hollow into which a similar but smaller dome then intruded. The Montaña Rajada also has a scattering of fine pumice on its surface, which could be derived from the Montaña Blanca nearby. The northwestern flank of the Montaña Rajada also burst open and extruded a very viscous phonolitic flow that congealed before it reached the foot of the 20° slope.

The pyroclastic rocks of Tenerife

Explosive eruptions have littered the volcanic past of the island of Tenerife. Its pyroclastic rocks are worthy of a specific mention, as the island contains some spectacular examples and the deposits from its explosive eruptions have been used to inspire our modern-day understanding of the pyroclastic density currents that result from such volcanic explosions (Fig. 4.11). Numerous widespread, and in some cases large-volume, pyroclastic density currents have swept radially across the island over the last 2Ma. The resultant deposits (ignimbrites) are variously preserved in and around the island, with a good concentration of very good outcrops to be found on the slopes and the coastline around the southern parts of the island, known as the Bandas del Sur ('Band of the Sun'). One ignimbrite of note is the 273,000-year-old Poris Formation, which fills valleys and is exposed along the coast (see Fig. 4.11A). The ignimbrites of Tenerife display many of the classic features associated with deposits formed from pyroclastic density currents, including cross- and massively-bedded deposits, pumice and lithic-rich layers, and accretionary lapilli; and in some cases the material was hot enough when it was emplaced that it started to plastically deform, stick back together and produce welded ignimbrite (Fig. 4.11C, D).

The coastline is also home to many shallow cones and ring-like formations, interpreted to be the result of hydrovolcanic eruptions, termed tuff rings, where water and magma interact explosively in the shallow subsurface. These structures have, in some instances, been dissected by erosion from the sea, and reveal the inside stratigraphy of the tuff rings in extraordinary detail. A classic example of one of these features is the Montaña Pelada tuff ring, near the airport on the southern coast, where complex layers of tuffs are exposed along the beach and slopes, with some showing fantastic impact structures from some of the blocks/bombs associated with their eruption (e.g. Fig. 4.12).

Rift-fissure eruptions

As the Cañadas stratovolcanoes were growing up, Tenerife also resumed, or continued, its old pattern of rift-fissure eruptions that have given the island its characteristic Y-shape, and they have persisted into historical times. The emissions seem to have come from blade-like insertions of magmas that could rise fairly quickly from a deep source, and which formed basaltic cones about 100m high and flows up to 10km long.

Northwest of the Caldera de las Cañadas, for example, a set of rift fissures trending from northwest to southeast gave rise to the Santiago volcanic field of recent cones and basaltic flows. Where these fissures extend into the caldera, they formed the Montaña de Samara, the Volcán Botija, the Montaña de la Cruz de Tea, and the Montaña Reventada ('breached'), which has been dated to AD1000–1300. This set of fissures also produced eruptions in 1706 and 1909. At the northwestern end of these rift fissures, the large Taco cone erupted near the sea at Buenavista. Its unusual size – it is almost 1km across and 200m high – and its large, deep crater, as well as its fine tuff layers, all indicate that it was formed by Surtseyan eruptions in shallow water, during a recent period of higher sea level.

A companion set of fissures erupted aligned cinder cones and fluid basaltic flows that built the backbone of northeastern Tenerife, the Cumbre Ridge. It is 25km long, 18km wide and 1600m high, and probably represents one of the fastest volcanic accumulations experienced on the island. Such rapid construction helped make its flanks unstable and eventually led to the landslides by which the Orotava Valley in the north and the Güímar Valley in the south collapsed between 830,000 and 560,000 years ago (see map in Fig. 4.3).

Both the Orotava and Güímar valleys are 10km wide, with smooth floors sloping gently seawards and bounded by crisply outlined inward-facing straight cliffs extending at right angles to the Cumbre Ridge. They look like two piano keys pressed down on either side of the Ridge. They have been recently interpreted as great landslides displacing more than

Figure 4.11 Pyroclastic rocks of Tenerife. **A)** Students working on exposures of the Poris Formation on the South Coast of Tenerife. **B)** Different layers within ignimbrite deposits made from pyroclastic density currents. **C & D)** Thin welded ignimbrite showing streaked out textures. **E)** Dougal Jerram highlighting a fossil tree mould in an ignimbrite. **F)** Rounded accretionary lapilli showing concentric rings.

$100km^3$ of lavas with a total thickness of 150–600m. They now lie upon a layer of chaotic breccia and sandy clay that was apparently crushed when the volcanic materials slid over them.

One of the most notable fissures recently erupted Montaña Colmenar and the Siete Fuentes cones on the Cumbre Ridge. It extends into the eastern part of the Caldera de las Cañadas, where it has given rise to

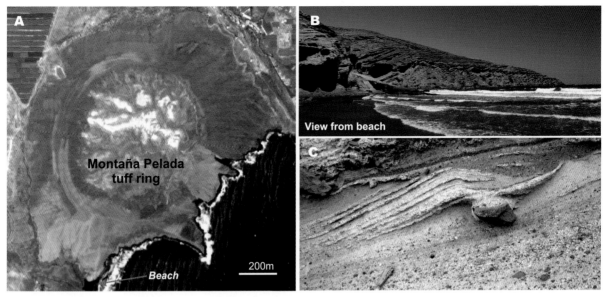

Figure 4.12 The example of the Montaña Pelada tuff ring, southern Tenerife; **A)** aerial view of circular tuff ring; **B)** view of deposits from beach (see location on **A**); **C)** close-up of volcanic bomb and sag on the slopes of the tuff ring (aerial view – NASA World Wind).

The longest lava tube in Europe

Tenerife is home to the largest lava tube in Europe, and offers some fantastic opportunities to explore the underground world of a volcano. Located in the northeast of the island, the Cueva del Viento-Sobrado ('Cave of the Wind') is an underground lava tube complex that extends over 17km in length. The name of the tubes comes from the powerful winds and draughts which rattle through the extensive cave system. Located near the town of Icod de los Vinos, the lava tubes are the result of basalt lava flows sourced from the Pico Vejo area some 27,000 years ago. The tubes display a number of morphological features that help decipher how they formed. Including lava stalagtites, flow patterns and lava ledges mark different levels of lava flows through the tunnels (e.g. Fig. 4.13).

This geological beauty is also of special interest from a biological and anthropological standpoint. The cave is home to a total of 190 known species, most of which are invertebrates. Including some 44 troglobites (animals adapted to underground environments), and specialist mosses, lichens and cyanobacteria. Of these species condemned to live in the darkness, 15 are new to science, such as the eyeless cockroach (*Loboptera subterránea*). Fossils have also been found in the cave system, including the giant rat (*Canariomys bravoi*), and the giant lizard (*Gallotia goliath*). Anthropological

findings include artefacts and burial remains of the original 'Guanches' settlers from as far back as 2000 years ago. Today small tourist tours are allowed through parts of the lava tube network, and in 2014 the Cueva del Viento Natural Resource Management Plan was adapted to European legislation for Special Areas of Conservation. Exploration and discovery of the caves on this island and elsewhere on the Canaries is ongoing by scientists and members of the Canary Islands School of Speleology

Figure 4.13 View within the Cueva del Viento-Sobrado ('Cave of the Wind'), the longest lava tube in Europe (photo courtesy and © of Alfredo Lainez Conception on behalf of Speleology of Tenerife Benisahare, and La Cueva del Viento, Icod, Tenerife).

the Montañas Negra, Los Tomillos, Los Corrales and the highest, Mostaza, which rises over 100m. Montaña Mostaza was formed before the Montaña Rajada, whose lavas are wrapped around its base.

Historical eruptions on Tenerife

Most of the activity on Tenerife during historical times took place along rift fissures. The early references to eruptions on the island are so vague that it is uncertain when, or even if, they occurred. However, there can be little doubt that, before the Spanish conquest, the local Guanche peoples must have witnessed eruptions on the island, notably on Teide. For the Guanches, Teide was Echeide, the Inferno. As the Italian engineer Torriani wrote in 1592, 'because of the terrible fire, noise and tremors that come from it, they consider it to be the home of demons'.

Andalusian and Basque sailors might have seen an eruption on Teide in 1393 or 1399. Hum-boldt reported that the Guanches had told their Spanish conquer-ors of an eruption in about 1430, which might have formed the cinder cones of Las Arenas, Los Frailes and Gañañías, in the lower Orotava valley. The Venetian mariner, Cadamasto, approached Tenerife in 1455 and later declared that 'the mountain that rose above the clouds [Teide] ... was glowing incessantly.' Alonso de Palencia, chronicler of the conquest of Gran Canaria, wrote that, somewhere between 1478 and 1480, 'fire surged continually from an infernal mouth in the centre of the highest mountain [Teide] ... Small chips of stone were carried on the wind to the very edge of the sea'. Christopher Columbus called at the Canary Islands on his first voyage to the Indies. On 24 August 1492, his logbook recorded that, while La Pinta was anchored off La Gomera, 'they saw very large flames coming from the mountain, which filled the crew with wonder. [Columbus] explained the cause of such fire, saying that it was just like Etna.' It is possible that one of these fifteenth-century accounts could refer to the eruption that gave off the fresh-looking lavas that stream down from El Pitón. On the other hand, it is strange that no record survives, of what must have been a spectacular display, from the Spaniards who were established on nearby La Gomera from 1402.

Then, for over a century after the Spanish conquest of Tenerife in 1493, the only signs of volcanic activity were the fumes that issued from the crater of El Pitón. But the first decade of the eighteenth century saw four basaltic eruptions and the destruction of half a town.

The crisis began on Christmas Eve in 1704. Twenty-nine earthquakes had been counted before dawn on Christmas Day. During the next few days, more and more earthquakes and rumblings were felt throughout the island. Calm returned on 29 and 30 December. Then, on 31 December 1704, a fissure opened on the Cumbre Ridge and began the little eruption that built the Siete Fuentes cone. A string of vents expelled lava fountains, ash and cinders, which constructed three cones – the largest of which reached only 22m high – and sent an olivine basalt lava flow, 1km long, into the neighbouring valley. Siete Fuentes had probably stopped erupting when the next episode began about 900m to the northeast.

At 08.00 on Monday 5 January 1705, a new set of earthquakes started and, that afternoon, a fissure opened and the Fasnia volcano began to form. Over a length of about 1400m, 30 initial vents formed an array of cones, explosion pits, hornitos, lava fountains and lava flows. The number of active vents fell to eight on 7 January, but they still 'made as much noise as the artillery', and fumes and ash descended on the nearby towns of La Orotava and Güímar. When the eruption ended, on 16 January 1705, the largest cone was 550m long but only 37m high, and half of it had been destroyed as basalts had flowed out into the valley nearby. This aa flow has some fine lava moraines and flow channels. No more lava emerged during the rest of January, but the earthquakes continued, the ground rumbled, the surface cracked and fumes escaped. It is said that one shock threw down 70 houses and killed 16 people in Güímar. If this tale is to be believed, this volcano-seismic event was the most lethal of all the recorded eruptions in the Canary Islands.

The next eruption arose on the same fissure, about 7km northeast of Fasnia volcano. The Volcán de Arafo began to erupt between 16.00 and 17.00 on Monday 2 February 1705, at the very head of the Güímar valley, whose upper walls tower 400m above it. Fountains of lava spurted 30m into the air, and ash and cinders rose much higher. Arafo produced higher columns of fumes, more fragments, more lava, and more noise than its three predecessors – and it lasted at least until 27 February. Thus, Arafo reaches a height of 102m. One of its two main craters is breached by an aa flow that developed distinct flow channels and many accretionary balls, and almost reached the sea, 8km away.

These three eruptions were brief and small-scale, even by the modest standards of their type. All gave out

Figure 4.14 *Eruption 1706* by Ubaldo Bordanova Moreno in 1898 (it is actually a copy/interpretation of an earlier painting by an unidentified artist).

olivine basalts, and covered a total area of only 12km². However, earthquakes continued to shake Tenerife for more than a year afterwards, hinting that another eruption was yet to come. But it came on the northwestern rift fissure and severely damaged Garachico in 1706.

The only eruption that is certain to have occurred within the Caldera de las Cañadas in historical times took place in the summer of 1798. This time, at least, it happened well away from any settlements – 2300–2800m high on the southwestern flanks of the Pico Viejo. However, it is rather confusingly called the Chahorra, or Narices, de Teide eruption. It took place on a fissure 850m long, and it was unusual for such an eruption to occur on a stratovolcano, because the activity started at the lower end of the fissure and progressed up slope. If it did indeed last for 99 days, as some have supposed, it could have been one of the longest eruptions in the Canary Islands during historical times. Earthquakes had been felt as early as 15 June 1795, but they did not reach their climax until late April 1797. However, the Chahorra magma did not make its appearance until just after 09.00 on 9 June 1798. Fifteen vents opened in the next two days as the fissure spread up the slope, as if Pico Viejo were being unzipped. By 13 June, eruptions

The destruction of Garachico, 1706

The eruption that began at 03.00 on 5 May 1706 was the most destructive that has been clearly recorded on Tenerife. There was only one preliminary earthquake before fluid basalt started gushing from a fissure on the ridge rising behind the flourishing port of Garachico, on the north-western coast (depicted in the painting in Fig. 4.14). Activity tended to migrate from northwest to southeast along this fissure. In the northwest, a spatter rampart, 5m high and about 400m long, formed only along its southern edge, because continuous streams of lava prevented it from developing on the downslope side to the north. On the southeastern part of the fissure, on the other hand, a mixture of explosions and effusions formed a cinder cone 80m high. But there again, persistent lava emissions stopped its northern sector from forming. This cone has been called the Montaña Negra, or the Volcán de las Arenas Negras ('the black sands'), but most often as the Volcán de Garachico, because of the damage that its eruption caused in the little town.

Founded in 1505, Garachico had soon become the main port in northern Tenerife. But, in 1645, floodwaters from a burst dam had badly damaged part of the town, and in 1697, fire had ravaged some of the best buildings along the shore. The new volcano was to bring yet another threat.

At 03.00 in the morning of 5 May 1706, the basalt flowed into the hamlet of Tanque, burned the church and several houses, and destroyed the vineyards nearby. At about 21.00 that same evening, a larger flow swept across the plateau and poured in seven separate cascades down the steep scarp dominating Garachico. It entered the town and filled the harbour to such an extent that little more than a creek remained. At 08.00 on 13 May, an even stronger flow rushed down the scarp, burned watermills and windmills, buried orchards and blocked springs. It reduced the San Francisco monastery and the Santa Clara convent to ashes, overwhelmed the finest quarter of Garachico, set fire to many houses and warehouses along the shore, and then spread in a fuming lobe out to sea. It is said that the basalts glowed for 40 days. But the citizens had time to flee with their goods towards the neighbouring town of Icod, often accompanied by – now equally homeless – members of the religious orders, who sang psalms to maintain morale. Many people in both the town and the surrounding countryside lost their livelihoods, but nobody was killed. The eruption probably ended on 28 May 1706, when the cinder cone had reached a height of 80m and its flows had spread 8km from their vent. During the following decades, Garachico was rebuilt, and more or less the same town plan was retraced over the surface of the new lava flows and lava delta (Fig. 4.15).

Figure 4.15 The rebuilt town of Garachico, constructed on the lava flows and lava delta of the 1706 eruption that destroyed the original town (photo Shutterstock/Tatiana Popova).

had concentrated on three vents: the top vent giving off billowing fumes, the middle vent ejecting lava and fragments in hydrovolcanic and Strombolian activity, and the lowest delivering only lava flows. On 14 June, a large explosion joined the two upper vents together. The new vent gave off a column of snow-white steam before resuming the habits of the old middle vent, with noisy explosions and spurting lavas. It eventually built a cinder cone with two craters. The lower parts of the fissure produced a continuous wall of spatter, but the chief eruptive effort of the lower vent was to disgorge copious lava flows. These have both aa and pahoehoe surfaces that vary in aspect from strikingly rugged to smoothly shining. These basanitic lavas broadened out as they reached the floor of the Caldera de las Cañadas and came to a halt and solidified, with a slabby pahoehoe surface, in a big lobe at the foot of its great boundary scarp, some 8km from their source. On 16 June, one

observer saw such a flow reach the floor of the caldera in less than a day. Another observer later declared that the eruption ended in mid-September, but apparently little information about the course of events after June has survived, and it is thus not certain when the activity stopped.

The eruption of Chinyero was the latest to occur in Tenerife (Fig. 4.16). It was typical of the Canaries and lasted for only ten days. It started between 13.00 and 15.00 on 18 November 1909 in the Abeque plateau: not far, in fact, from the Volcán de Garachico. Chinyero caused nothing like the same amount of damage. Although many small earthquakes had been recorded during the previous year, they became more intense and frequent in the autumn of 1909. In the week before the eruption began, they were even accompanied by underground noises, and the ground became warm where the lava was eventually to reach the surface. The

Figure 4.16 Chinyero volcano. **A**) View of the Chinyero volcano with the ash and volcanic bombs that surround it. The common path to the top can be seen winding up its cone. **B**) View back to Chinyero from its rubbly lava flow surface (photo **B** courtesy of Francis Abbott).

eruption started with explosions of ash from three or more vents on a fissure some 650m long that opened on the flanks of an older cinder cone also called Chinyero. Activity migrated towards the northwestern end of the fissure as the new cone grew. Windblown ash fell at least 25km from the vent, at La Orotava, for instance, before the first lava emerged late on 18 November. Next day, four craters were operating, bubbling out frothing lava fountains 50m high and throwing columns of ash 500m into the air, often to the accompaniment of loud explosions. For several days, each crater seemed to take up the main role in turn, but from 25 November activity began to decline, and from noon on 27 November 1909 only fumes issued from the vents. The new cone of Chinyero was of crescent shape and about 50m high, and its olivine basalt flow had wrapped itself halfway around the base of the cinder cone of the Montaña de Bilma, which had, no doubt, grown up in a similar fashion a few centuries or millennia before.

All the historical eruptions of Tenerife covered an area of only about 25.3km², which represents a low rate of production. On average, another eruption

on one of the rift fissures may be expected to occur within the next few decades, and add a little more to this modest total.

El Hierro

The site of some of the most recent volcanic action in the Canary Islands is the island of El Hierro. This is the smallest of the Canary Islands, covering an area of 278km², but it is quite mountainous, reaching 1051m at Malpasso (Fig. 4.17). Rising from the ocean floor at a depth of 3000m, El Hierro was the last island to emerge in the archipelago, and two consecutive, mainly basaltic, edifices quickly formed. The recent

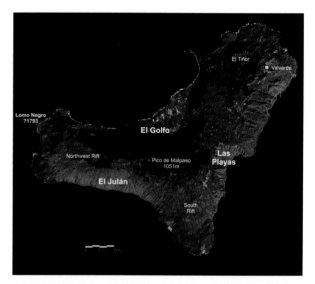

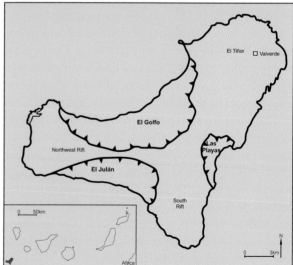

Figure 4.17 Location map, features and outline of recent activity on El Hierro (Satellite Image – USGS).

activity off the coast of El Hierro is testament to the continuing island-building activity at El Hierro (see section below). The oldest dated rocks are about 1.12 million years old and they come from the El Tiñor volcanic complex, which was active for at least about 250,000 years. The second volcanic edifice, El Golfo, grew up on the eroded western flanks of El Tiñor and was active about 545,000–176,000 years ago. It began about 545,000 years ago with dyke-intruded layers of basaltic fragments and ended, about 176,000 years ago, with a predominance of lava flows. Basaltic eruptions from rift fissures then began about 158,000 years ago, virtually as the activity of El Golfo ceased.

El Hierro is a trilobate island, and each arm is dominated by a rift zone. These three arms dominate the scenery of the island, and it is here that magma has been inserted, blade-like, into long parallel feeder dykes, which have thus built up the ridges by frequent and largely basaltic eruptions. The island most probably now lies over an active branch of the Canary Island hotspot. El Hierro, indeed, has the greatest concentrations of recent and well-preserved emission vents in the whole archipelago. Many small cones and lava flows, such as Julán and Orchilla volcanoes, are so little weathered that they can scarcely be more than a few thousand years old at the most. For example, near San Andrés on the central plateau the eruptions of the Montaña Chamuscada and the Montaña Entremontañas took place about 2500 years ago, and many vents near the end of the northwesternmost rift of El Hierro seem to have been active even more recently. But, in spite of its youth, El Hierro has had only one possible onland eruption during historical times. The Lomo Negro, at the western end of the island, suffered many earthquakes between 27 March and the end of June 1793, and a small eruption is thought to have taken place at that time.

Coastal embayments are prominent features of El Hierro. The northern coast of the island is scalloped by the impressive embayment of El Golfo, which is 5km across and bounded by cliffs 1100m high. It was apparently caused by an enormous landslide less than 158,000 years ago. Lava eruptions had built up the rift ridge to such a height that its unsupported seaward flank became unstable and slipped into the sea. On the southeastern coast, a similar landslide, 3km across, formed the Las Playas embayment, the bounding cliffs of which rise to 900m. A third landslide forms the Julán embayment on the southwestern coast. It is still uncertain whether these landslides occurred rapidly or over many centuries, as marine erosion progressively undermined the seaward buttress of the accumulated lavas. However, although El Hierro already has a triple rift system like Tenerife, it is apparently still too young to have developed the massive central vent complex of its larger neighbours.

The 2011–2012 submarine eruption at El Hierro

The offshore area west of El Hierro has seen the most recent volcanic activity at the time of writing this book, and warrants its rather elevated position in this chapter for such a small island. The eruption started in October 2011 along the southern submarine rift zone offshore from El Hierro, and was intermittently active until March 2012. The submarine vent that caused the eruptions was situated between 350 and 100 metres below sea level (fluctuating in depth as the eruption episode continued), and was located about 2km offshore. The activity sparked much interest in the press, as it was speculated that a new island or part of El Hierro might rise from the depths as an emergent volcano. This was not realized, but the activity itself produced some rather remarkable materials brought up from the deep.

The eruption was associated with discoloured murky plumes of water (termed 'la mancha'), which stained the sea water around (see Fig. 4.18). Additionally, early in the eruptions material was brought to the surface as frothy and bubbly fragments of lava, which were often still steaming hot when collected (Fig. 4.18C–F). Many of the fragments also contained a very light-coloured material. The latter period of eruptions was characterized by large gas escapes. It was these light-coloured fragments that were to prove most interesting. On closer inspection they were found to contain fragments of the pre-island sedimentary strata, upon which the volcanic island had grown. The rising magma that fed the eruption had plucked off fragments of this material as it rose, and brought them to the surface (e.g. Fig. 4.19). The sediments contained key microfossils (coccolithophores), which can be dated, and the youngest of these turned out to be 2.5 million years old. With sediments from the easternmost islands being around 20 million years old, these newly discovered younger sediments indicate an age progression of old to young from east to west. This age progression of the island to the youngest at El Hierro has been used to help support the idea of a hotspot/plume origin of the volcanic islands of the Canaries (see Fig. 4.2). It is not clear what additional materials El

Figure 4.18 Images of the 2011–2012 submarine eruption off El Hierro (images courtesy of IGN, Instituto Geográfico Nacional).

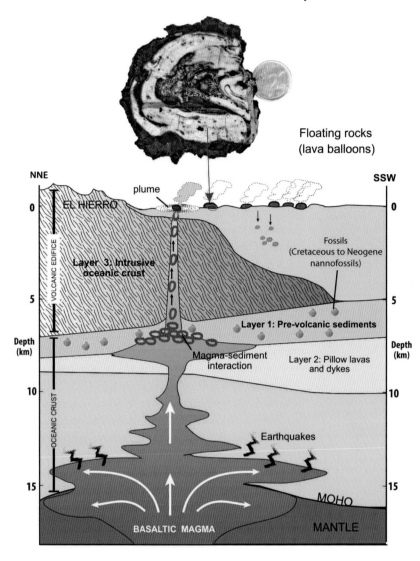

Floating rocks
(lava balloons)

Figure 4.19 Schematic cross-section highlighting the eruptive processes at El Hierro. Magma interacts with the sediment beneath the volcano, bringing sediment inclusions containing microfossils to the surface as floating stones. An example of one of these in section is also highlighted (cross-section courtesy of Val Troll and Frances Deegan; image courtesy of IGN, Instituto Geográfico Nacional)

Hierro will muster from the deep, but it is likely that the island itself will emerge and grow further as the youngest addition to the Canary chain.

Lanzarote

Lanzarote is the northeasternmost of the Canary Islands, covering an area of about 795km². It is 56km long, has a maximum width of 21km and reaches a height of 671m (Fig. 4.20). The higher parts of Lanzarote form the Famara Plateau in the north and the Los Ajaches Plateau in the south, both of which are composed of piles of basaltic lavas and bordered by steep, straight cliffs. Between these plateaux lies the main axis of Lanzarote, an area often below 300m, where lava flows form aprons around cinder cones that have erupted along fissures. The products of the older eruptions are covered with pale ochre caliche, but, in the northwest,

black or reddish basalts form a grim wilderness of cones and flows that erupted between 1730 and 1736, and briefly again in 1824.

The volcanic forms of Lanzarote are all the more striking because the rainfall averages less than 200mm per year and the natural vegetation is often limited to xerophytic plants ranging from *Sempervivum* to prickly pear. There are thus no permanent streams on Lanzarote and, although many older cones are ribbed by gullying, there are very few *barrancos*. However, much of the centre of Lanzarote is blanketed with layers of black ash and lapilli, or *picón*, in which the farmers have planted vines or vegetables (Fig. 4.21).

Practically all the volcanic activity on Lanzarote sprang from fissures that generally follow the overall trend of the island. Eruptions from similar fissures also, no doubt, built up the whole mass of Lanzarote from

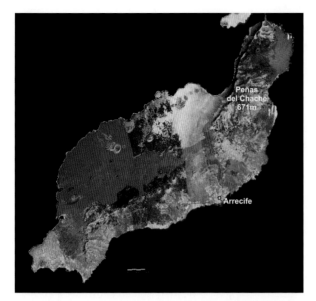

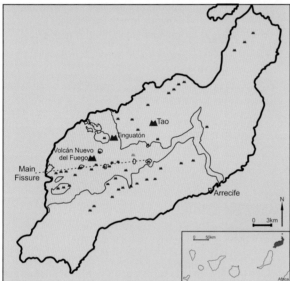

Figure 4.20 Location map, features and outline of recent activity on Lanzarote (Satellite Image – USGS).

Figure 4.21 Vines protected by semi-circular dry stone walls in the La Geria region of Lanzarote (photo by Yummifruitbat – Wikimedia Commons).

a depth of at least 2700m on the floor of the Atlantic Ocean. The first eruptions above sea level in Lanzarote have been dated to about 15.5 million years ago, but three quarters of the surface area of the island erupted from more or less parallel fissures probably less than 500,000 years ago – including one quarter that erupted between 1730 and 1736. Almost all of the lavas found on Lanzarote are basalts of one description or another.

The oldest basalts form the Famara and Los Ajaches plateaux, which are similar in age, origin and nature, and are covered in caliche. Their olivine basalts are at least 670m thick and they came from fissures that first produced thin fluid flows, then cinder cones, and then yet more thin flows. The emissions took place between 15.5 million and 5 million years ago, but were often separated by long intervals when thick fossil soils developed. The plateaux were then tilted about 15° down to the east, so that fault scarps or marine cliffs formed on their western edges. The cliffs reach a height of 500m on the west coast of Famara, but have been widely masked by later eruptions on the west of Los Ajaches.

After over 3 million years of rest, basaltic emissions resumed 1–2 million years ago. Some 25 vents, aligned on east-northeast to west-southwest fissures, formed aa lava flows and cones that now have eroded craters and flanks.

Less than about 10,000 years ago, more basaltic eruptions took place on two main fissures, one following the north coast and the other along the south coast. In all, more than a hundred vents can be identified and their lava flows are the most extensive in Lanzarote at present. These eruptions began with an explosive phase, forming cinder cones and ash layers, but later gave off abundant lava flows, which have now weathered enough to be extensively cultivated. The northern fissures include the Sóo volcanoes, near the north coast, which show a sequence of Surtseyan eruptions from aligned vents that were so close together that each cone was partly destroyed when its successor erupted. At the same time, several fissures inland produced, for example, the Pico del Cuchillo and the Caldera Blanca. The latter is not a true caldera but one of the largest cones on the island, rising 175m, whose vast lava-breached crater is 150m deep and 1200m across. Although it erupted inland, it has the typical dimensions of a Surtseyan cone and owes its name to the pale caliche covering its surface. The beautiful shallow ribbing on its outer flanks displays the typical dissection of the cones of this period (see Fig. 4.22)

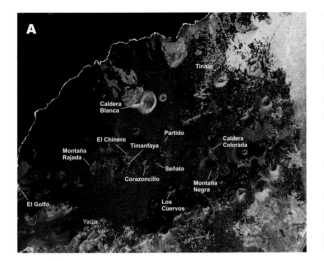

Figure 4.22 The young volcanic area of central Lanzarote. **A**) Satellite view with some of the volcanic cones and features labelled. **B**) Timanfaya volcano is the largest of a number of volcanic cones to be found in the National Park of the same name (photo courtesy of Francis Abbott). **C**) Panoramic view of Caldera Blanca (Shutterstock/FlorianKunde).

The southern bunch of closely parallel fissures produced essentially similar forms. Its cones include Tahiche, Zonzomas and the Montaña Blanca, and two notable cones in the southwest: the Montaña Roja (red) and the Atalaya de Femés ('lookout'). Both erupted on the same fissure. Montaña Roja is the older of the pair and it dominates the Rubicón plain in the far west of Lanzarote. As its name indicates, it forms a reddish cinder cone, 130m high, which is thickly mantled with caliche and many thin, reddish basaltic flows covering about 7km² on the western Rubicón plain. At 608m, the Atalaya de Femés lives up to its name as one of the highest points in Lanzarote, and it forms a cone 100m high. It grew up in three stages: the smaller northern crater probably formed before the larger southern crater, and a minor cone then erupted on the rim between them. Its blocky lava flow reaches the sea and covers about 20km² of the eastern Rubicón plain. This flow is quite young because it covers a recent beach now raised about 5m above the present shore.

At about the same time, the seaward ends of these fissures gave rise to notable Surtseyan cones. At El Golfo, in the far west, the sea has gouged out the deep crater to form the gulf that gave the volcano its name, and exposed innumerable multi-coloured layers of tuffs in the cliffs enclosing the green lagoon (Fig. 4.23). Originally, El

Figure 4.23 The crater and green lagoon of El Golfo, Lanzarote (photo by Gernot Keller– Wikimedia commons).

Golfo may have resembled the five large cones composing the islet of Graciosa that lie at the opposite end of this northern group of fissures. Each cone is almost 1km across, with a typical deep Surtseyan crater that has been protected from marine erosion because the eruptions took place some 5m or so above sea level.

The latest eruptions before the settlement occurred after the latest raised beaches had formed. They include the six cones aligned on a single fissure that crosses the Famara Plateau from northeast to southwest. They probably date from just before the Spanish settlement, because their forms are remarkably fresh, just as *quemado* (burnt) describes their appearance. Their cones include Quemado de Orzola, the Montaña de los Helechos, La Quemada, and especially the Montaña Corona. These vents have erupted a *malpaís* of olivine basalts that extends mainly eastwards to the coast and covers about 50km². The main contributor was the Montaña Corona, which was probably the youngest cone in the series, 609m high and with a crater 418m deep, which was breached on the downslope side by a lava flow that built a broad bulge out into the ocean. This flow contains some well-developed lava tunnels, here called *jameos* (The word used to refer to the large openings in the tube due to collapse). One of these exceeds 6km in length, is 35m high in places and wide enough to have been converted into a concert venue and nightclub. It is also host to the Cueva de los Verdes, a spectacular section of the lava tube system and one of the island's great volcanological tourist attractions (see Fig. 4.24).

The eruptions in 1730–36

The historical eruptions of Lanzarote took place from 1730 to 1736, after at least three centuries of calm. Between 1 September 1730 and 16 April 1736, the island was the scene of one of the longest eruptions in Europe during historical times. One quarter of Lanzarote was given an entirely new landscape of cones and lava flows, and farms and villages in the west and centre were buried completely. Access to part of this area is restricted to protect the landscape as an educational and ecological volcanic reserve in the Timanfaya National Park. But the volcanic forms are equally fine in the rest of the area, where access to this spectacular tourist attraction is free.

During the five and a half years, the focus of activity switched from place to place, and eruptions may not, in fact, have been incessant. However, the magma probably came from the same deep source, and the olivine basalts with a tholeiitic tendency erupted from fissures forming a belt, about 18km long and some 4km broad, running east-northeast to west-southwest. The eruptions produced an extensive *malpaís* decorated with a variety of cones, often with evocative names, such as Calderas Quemadas (burnt), Montaña Rajada (split), Pico Partido (cloven). The cloven profile of Pico Partido, for example, makes it the most distinctive volcano on the island, and its jagged black outline, rising 230m above the *malpaís*, dominates the skyline of central Lanzarote. It is made all the more distinctive by a lava flow, outlined by pale-green lichen, which has spilt like paint down its black lapilli-strewn northern flanks. Pico Partido was, in fact, formed from a cluster of vents (see

Figure 4.24 Cueva de los Verdes. **A)** Concert theatre and night club venue in the lava tunnel (Shutterstock/Jorg Hackemann). **B)** Stunning reflection along section of the lava tube in the Cueva de los Verdes (photo by Luc Viatour – Wikimedia Commons – see also **www.Lucnix.be**)

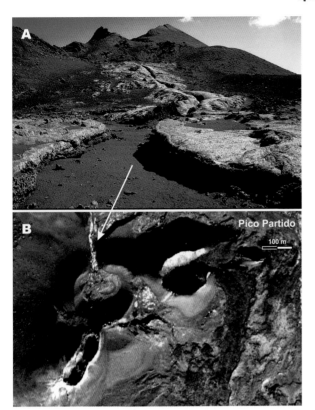

Figure 4.25 Pico Partido. **A)** Its craters form a distinctive cloven summit. An incompletely covered lava tunnel exists from the now solidified lava lake just below the summit. **B)** Pico Partido was formed from a cluster of vents that can be seen from an aerial view (image – NASA World Wind).

Figure 4.26 Volcanoes and vents from the eruptions in 1730–36. **A)** Montaña del Señalo, Lanzarote, Canary islands. **B)** Caldera del Corazoncillo, Lanzarote, Canary islands. The Caldera del Corazoncillo is located close to the Timanfaya volcano and is one of a number of cones to have been formed during the 1730 to 1736 eruptions (photos courtesy of Francis Abbott).

Fig. 4.25) – Mt. Colorada ((red)-coloured), Mt. Roja del Fuego (fire-red), Caldera Fuencaliente (hot spring) and Volcán Negro (black). They make the starkest and most brutal landscape in the Canary Islands, where scarcely a bush has yet taken root after more than two centuries.

During the first 16 months of the eruption, Father Andrés-Lorenzo Curbelo, parish priest of the village of Yaiza, kept a diary of events. The original is lost, but the German scientist Leopold von Buch published a summary, albeit spiced with his own interpretations of events. Other accounts have emerged recently in the Spanish National Archives at Simancas, which describe some of the first efforts ever made to manage a volcanic crisis.

It is uncertain whether any earthquakes preceded the eruption on 1 September 1730. It began modestly with the formation of the Los Cuervos cone over the next 18 days. Calm then returned until 10 October, when the Santa Catalina and Pico Partido cones and flows began to erupt on top of the villages of Santa Catalina

and Mazo respectively. The eruption could now be seen from Gran Canaria. An ad hoc committee was set up in Lanzarote to deal with the crisis, but it could do little to alleviate the ensuing distress. Santa Catalina and Pico Partido stopped erupting on 30 October, but Pico Partido resumed with even greater violence on 10 November, and probably continued for the rest of the winter. By 20 March 1731, a new series of vents had begun more than three months of eruptions that built the Montañas del Señalo (Fig. 4.26A). In June, the focus of activity suddenly switched to the sea off the west coast, where Surtseyan eruptions occurred for a time before the fissures extended onto the land nearby. For the rest of the year, successive vents exploded the cones of El Quemado, in June, Montaña Rajada, in July, and the four Montañas Quemadas from October 1731 to January 1732.

The construction of Timanfaya, a little to the east, seems to have occupied most of 1732 and probably the

following year, for this is one of the largest in the area. Eruptions may have been less vigorous in 1734 and 1735, but in 1736 they found a new site in the east, in March, when the Las Nueces cone was formed; and they reached their last abode in early April when the Colorada cone erupted. The whole episode finished on 16 April 1736.

These eruptions gave rise to some impressive and varied landforms. Timanfaya is a complex accumulation of reddish-brown lapilli and cinders forming a sharp crescent-shaped ridge, 190m high, surrounding a crater about 80m deep. The chief lava emissions came from small vents on its western flanks and made a major contribution to an extensive *malpaís*, decorated with a sinuous lava tunnel that has partly collapsed. East of Timanfaya, Corazoncillo is one of the most striking cones, attracting attention as much by its colour as by its form (Fig. 4.26B). The squat cone, 65m high, forms a circular rim 500m across, enclosing a deep funnel-shaped crater. Its steep and perfectly smooth slopes are clothed entirely in small pink tuffs. Corazoncillo is reminiscent of features formed from a Surtseyan eruption, although it is some distance from any visible water bodies. In direct contrast, there is no evidence of the slightest influence of water about 500m to the west, where the same fissure erupted three spatter cones, several hornitos, and – most notably of all – the copious basaltic flows that threatened Yaiza. These could have been the eruptions on 28 December 1731 that caused most of the people of Yaiza to abandon their homes. However, a few people must have stayed behind, because the whole township registered 93 births and 71 deaths between 1732 and 1736. Numbers reached their lowest in 1733, when only 9 births and 6 deaths were recorded, but there was a return to normal by 1737. It may be inferred from these statistics that the volcanic activity waned after 1733, which could, therefore, have encouraged the refugees to return home. In fact, Yaiza was spared and remained on the edge of the new *malpaís*.

When the eruptions ended, 11 villages had been overwhelmed and 400 houses destroyed. The fields made fertile by careful husbandry had been blanketed with rough lava, livestock had been killed, and over 30 cones formed a threatening assembly 18km long. But although many people had to abandon their goods and their homes, not a single human life had been lost.

Although lavas undoubtedly entered the sea, and some eruptions occurred within it, most of the vents were on land. Thus, probably only a small proportion of the 200km² that was resurfaced by the eruptions

Figure 4.27 Timanfaya National Park. **A)** View of the park with its barren volcanic landscape (photo by Gernot Keller –Creative Commons, Wikimedia). **B)** Man-induced geyser eruption for tourists at the Islote de Hilario in the national park (photo by Andreas Tusche – Creative Commons, Wikimedia).

represented additions to the area of the island. The area of Timanfaya now forms the impressive national park between the municipalities of Tinajo and Yaiza (Fig. 4.27).

The eruptions of 1824

The calm that returned to Lanzarote in April 1736 has been broken only by three eruptions of olivine basalt from the same fissure in 1824. These eruptions were brief, mild and covered only 3 km². They were described by Father Baltasar Perdomo, the parish priest of San Bartolomé. Although earthquakes had been felt in Lanzarote as early as 1822, stronger shocks occurred in the centre of the island during July 1824. Earthquakes were so strong in the early hours of both 29 and 30 July that people fled from their homes. Near Tao, the ground vibrated and rumbled 'as if it was boiling' and fumes escaped from newly formed cracks. The eruption started from a fissure near Tao at 07.00 on 31 July 1824, on land belonging to the priest, Luis Duarte. Hence the cone is often known as the Volcán del Clerigo Duarte, although its more official name is the Volcán de Tao. It spent most of its life giving off nothing but fumes. The actual basaltic eruption proved to be one of the shortest

The early months of the eruption seen from Yaiza

'On the first of September 1730 between nine and ten in the evening, the earth suddenly opened up near the village of Timanfaya, two leagues [in fact, 8km] from Yaiza. During the first night an enormous mountain [Los Cuervos] rose up from the bosom of the Earth and it gave out flames from its summit for 19 days. A few days later, a fissure opened up … and a lava flow quickly reached the villages of Timanfaya, Rodeo and part of Mancha Blanca. This first eruption took place east of the Montaña del Fuego, half way between that mountain and Sobaco. The lava flowed northwards over the villages, at first as fast as running water, then it slowed down until it was flowing no faster than honey. A large rock arose from the bosom of the Earth on 7 September with a noise like thunder, and it diverted the lava flow from the north towards the northwest. In a trice, the great volume of lava destroyed the villages of Maretas and Santa Catalina lying in the valley. On 11 September, the eruption started again with renewed violence. The lavas began to flow again, setting Mazo on fire and then overwhelmed it before continuing on its way to the sea. There, large quantities of dead fish soon floated to the surface of the sea or came to die on the shore. The lavas kept flowing for six days altogether, forming huge cataracts and making a terrifying din. Then everything calmed down for a while, as if the eruption had stopped altogether. But on 18 [in fact, 10] October, three new openings formed just above Santa Catalina, which was still burning, and gave off great quantities of sand and cinders that spread all around, as well as thick masses of smoke that belched forth from these orifices [Santa Catalina and Pico Partido] and covered the whole island. More than once, the people of Yaiza and neighbouring villages were obliged to flee for a while from the ash and cinders and the drops of water that rained down, and the thunder and explosions that the eruptions provoked, as well as the darkness produced by the volumes of ash and smoke that enveloped the island. On 28 October, the livestock all over the area nearby suddenly dropped dead, suffocated by an emission of noxious volcanic gases that had condensed and rained down in fine droplets over the whole district. Calm returned on 30 October.

Ash and smoke started to be seen again on 1 November 1730 and they erupted continually until 10 November, when a new lava flow appeared, but it covered only those areas that had already been buried by previous flows. On 27 November, another lava flow [from Pico Partido] rushed down to the coast at an incredible speed. It formed a small islet that was soon surrounded by masses of dead fish. On 16 December, the lavas changed direction and reached Chupadero, which was soon transformed into what was no more than an enormous fire. These lavas then ravaged the fertile croplands of the Vega de Ugo [1 km east of Yaiza]. On 17 January 1731, new eruptions [from Pico Partido] completely altered the features formed before. Incandescent flows and thick smoke were often traversed by bright blue or red flashes of lightning, followed by thunder as if it were a storm. On 10 January 1731, we saw an immense mountain rise up, which then foundered with a fearsome racket into its own crater the self-same day, covering the island in ash and stones. Burning lava flows descended like streams across the malpaís as far as the sea. This eruption ended on 27 January. On 3 February, a new cone grew up [Montaña Rodeo] and burned the village of Rodeo. The lavas from this cone … reached the sea. On 7 March, still further cones were formed, and their lavas completely destroyed the village of Tíngafa. New cones with craters arose on 20 March [Montañas del Señalo] and continued to erupt until 31 March. On 6 April, they started up again with even greater violence and ejected a glowing current that extended obliquely across a previously formed lava field near Yaiza. On 13 April, two [of the Montañas del Señalo] collapsed with a terrible noise. On 1 May, the eruption seemed to have ceased, but on 2 May a quarter of a league further away, a new hill arose and a lava flow threatened Yaiza. This activity ended on 6 May and, for the rest of the month, this immense eruption seemed to have stopped completely. But, on 4 June, three openings occurred at the same time, accompanied by violent earthquakes and flames that poured forth with a terrifying noise, and once again plunged the inhabitants of the island into great consternation. The orifices soon joined up into a single cone of great height, from which exited a lava flow that rushed down as far as the sea. On 18 June, a new cone was built up between those that already masked the ruins of the villages of Mazo, Santa Catalina and Timanfaya. A crater opened up on the flanks of this new cone, which started to flash and expel ash. The cone that had formed over the village of Mazo then gave off a white gas, the like of which nobody had ever seen before. More lava flows also reached the sea. Then, about the end of June 1731, the whole west coast was covered by enormous quantities of dead fish of all kinds, including some that had never been seen before. These eruptions took place under the sea. A great mass of smoke and flames, which could be seen from Yaiza, burst out with violent detonations from many places in the sea off the whole west coast. In October and December, further eruptions [of the Montañas Quemadas] renewed the anguish of the people. On Christmas Day 1731 the whole island was affected by the most violent of all the earthquakes felt during the previous two [sic] years of disasters. On 28 December, a new cone [in the Montañas Quemadas] was formed and a lava flow was expelled from it southwards towards Jaretas. That village was burned and the Chapel of San Juan Bautista near Yaiza was destroyed.'

Many of the panic-stricken villagers of Yaiza then decided to take refuge in Gran Canaria. (excerpt from Andrés-Lorenzo Curbelo, in von Buch 1825).

ever recorded in the Canary Islands, for it had only one day of glory. Ash, spatter and several lava flows issued from as many as 25 vents that formed three small cones, one reaching 38m high. Lines of craters developed on their summits, which followed the northeast–southwest trend of the fissure. But, from 04.00 on 1 August, activity was reduced to emissions of fumes, with rare explosions of ash.

At length, on 21 August, new cracks opened in the ground after an earthquake and precipitated the most notable event of the eruption at 07.00 on the following morning. Brackish water suddenly spurted from the ground and great columns of steam and fumes soared into the air, accompanied, from time to time, by fine ash. Such hydrothermal activity had never been seen before in the Canary Islands. The Alcalde Mayor had some of the water collected and sent to Santa Cruz de Tenerife, where analysis revealed that it contained various soda salts and sulphuric acid. This hydrothermal activity lasted until 25 August, when it stopped as suddenly as it had started. Thereafter, Clerigo Duarte's volcano returned to its fumarolic somnolence until the eruption ceased altogether on 29 September.

By then the next eruption was already under way and, in contrast, was almost entirely magmatic. It began on the western end of the fissure, 13km from its predecessor, and formed the Volcán Nuevo del Fuego. Lava in both fountains and flows immediately emerged from the fissure, but ash explosions soon followed. The sky glowed like the Aurora Borealis as the eruption entered its climax between 2 and 4 October. The ash and the sulphurous fumes made breathing difficult, even 20km away in Arrecife. The cone attained a height of 60m, with two craters breached on their northern sides by lava flows. The basalts arrived on the north coast, 6km away, at 09.00 on 3 October, after having travelled at a speed of about 65m an hour. But they did little damage because they flowed mainly across the *malpaís* that had formed in 1730–36. Activity came to an end with a bang. In the early hours of 5 October, there was a loud explosion and Volcán Nuevo del Fuego gave off nothing but fumes thereafter. The baton was soon taken up by the final eruption of 1824. It formed the Volcán de Tinguatón, about 4km along the fissure from its predecessor. It started at about 06.15 on the morning of 16 October with explosions, which created a terrifying din and seemed to light up the whole island. An hour later, three lava flows emerged. This mixed explosive and effusive activity lasted until noon on 17 October, when the flows were hardly more than 1km long, and the cone of cinders and spatter, elongated along the fissure, was scarcely 30m high. No further magmatic materials erupted.

However, then began perhaps the oddest episode of any eruption in the Canaries since the Spanish settlement. That afternoon, vigorous explosions sent black columns of gas, steam, and old volcanic fragments soaring skywards. Then, at 16.30, hot water suddenly gushed from the crater in powerful jets reaching 16m high. With them came billowing columns of steam, fumes and old ash – sometimes white, sometimes dark brown, and sometimes like cypress fronds. The eruption seemed to combine the hydrothermal and Surtseyan styles, although the source of the water is not clear in such an arid area, which is both 200m above and 8km from the sea. The water developed enough erosive power to break through the walls of the crater and then deposit a delta of fragments on the flows to the north. This extraordinary eruption ceased on Sunday 24 October 1824. Now the Volcán de Tinguatón stands like a squat ruined castle in the midst of the flows of central Lanzarote. Six deep and narrow pits on the floor of the crater mark the vents that disgorged the water: the Cuevas del Diablo or the Simas del Diablo.

The respite for Lanzarote can only be temporary. Molten magma, with a temperature probably between 900°C and 1100°C, still lies only about 4km below the surface. A maximum temperature of 600°C has been measured at a depth of only 13m, and over 100°C has been recorded beneath several craters and fissures. Steam emissions, and temperatures high enough to kindle straw bundles and even to cook food, continue at the Islote de Hilario in the national park, and an artificially-induced geyser is erupted for tourist parties (Fig. 4.27B).

Fuerteventura

Fuerteventura is the second largest of the Canary Islands, 100km long, less than 30km wide, and 1725km² in area (Fig. 4.28). It is separated from Lanzarote only by the La Bocaina channel, which is less than 40m deep. Lying less than 115km from Africa, Fuerteventura is arid, and its climate bears more of the stamp of the Sahara than of the northeast trade winds. Thus, many areas have developed a caliche that has given an unusual buff colour to the lavas, although many older volcanic formations have been laid bare in the *barrancos* cut by ephemeral streams. The relief of Fuerteventura falls

naturally into three longitudinal zones. In the west, the rolling hills of the Betancuria massif form the area where the old basement complexes are best exposed. They are transected by one of the most remarkable dyke swarms in the world, which trends from north-northeast to south-southwest. Fissures running in the same direction erupted thin, but widespreading flows, often exceeding 700m in total thickness, which eventually buried most of their associated cinder cones. They now form an eroded tableland of horizontal basalts in the east, which rises to 807m at the Pico de la Zorza, the highest point of the island. The possibly downfaulted central depression between them is more than 25km long and between 5km and 10km broad.

Fuerteventura rises in a long hump from a depth of about 3000m on the Atlantic Ocean floor, where the oceanic crust is some 180 million years old. But the rocks of Fuerteventura are altogether much younger. The eroded remnants of three ancient volcanic edifices – the northern, central and southern volcanic complexes – erupted between 22.5 and 13.2 million years ago. They now form much of the plateaux, mesas and *cuchillos* of the east and southeast, and in the Jandía Peninsula in the south, and they have developed a thick and widespread caliche.

After a long period of calm, eruptions resumed about 2 million years ago and gave rise to basalts that were concentrated chiefly in the northern and central parts of Fuerteventura. As time went on, the emissions became

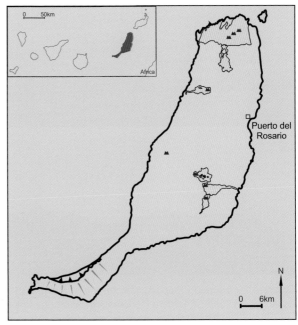

Figure 4.28 Location map, features and outline of recent activity on Fuerteventura (Satellite Image – USGS).

smaller and less frequent and less voluminous. Eruptions along fissures produced a wealth of cinder cones and lava flows, which have been dated chiefly according to the thickness of the caliche, the depth of gullying upon the cones, and in relation to raised beaches around the island. The lavas show some variations from alkali olivine basalts towards those approaching the nature of trachybasalts, although the newer basalts show tholeiitic rather than alkaline affinities.

The olivine basalts were the most extensive of these emissions. At first, they formed shield volcanoes, such as La Ventosilla, but more explosive fissure eruptions later formed large caliche-covered cinder cones and widespread aa flows, such as Piedra Sal, Temejereque, Montaña San Andreas and La Caldereta near Tetir. The more recent eruptions of olivine basalts have given rise to many cinder cones, spatter cones and rugged *malpaís* (Fig. 4.29), which are so well preserved that they cannot have long preceded the Spanish settlement of Fuerteventura in 1402. Devoid of caliche, vegetation or gullying, they have retained their original form and colours so that they stand out readily in the landscape. They form, for instance, the Pajara cones, the Malpaís de la Arena, the Malpaís Chico, in the centre, and the Malpaís Grande, in the south of the island. The Volcano Arena, which rises to 420m with its surrounding *malpaís*, provides an impressive example between the villages of La Oliva, Lajares and Villaverde (Fig. 4.29A).

But their greatest volcanic contribution to Fuerteventura occurred when the northern promontory of the island was formed. Here, fissures trending from east-northeast to west-southwest gave rise to a dozen cinder cones and thin fluid flows, which pushed seawards and reduced the width of the La Bocaina straits between Fuerteventura and Lanzarote. At this time, alternating eruptions of cinders and tuffs formed the Isla de Lobos in the straits themselves. The south-westernmost of the volcanoes is Montaña Colorada with the impressive Calderon Hondo (Fig. 4.29C, D). Eruptions along the same fissure gave rise to the cinder cones of Rebenoda, Encantada, Las Calderas and Bayuyo (see Fig. 4.29B), which dominate the landscape of the northern promontory. Contemporaneous emissions of voluminous lava flows breached each cone down to its base. In all, the most recent episode of volcanic activity on Fuerteventura probably added some 50km² to the island. Although no eruptions have been witnessed in Fuerteventura in historical times, activity may be expected to resume, most probably in the north, perhaps within the next few millennia.

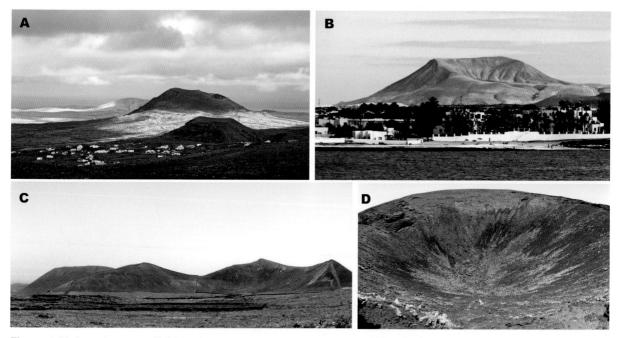

Figure 4.29 Location map. **A)** Malpaís de la Arena and Montana de la Arena (Volcano Arena), as seen from Montana Escanfraga (photo by Tamara K – Creative Commons Wikimedia). **B)** Corralejo with cone of Bayuyo (seen from the ferry to Lanzarote). **C)** *Malpaís* and agricultural walls around Montana Colorada near Lajares. **D)** Calderon Hondo volcano, Montana Colorada near Lajares (photos **B**, **C** & **D** by Norbert Nagel – Creative Commons, Wikimedia).

La Palma

La Palma is mountainous, rising to 2426m at the Roque de las Muchados and covering 728km². It rose above sea level during the past two million years from a base on the ocean floor about 3000m deep. La Palma is pear-shaped in outline, 47km long. It is broadest, at 28km across, in the north, where a high and complex shield has accumulated (Fig. 4.30). Occupying the heart of this shield is the most famous landform on the island, the Caldera de Taburiente, which is 7km wide, 15km long and nearly 2000m deep, and surrounded by the highest peaks in the island. In the south, it is joined to a shallower companion, the Cumbre Nueva caldera. La Palma has a volcanic sting in its tail: the modern eruptions have been concentrated on the series of parallel fissures stretching southwards along the Cumbre Vieja ridge.

In the north, the Taburiente stratovolcano grew up above a basal complex of altered basic and ultrabasic rocks about 2.0 million years ago. It is composed mainly of thick accumulations of basaltic flows interspersed with cinders and transected by innumerable basaltic dykes, whose upper layers have been dated to between 853,000 and 566,000 years old. A rift developed southwards from the Taburiente volcano and quickly grew into the steep, unstable Cumbre Nueva ridge, which may have reached a height of 2500m. About 180– 200km³ of this area then collapsed about 560,000 years ago, and another volcano, Bejenado, erupted within the collapsed area during the following 60,000 years. There seems then to have been a quiescent period of some 370,000 years, during which the Caldera de Taburiente developed. A straight, fault-guided stream, the Barranco de las Angustias, formed on the southwestern flanks of the Taburiente volcano in a particularly favoured position between the faulted edge of the Cumbre Nueva collapse structure and the new Bejanedo volcano. Vigorous headward erosion thus cut the vast hollow of the Caldera de Taburiente deep into the heart of the volcano. Consequently, fluvial erosion continued what gravitational collapse had started.

From about 125,000 years ago, the eruptive emphasis on La Palma pushed further southwards and began to develop the present rift-fissure activity that

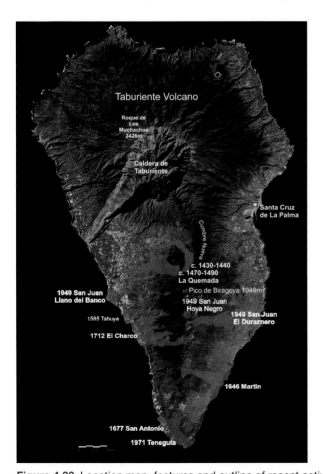

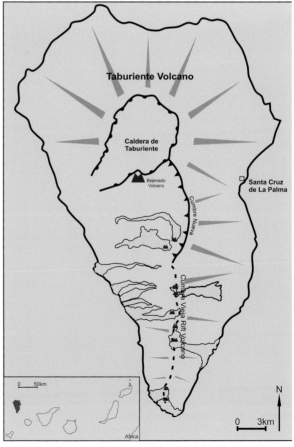

Figure 4.30 Location map, features and outline of recent activity on La Palma (Satellite Image – USGS).

Figure 4.31 Examples along the Cumbre Viejo ridge, La Palma. **A**) The crest of the Cumbre Viejo Volcano with the El Duraznero summit and Hoyo Negro explosion crater in the centre (photo courtesy of Juan Carlos Carracedo). **B**) Panorama at the top of San Antonio cone (photo courtesy of Helen Robinson). **C**) El Duraznero volcano. **D**) Detail within The Hoyo Negro (Black Hole) (photos **C** & **D** courtesy of Francis Abbott).

is still forming the Cumbre Vieja volcano. However, the eruptions were much reduced between 80,000 and 20,000 years ago, during which time marine erosion pared the flanks of the ridge and formed imposing cliffs around the island. Then, some 20,000 years ago, activity increased both on the north–south rift and also on a northeastward trending rift. For the next 13,000 years, this renewed activity lengthened and widened La Palma as lava flows cascaded down the old cliffs and formed lava deltas at their bases. But 7000 years ago, activity became almost exclusively concentrated on the north–south rift. These eruptions added the latest touches to the Cumbre Vieja. It was on this volcano, too, that all the historic eruptions of La Palma have occurred (Fig. 4.31 and maps in Fig. 4.30).

Historical eruptions on La Palma

All the historical eruptions on La Palma were explosions of ash and cinders and emissions of fluid, basaltic lavas. There was an eruption on the northern axis of the Cumbre Vieja ridge just before the island was settled in 1493, but its exact date, location, and even its name, have been subject to some discussion. It seems to have occurred either between 1430 and 1440 or about 1470–90. It probably formed the Montaña Quemada, a cinder cone rising 118m high, with a breach on its northern side from which issued a lava flow about 6km long.

An eye-witness, Father Alonso de Espinosa, and an Italian engineer, Leonardo Torriani, recounted the events of the next eruption in 1585. It was one of the most complex eruptions in the Canary Islands since the Spanish settlement. From late on Sunday 19 May, earthquakes of increasing vigour and frequency were felt, especially at Los Lanos, on the northern side of the Cumbre Vieja ridge. The ground began to swell up and became riddled with fissures. The magma duly reached the surface on the night of 26/27 May. Within the next two days, explosions produced masses of ash, cinders and high columns of fumes, quickly formed the Tahuya (or Teguso) cinder cone, and emitted a lava flow that eventually reached the west coast. There, said Father Espinosa, 'the lava extended half a league into the sea, warmed the waters, boiled the fish therein, and melted the tar of the boats.' The activity weakened considerably during the ensuing month. Then came the most extraordinary event of the eruption. The fissures began to exude huge masses of old phonolite from the substratum, which now form the Companions of Jedey, named from the hamlet nearby. At the same time, about 1km north of the main cone, more than half a dozen vents started exploding, which may have continued until 10 August 1585.

In the autumn of 1646, a much smaller eruption formed the Volcán Martín at Tigalate, on the southeastern flanks of the Cumbre Vieja ridge. On 30 September, earthquakes rumbled all over La Palma and the islands nearby. The eruption started on 1 October, and ash and cinders built a prominent cone during the next three days. The explosions could be heard in Tenerife, and, at night, the volcano 'glowed like a candle'. Renewed earthquakes on 4 October preceded the opening of new vents along a fissure that spurted out ash and lava flows for the next six weeks. On 15 November, yet another fissure opened close to the coast, and the flows reached the shore in a band 4km wide. One day, the people processed to the eruption with a miraculous image of the Virgin. The following dawn, the volcano was covered in snow. On 21 December 1646, a few days later, activity ceased.

Activity resumed in 1677 in the far south of the Cumbre Vieja, between the coastal spa of Fuente Santa and the town of Fuencaliente, 5km inland. It was long believed to have formed the large, often hydrovolcanic, San Antonio cone (Fig. 4.31B), which in fact is probably over 10,000 years old. The scope of the events in 1677 was much more limited: a small vent opened on the northern rim of San Antonio, and four rows of little vents erupted on its steep southwestern flanks. On 13 November, there was 'a pestilential odour of sulphur in the air' and earthquakes split open fissures 'that were difficult to jump across'. The magma burst from the fissures at sunset on 17 November. The vent on San Antonio volcano exploded ash that formed a small cone, 30m high, perched on its crest, but accompanying earthquakes shook most of it into the larger crater – and, no doubt, also destroyed the church tower in Fuencaliente. In December, the small cone gave off carbon dioxide, which sank into the valley near Fuencaliente and killed many head of cattle, rabbits and birds, and a peasant. In January 1678, 27 goats succumbed to the invisible gas. While the upper vent was exploding fragments, the lower vents were emitting spatter and fluid basaltic lavas. These flows buried the spa of Fuente Santa on 26 November 1677. Indeed, basalts had covered much of the southernmost toe of La Palma when the eruption stopped on 21 January 1678. At this date, the historic eruptions on La Palma had migrated to the southern end of the island in a little over two centuries.

The following eruption paid no heed to this trend, and broke out at El Charco in 1712. It was only 2.5km from the Volcán Martín, but it occurred this time on the north-western flanks of the Cumbre Vieja ridge. Earthquakes began to offer their usual warnings on 4 October, and ash explosions duly started on Sunday 9 October from a fissure 2.5km long. They soon built up a cinder cone that perched rather uneasily on the steep flanks of the ridge: its downslope side now rises 361m, but the upslope side reaches no more than 25m high. It is crowned by three craters, one of which is 135m deep. After the cone formed, activity concentrated on the lower parts of the fissure, where fluid lava gushed out and reached the coast in a broad apron about 5km wide, before the eruption stopped on 3 December 1712.

After this minor flurry of activity, which had seen four eruptions in 128 years, La Palma rested. Nearly 237 years later, the fissures sprang back to life on the Feast of St John the Baptist, 24 June 1949. This is why this episode of widespread activity is named after San Juan. The main focus of explosive operations lay near the crest of the Cumbre Vieja ridge, between the older volcanoes of Nambroque and El Duraznero. Notable underground rumblings and earthquakes started on 21 June. At 08.30 on 24 June, a hydrovolcanic explosion burst out north of El Duraznero. Soon ash was raining down all over southwestern La Palma, and similar pulsating emissions went on until 7 July. At 04.30 on 8 July, a more vigorous hydrovolcanic eruption heralded the opening of another fissure, about 3km away, across the Llano del Banco. At once it became the new focus of activity and brought with it a new eruptive style: emissions of fluid basalts. They cascaded down the coastal cliffs, reached the sea, 7km away, in a wide delta by 10 July, and continued to gush out until 26 July.

Meanwhile, the El Duraznero fissure (Fig. 4.31C) had two last flings. On 12 July, hydrovolcanic explosions began blasting out the Hoyo Negro (Fig. 4.31D), which, by the time they had finished on 30 July, had created a 'Black Hole' 192m deep, which was surrounded by a small tuff ring. At noon on 30 July, the El Duraznero fissure started spewing out fluid lavas. This phase lasted for eleven hours and eventually formed a flow 8km long. The San Juan eruption had an unusual epilogue for a Canarian eruption. The first autumn rains on 28 November 1949 produced several volcanic mudflows, or lahars, that destroyed several bridges and a stretch of the coastal road. Luckily, they hit only a thinly populated area, and no lives were lost. The San Juan eruption

also produced unusually strong seismic activity, and the fissures developed during this eruption may mark the inception of another great landslide like those of the Caldera del Taburiente and the Cumbre Nueva caldera just to the north. The latest eruption in La Palma was in 1971 at Teneguía at the southern end of the Cumbre Viejo volcano (Fig. 4.32). In all, La Palma is presently one the most active of the Canary Islands, and its historical eruptions have covered a total area of 37km².

The eruption of Teneguía in 1971: the case of the migrating breach

The latest on-land eruption to be recorded in La Palma – and in all the Canary Islands – built the Teneguía cone and surrounding flows between 26 October and 18 November 1971 (Fig. 4.33). It occurred close to where the eruptions had taken place near the San Antonio cone almost 300 years before. Earthquakes warning of the impending eruption began on 15 October 1971. They made up in frequency what they lacked in power, for over a thousand were recorded between 21 and 24 October. Then two days of relative calm ensued. But at 15.00 on 26 October, a fissure opened and lava fountains spurted from many vents along it. Basaltic flows soon emerged and reached the sea near Fuencaliente lighthouse the following dawn. On 27 and 28 October, explosions formed two successive cinder cones. The second cone to form was, in fact, known as Teneguía I because it soon became the main focus of activity. During the next two days, a lava flow breached the cone on the sector facing up slope to the north. The flow could not easily defy gravity for long, and, in the early hours of 31 October, the position of the breach began to migrate clockwise. The slope of the land tended to shift the flow sideways towards the east, and thus undermined the eastern arm of the cone. At the same time, ash and cinders could then pile up over the original exit and thus extend the western arm round over the original northern breach. This migration went on until 3 November, when the flow had shifted the breach to the south-southeast, where the land sloped most steeply away from the cone and offered the easiest exit for the lavas. Thereafter, of course, the breach stayed in this optimum position.

Meanwhile, at 12.39 on 31 October, an explosion 200m north of the main cone formed Teneguía III.

Figure 4.32 The Teneguía eruption of 1971. **A**) Teneguía from the air showing the eruption centres and flows. **B**) Lava entering the sea and extending the coastline. Inset stamp from special edition 'Sciences of the Earth and the Universe' stamp depicting the 1971 eruption of the Teneguía Volcano as well as a seismograph issued in 2006. **C** & **D**) Explosive eruptions at the Teneguía vents (photos courtesy of Juan Carlos Carracedo; stamp image sourced from UPU – Universal Postal Union).

This remained active for only two days and formed a little pit circled by a tuff ring. It was to revive briefly once more, on 5 November. However, the effusions of the main cone continued at an increasing pace, when sustained lava fountaining continued for eight days. On 8 November an explosion blasted out another pit, called Teneguía IV, on the northern flanks of the main cone. Two days later, a similar explosion alongside it created yet another pit, Teneguía V, whereupon the activity of the main cone began to wane. On 18 November, lava started to issue from the pit of Teneguía V. Like the effusions from El Duraznero in 1949, these flows heralded the end. At midday on 18 November, the whole eruption came to a halt, and the cone has produced nothing but fumaroles ever since. The horseshoe cone was 89m high and the lava flows had reached the sea in four broad and rugged tongues that covered 4km^2.

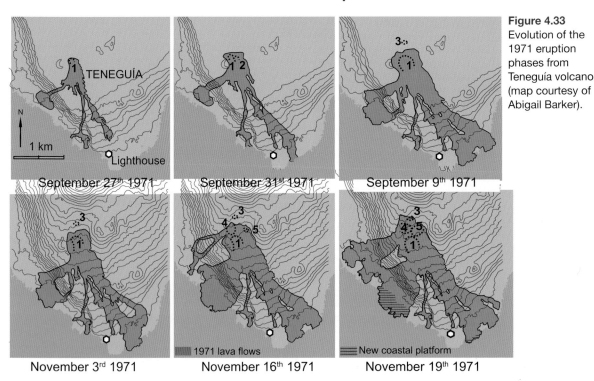

Figure 4.33 Evolution of the 1971 eruption phases from Teneguía volcano (map courtesy of Abigail Barker).

Meet the Scientist – Helen Robinson

At the time of writing Helen is finishing her PhD at the University of Glasgow. She has undergone a number of monitoring and geothermal projects in the Canaries, Africa and is branching out further looking at the geothermal potential of volcanoes. She sheds some light on the work on La Palma and on how gas monitoring can help volcanoes help us.

How did you first start getting into research on volcanoes?
When I was young my parents loved all the 'natural world' type of documentaries; it was a Sunday evening ritual. I became mesmerized by the things I saw on the TV. Then when I was about seven years old I saw a documentary on Mount St. Helens and that was it! I was hooked. My mum would probably agree that I had a slightly morbid obsession with natural disasters from then on in.

I then did some coursework on renewable energy; I chose geothermal as my subject, and found it really fascinating. Everything I have done since has snowballed from there.

When did you first visit the Canaries and which islands have you seen?
I only first visited the Canary Islands for the first time last year (2015). Mt. Teide has been on my 'must climb' list for a few years!

You took part in a programme of gas measurements on La Palma: what was it about?
This was working in collaboration with the Instituto Volcanológico de Canarias (INVOLCAN) and GeoTenerife. Every year small teams from INVOLCAN head out to locations across the Canary Islands and Cape Verde to monitor the volcanic activity. This involved collecting soil gas samples, spring samples and *in situ* CO_2 efflux data. The soil gases are used to analyse annual changes in the quantitative volumes of CO_2 and helium in soils on the flanks of the volcanoes; some of the samples were also used to evaluate $^{13}C/^{12}C$ ratios.

The spring samples are used for ratio analysis of ^{3}He/^{4}He and the CO_2 efflux measures how much is escaping into the atmosphere. After water vapour, CO_2 is the most abundant gas dissolved in magma and the main species released along tectonic structures. As the gas ascends it is modified by a range of processes; we can use the efflux data to monitor how much arrives at the surface. We use the ^{13}C/^{12}C ratios to determine how much of this gas has a magmatic source and how much has a crustal source. The helium ratios are also used for a similar analysis. Changes in this annually collected data can assist in identifying changes in the volcanoes that may lead to an eruption. This, coupled with a range of other data collected, can help develop education programmes for locals, tourists and governments, as well as highlight the need for evacuations in extreme circumstances and protection of key infrastructures.

Which other volcanoes are you working on and why?
At the moment I am working on a few volcanoes. My main area of study is Menengai Volcano and summit caldera in the Gregory Rift, East African Rift Valley, Kenya. My work here is largely associated with the geothermal power generation in the country. It is estimated that only ~30% of Kenyans have access to the national grid. The country relies on hydro-schemes for their electricity generation, yet it is a country under great water stress. This often leads to an inability to meet base load demands, the result of which is energy rationing, high prices, and an unreliable power source.

The geothermal project at Menengai (one of many planned across Kenya) is believed to have the potential to provide Kenya with ~17% of its electricity needs. The area is already well developed for power generation, so I have been developing comparatively cheap surface exploration methods and comparing the results from these to the data already collected, allowing me to support the methods and conclusions from this work.
I have also started doing some hazard monitoring of the volcano, as it has the potential to be dangerous in the future. The data collected from the hazard monitoring will hopefully provide support for a more long-time monitoring programme.

I am also working on Fantale Volcano in the Main Ethiopian Rift Valley, Ethiopia and the Butajiri-Silti Volcanic Field, also in Ethiopia. These two localities have been identified for geothermal surface exploration. I will be using techniques tried and tested at Menengai.
I have also started phase 1 studies of the geology and potential for geothermal resources in the Comoros Islands and Mauritius in the southern Indian Ocean.

What do you think can link monitoring gases on volcanoes with energy?
For the development of rational engineering designs and cost-effective drilling programmes in geothermal projects, it is of vital importance to target the fault and fracture networks in the sub-surface that are actively used as fluid and gas conduits. Collecting data relating to soil gas volumes and compositions will enable those developing the project to identify, from surface data, where the sub-surface structures are that need to be targeted. The soil gas and efflux methods are low-cost and quick to complete, resulting in a more efficient surface exploration survey and reducing the cost of 'trial and error' drilling that is often used in East Africa.

Figure 4.34 Helen Robinson soil gas sampling on La Palma.

Gran Canaria

Gran Canaria forms the emerged centre of an eroded shield, 45km across, which covers an area of 1532km². It rises from a depth of 3000m on the ocean floor to a height of 1949m above sea level at the Pozo de las Nieves (Fig. 4.35). This compact and symmetrical mass is deeply scarred by radiating *barrancos* and trimmed by marine cliffs where thick piles of lava are often exposed in sections approaching 100m high. Generally speaking, the older volcanic rocks are exposed in the southwestern part of the island, whereas the younger formations occur in the northeast, most notably in the peninsula of recent cones composing La Isleta north of the capital, Las Palmas.

The basal complex of Gran Canaria was masked when intense emissions occurred during a very short interval about 14.5–13.9 million years ago, at an average rate of about 5km³ per thousand years. They formed a basal shield of long, thin flows, ranging from alkaline basalts to basaltic trachyandesites, that now constitutes more than 90 per cent of Gran Canaria.

There was a short burst of violent eruptions of trachytes, trachyphonolites and peralkaline rhyolites about 13.5 million years ago. They form lava flows, tuffs and ignimbrites and ash that were expelled when the hub of the old shield collapsed to form a caldera 15km across and 1000m deep. Similar eruptions continued much more feebly until about 10.5 million years ago. Gran Canaria then experienced a long period of volcanic calm and much erosion.

The third major eruptive phase took place between 5 million and 3 million years ago with the formation of the Roque Nublo stratovolcano, which covers much of the centre and northeast of the island. At first, basanites, tephrites and phonolites erupted in thin lava flows, but, as the volcano grew to about 3000m, widespread pyroclastic density currents and a few phonolitic domes erupted from vents concentrated around Tejeda in the centre of Gran Canaria. The stratovolcano then collapsed to form the Roque Nublo caldera, and an avalanche over 3km³ in volume swept southwards for 28km down its flanks. Roque Nublo itself forms a prominent

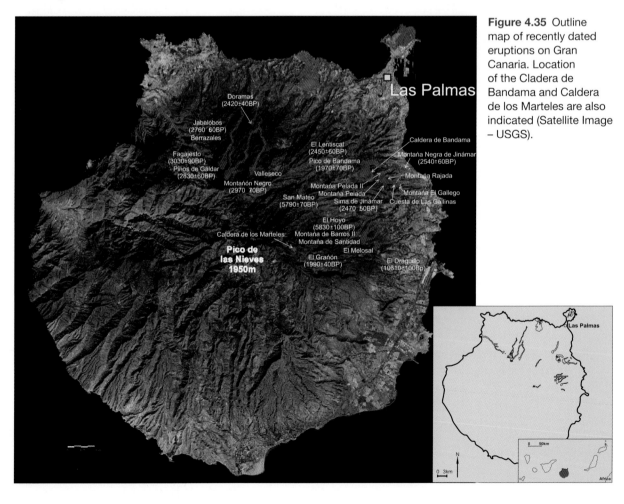

Figure 4.35 Outline map of recently dated eruptions on Gran Canaria. Location of the Cladera de Bandama and Caldera de los Marteles are also indicated (Satellite Image – USGS).

Figure 4.36 A) Roque Nublo, Gran Canaria. The Roque Nublo, once revered by the ancient Guanche population and today visited by thousands of tourists each year, is the emblematic symbol of Gran Canaria. **B)** Roque Nublo, Gran Canaria in the forground with Mt. Teide and the outline of Tenerife in the background (photos courtesy of Francis Abbott).

Figure 4.37 A) Caldera de Bandama, Gran Canaria. The Caldera de Bandama is one of a number of small calderas to be found on Gran Canaria. This caldera of 1km in diameter and 200 metres depth has been dated at 20**BC**. Originally a double cinder cone, the caldera was formed later. **B)** Caldera de los Marteles, Gran Canaria. Situated close to the summits of Gran Canaria (Pico de Las Nieves) and at the head of the Guayadeque canyon, the Marteles caldera was formed some 100,000 years ago by a phreatomagmatic explosion (photos courtesy of Francis Abbott).

pinnacle 60m high, an attractive tourist site on the island (Fig. 4.36).

The latest phase of activity on Gran Canaria was different. Eruptions occurred from vents, often aligned on fissures, which were overwhelmingly concentrated in the northeastern half of the island. The eruptions took place in brief spurts of activity from 2.85 million to 1.5 million years ago, and they generally became less voluminous and more scattered as time went on. They mark much the smallest major volcanic episode on the island, and produced fresh-looking cinder cones and lava flows. The older eruptions covered much the widest area and came from scattered vents. For example, cinder cones such as the Osorio volcano have lost their original sharp outlines, and erosion has exposed the columnar cores of their thick flows. Later eruptions, such as those forming the Pico Gáldar, were concentrated in the northeast of Gran Canaria, and the most recent have preserved most of their original freshness. Their basanitic cinder cones and rugged aa flows, for example, constitute much of the La Isleta peninsula near Las Palmas, the Montaña de Arucas and Las Montañetas. Hydrovolcanic eruptions also formed the well-preserved maars and tuff rings of the 'Calderas' of Bandama, Las Piños de Gáldar and Los Marteles (see Fig. 4.37). No eruptions have been recorded on Gran Canaria during historical times, but the fine state of preservation of many of its most recent features indicates that they cannot have erupted long before the Spanish Conquest of the island in 1483.

La Gomera

La Gomera, 380km^2 in area, is the smallest of the Canary Islands apart from El Hierro and rises to a height of 1487m at Garonjay. La Gomera is a round and broadly symmetrical island rising from a depth of some 3000m on the Atlantic Ocean floor. It forms the summit of a vast shield volcano that has been constructed in eruptive bursts separated by long periods of erosion over several million years. La Gomera has witnessed no historical eruptions, and perhaps none for 4 million years.

In proportion to its size, La Gomera has the largest exposed surface area of the basal complex, because the island has subsequently undergone marked uplift. As elsewhere, it is composed of peridotites, gabbros and dolerites, and it is riddled with many dykes that make up two-thirds of its total volume. The basal complex underwent a long period of erosion lasting over a million years before activity resumed with eruptions of lavas some 12.5 million years ago. These flows now cover much of the island, and were erupted in two main bursts: about 11 million years ago and between 8 million and 6 million years ago. After another long dormant episode, the fluid Younger Basalts covered 200km^2 between 4.6 million and 4.0 million years ago. During the subsequent dormant period, La Gomera has undergone considerable erosion, so that clearly displayed volcanic features are rare, with the notable exception of cliffs revealing fine basaltic colonnades.

Chapter 5

Portugal

Introduction

A mixture of green, fertile, and sometimes misty islands rising from the Atlantic Ocean floor, the Azores are an autonomous region of Portugal situated in the Atlantic Ocean 1500km west of Lisbon and about 880km northwest of Madeira. The archipelago of nine islands is scattered over 600km and falls into three groups: Flores and Corvo in the west, Santa Maria and São Miguel in the east, and a central cluster of five islands – Terceira, Graciosa, São Jorge, Pico and Faial (Fig. 5.1). Seven of the islands rise from the Azores Platform on the eastern flanks of the Mid-Atlantic Ridge. Only Flores and Corvo rise on its western flanks. All the islands are active except Santa Maria, which is the furthest from the Ridge. All except São Jorge have stratovolcanoes, and all except one have been decapitated by calderas. The Pico do Pico, soaring to 2351m above sea level, is the only stratovolcano that still retains the pristine glory that makes it the incomparable landmark of the Azores.

Although the Azores are almost entirely volcanic, they are far from stark and bare: laurels, hydrangeas and azaleas dominate a floral extravaganza that is rarely seen in temperate climates. The Azores were uninhabited when the Portuguese explorers were led to them by the goshawks, the Açores, which were flying about the islands. Settlement began first on Santa Maria and São Miguel in 1439 and then on Terceira in 1450, on Pico and Faial in 1466, on Graciosa and São Jorge in 1480, and on Flores and Corvo at the start of the next century. A map from 1584 higlights the extent to which the islands were known even at this early stage (Fig. 5.2). Historical records of eruptions therefore extend back to between 500 and 600 years, with over 30 eruptions that have taken place during this period, either on the islands or offshore. However, the Azores are not growing rapidly: Iceland has a much higher output of lava per visible eruption; eruptions occur four times more often in Iceland; and eruptions of similar composition have built the Canary Islands five to ten times faster. But, of course, Iceland is 40 times larger

than the Azores islands, and many submarine eruptions must have passed unnoticed on the Azores Platform.

The Azores and their submarine plinth grew up on the Mid-Atlantic Ridge near the triple junction of the North American, Eurasian and African plates. Their activity is probably related to a weak hotspot, and perhaps also to a secondary band of seafloor spreading. The Mid-Atlantic Ridge, forming the boundary between the North American and Eurasian plates, passes through the Azores. The western islands of Flores and Corvo belong to the North American plate, but the location of the boundary between the Eurasian and African plates is not at all clear (see Fig. 5.1). The East Azores fracture zone runs from the Mid-Atlantic Ridge along the southern edge of the Azores Platform. Thus, if the fracture zone forms the main plate boundary, then the central and eastern Azores must belong to the Eurasian plate. However, an axis of secondary seafloor spreading runs through the central Azores along the Terceira Rift, which probably passes through Graciosa, Terceira and São Miguel. In either case, parts of the central and eastern Azores could therefore belong to the African plate or to an Azores microplate. The volcanic activity in the Azores was most probably also intensified by one large, or several small, hotspot plumes. Whatever the reasons behind the growth of the central and eastern Azores, all display a common and impressive predominance of stratovolcanoes, rifts, faults, fissures and volcanic alignments, running parallel to the spreading axis, that have been the elements in the development of their scenery.

The stratovolcanoes in the Azores are mostly gently sloping cones, usually more than 10km in diameter and rising about 1000m above sea level. They are crowned by beautiful deep calderas, which so impressed the early settlers that they also rather confusingly gave the name Caldeira (cauldron) to the whole mountain. These calderas are the hallmarks of the Azores.

The eruptions that have taken place along fissures running from northwest to southeast are the second

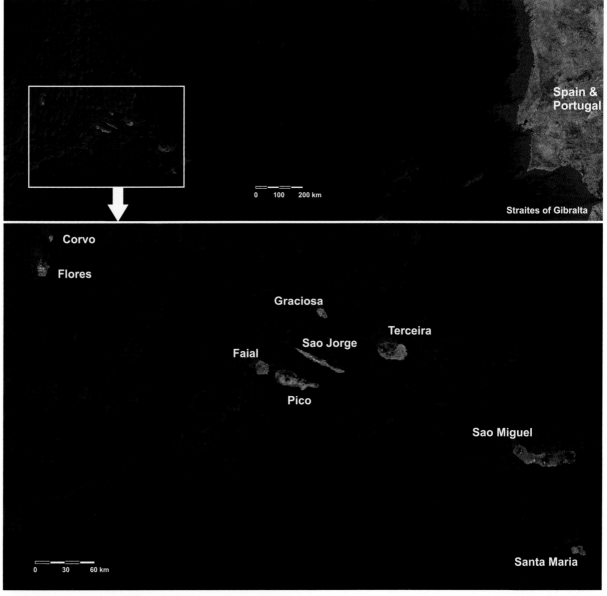

Figure 5.1 Location satellite maps of the Azores islands (USGS).

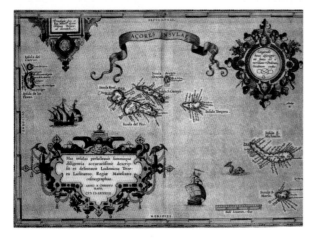

Figure 5.2 Historic map of Azores 'Açores Insulae' ('Ilhas dos Açores'), Luís Teixeira, c.1584 (Wikimedia Commons).

most striking characteristic of the Azores, and they completely dominate the landscape of São Jorge. What these eruptions lack in volume they make up for in number, for there are over a thousand cinder cones in the Azores. They occur on the flanks of stratovolcanoes and dominate the scenery on the plains and plateaux, and they are commonly associated with lava flows. The younger flows often have rugged black aa surfaces, which the early settlers called *mistérios*, because they had frightening and mysterious associations, which the eruptions during historical times did nothing to dispel. Although they have now often been planted with woodlands, they still stand out among the meadowlands of the Azores, and they reach their finest expression on Pico.

The fissure eruptions in the Azores extend well below sea level. Deeper eruptions, stifled by the water pressure, reached the surface as bubbling gas emissions and hot discoloured seas. Such eruptions have marked the activity of the Banco de Don João de Castro. But, in the shallower coastal waters, Surtseyan eruptions built bulky tuff cones such as those protecting both Angra do Heroísmo in Terceira, and Horta in Faial. Older Surtseyan tuff cones are dotted about the coasts of the Azores, but, because they formed in such vulnerable positions, marine erosion soon reduced them to picturesque islets such as the Ilhéus das Cabras, off Terceira, and the Ilhéus dos Mosteiros off São Miguel.

Several islands have been marked by faulting and rifting, which was probably associated with the zone of secondary spreading branching from the Mid-Atlantic Ridge. The Terceira Rift transects that island and São Miguel, and another rift in eastern Faial forms a distinct fault trough. On the other hand, uplifted blocks seem to delimit most of both São Jorge and eastern Pico. The archipelago still suffers from typical mid-ocean ridge earthquakes that are less than magnitude 5.0 on the Richter scale and have shallow epicentres, less than 30km deep.

Although six of the islands are still active, fumaroles are their only persistent manifestation. They make their best displays in the Caldeira das Furnas in São Miguel, where tourists can bathe or have meals specially cooked – in different vents, of course. Weak fumes usually issue from the summit of Pico, and the earthquake in May 1958 briefly revived those in the caldera of Faial. Magmatic emissions are much less frequent; the latest formed Capelinhos in Faial in 1957–8, although other eruptions have since occurred below sea level. At all events, volcanic features are never out of sight in the Azores, and volcanic activity plays a role in the local place names, exceeded only by religion. Mistério and caldeira have already been mentioned, but other common names include *biscoitos* (rugged aa lava flow), *bagacina* (cinders), *queimado* (burnt), *fogo* (fire), *furna* (oven), *timão* (arched yoke) and *cabeço* (head). Thus, volcanic activity is never far from the consciousness of the Azoreans.

The Azores are young. The eastern group contains the oldest rocks, which reach about 5 million years old

in Santa Maria and 4 million years old in eastern São Miguel. No lavas on any of the remaining islands are apparently more than a million years old. The oldest rocks commonly outcrop in the southeast of each island, and the more recent eruptions have tended to occur in the northwest, nearest the Mid-Atlantic Ridge. Nevertheless, there was no regular progression of eruptions from island to island towards this ridge. Thus, for instance, Faial, in the west, and São Miguel, in the east: both have some of the oldest and some of the youngest lavas in the archipelago.

The eruptions in the Azores produced a predominance of basaltic compositions of various types, but some evolution to more silica-rich compositions was associated with the violent explosions that formed the calderas. The basaltic rocks, commonly erupted from fissures, formed many cinder cones and lava flows, and Surtseyan eruptions took place where these fissures extended into shallow water.

São Miguel

São Miguel is the largest and most varied island in the Azores. It covers an area of 747km² and is 65km long, with a maximum width of 15km, culminating at a peak of 1103m at the Pico da Vara (Fig. 5.3). It is notable for its rich vegetation, fertile soils, four stratovolcanoes with majestic calderas, dozens of cinder cones, aligned for the most part on fissures, thermal springs, and a beautiful coastline dominated in many places by high cliffs. However, large tracts of bare lava are rare, for most of the flows are either weathered or have been blanketed by pumice exploded from the stratovolcanoes.

The eruptions that gave rise to São Miguel began about 4 million years ago at a depth of 2000m on the floor of the Atlantic Ocean, and the island has some of the oldest exposed lavas in the archipelago. However, five eruptions on land, and a further seven offshore, have been recorded since the island was first settled in 1439. In general, activity on São Miguel spread westwards and has now probably ceased in the east.

Four stratovolcanoes form the backbone of the island. Povoação and Furnas are contiguous in the east; the Água de Pau, or Fogo volcano, occupies the centre; and the Sete Cidades volcano forms much of the northwestern part of the island. Sete Cidades was a separate island until it was joined to the rest of São Miguel by the eruptions of the Região dos Picos.

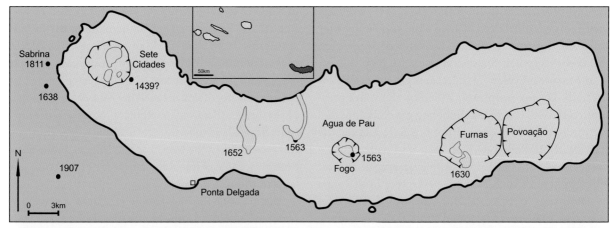

Figure 5.3 The calderas and recent eruptions on and around São Miguel (satellite map – USGS).

Povoação and the Nordeste

The northeastern region contains the oldest rocks exposed on São Miguel, 4.00–0.95 million years old. It is a broad upland, composed of thick piles of basaltic flows, which are probably the remains of a shield that culminated at the Pico da Vara. However, this upland has been slashed by narrow gorges and pared by marine erosion into cliffs commonly over 100m high. Povoação, the oldest stratovolcano on São Miguel, grew up on the southern fringe of these uplands. The eruptions started with basaltic, and ended with evolved (trachytic) emissions. Plinian explosions of trachytic pumice accompanied the formation of the Povoação caldera on the southern flanks of the stratovolcano. This seems to have occurred in two stages about 820,000 and 700,000 years ago. The southern wall of the caldera is missing. It could have foundered along the fault that seems to delimit the adjacent coast, but is more likely to have been swept seawards by a massive landslide. Neither the northeastern uplands nor Povoação volcano has

experienced any eruptions during the past few thousand years.

Furnas stratovolcano

Furnas is an indistinct stratovolcano with an imposing caldera. It emerged from the Atlantic Ocean more than 100,000 years ago and its products cover some 75km^2 and have a volume of about 60km^3. Most of the rocks are trachytes that occur in lava flows, pumice, ash and cinders, as well as in pyroclastic density currents and mudflows.

The best view of Furnas caldera can be obtained from the trachytic dome of the Pico do Ferro, perched on its northern rim (Fig. 5.4). It is the youngest caldera on São Miguel and, when it collapsed about 12,000 years ago, massive eruptions scattered trachytic pumice over much of the island. The caldera covers an area of 35km^2 and the southern rim is low, but the remaining walls rise 100–250m almost vertically from its floor. Its

COVA DA BURRA

Hot springs

Figure 5.4 Furnas caldera, the trachytic dome of the Pico do Ferro, location of the hot springs, and the Cova da Burra are indicated (Shutterstock/Oliver Hoffmann).

The eruption in Furnas caldera in 1630

At about 20.00 on 2 September 1630, earthquakes began to shake the whole of São Miguel, virtually destroying the villages of Ponta Garça and Povoação, near Furnas, causing landslides on the south coast at the mouth of the Ribeira Quente and making the church bells ring 30km away in Ponta Delgada. The magma duly erupted inside the caldera between 02.00 and 03.00 on 3 September; a small column of ash and fumes rose skywards, and the wind blew the ash southwestwards. Soon, more violent hydrovolcanic eruptions and pyroclastic density currents blasted out from the caldera lakes, and an eye-witness said that two of the lakes gave off 'clouds of fire'. One of these pyroclastic density currents swept down the gorge of the Ribeira Quente and surged seawards to form a small delta. Another formed 'a burning stream' that spread through the woods to the village of Ponta Garça. On 4 September, a subPlinian climax started three days of full fury, sending a column of ash and steam 14km into the air, spreading pumice, ash and lapilli over a wide area, burning the forests, and plunging the whole island into darkness. Ash even fell 550km away to the west on the island of Corvo. The wind changed direction for the last, most violent and purely magmatic phase of the eruption. It lasted for 15 hours and expelled nearly one-third of all the magma ejected. Ash and lapilli showered down mainly to the northeast of the vent, and deposited a blanket of fragments that reached at least 2.5m thick in Ponta Garça and even 10cm in Ponta Delgada. When the climax drew to a close on 6 September 1630, the Lagoa das Furnas was the only lake left in the caldera. The eruption had built a distinctly asymmetrical tuff ring, the Cova da Burra, inside which the trachytic dome now forming Monte Areia extruded during the next two months. The eruption expelled about 2.5 km^3 of fragments and killed 195 people. Between 80 and 115 people – half the population of the village – died in Ponta Garça; some 30 people were killed who had lived in huts beside the lakes; and the remainder met their deaths in and around the villages of Furnas and Povoação.

central part later foundered even further, perhaps about 11,000 years ago, to form an inner caldera that has a diameter of 5km, and its western parts are occupied by the Lagoa das Furnas. Plentiful hot springs are still concentrated on the northern lakeshore at the foot of the Pico do Ferro, and especially in the spas at Terra Nostra and Caldeiras, on the outskirts of Furnas (see map in Figs 5.5 and 5.6).

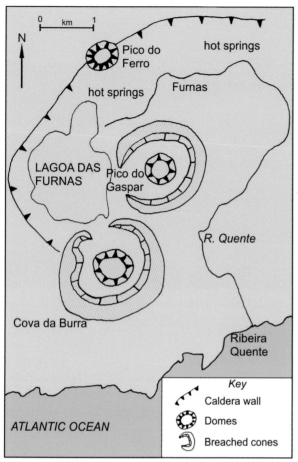

Figure 5.5 Map of the main volcanic features in Furnas Caldera.

Figure 5.6 The hot springs are a popular feature of the Furnas caldera (photo by José Luís Ávila Silveira/Pedro Noronha e Costa).

Trachytic eruptions have already started to fill the caldera. The Pico do Ferro itself is about 11,200 years old and it could have extruded before the caldera collapsed completely. But the main volcanic events in the caldera were the ten explosive eruptions, designated Furnas A to Furnas J, which took place during the past 5000 years. Most of these eruptions were hydrovolcanic and related, no doubt, to the three lakes lying within the caldera at that time. About 2900 years ago, the third and largest eruption (Furnas C) expelled a widespread blanket of 1.5km³ of pumice from a vent located on the flat plain where the village of Furnas now stands. This plain may therefore be the filled remains of a maar. In fact, the subsequent eruptions occurred on a ring fracture, and they formed a crescent of tuff rings into which trachytic domes have extruded – and where many of the chief fumaroles are found. The twin domes and tuff rings of the Pico das Marcondas, for instance, first formed during the fifth (Furnas E)

and eighth (Furnas H) episodes. About 1100 years ago, the dome and tuff rings of the Pico do Gaspar erupted and also produced the extensive pumice deposits of Furnas F, G and I. Pico do Gaspar also had the important morphological role of impounding the Lagoa das Furnas, the sole successor of the three former lakes. Thus, the Lagoa das Furnas has the unusual distinction of not occupying the lowest parts of the Furnas caldera, for the plain of Furnas lies almost 100m below it. The lake is only 14m deep and could itself be an old maar, now largely filled by pumice.

About 1445, there was a hydrovolcanic eruption within the caldera, but the only magmatic eruption recorded here during historical times occurred near its southern edge in 1630. Furnas caldera is clearly not extinct. The average interval between eruptions during the past 2900 years has been 362 years. The caldera has only to continue that pattern for another eruption to happen there within the next few decades.

Achada das Furnas

The plateau of Achada das Furnas abuts onto both the Água de Pau and Furnas volcanoes. It forms a moorland saddle of basaltic lava flows, between about 10,000 and 26,000 years old, which have been blanketed with trachytic pumice expelled mainly from Água de Pau stratovolcano. On the plateau there are also cinder cones about 100m high, such as the Pico do Meirim, the Pico das Tres Lagoas and the Pico de el Rei. They were formed along fissures at the same time as the later eruptions of Água de Pau and Furnas volcanoes. About 3800 years ago, a hydrovolcanic explosion of pumice and lithic debris also formed a tuff ring enclosing a maar, inside which a trachytic dome later arose. The Lagoa do Congro now occupies the maar.

Água de Pau stratovolcano

The low pyramidal outline of the Água de Pau strato-volcano dominates central São Miguel. It is 15km in diameter and reaches 947m at the Pico da Barrosa. Its elliptical crest rises in a wild, isolated moorland that is quite unlike any other on the island. It surrounds a caldera, with a maximum width of 3.5km from north-west to southeast, which is the smallest in São Miguel.

Some 250m below its rim lies the green Lagoa do Fogo ('lake of fire'), which has been the source of five major eruptions in the past 5000 years (Fig. 5.7).

Água de Pau began erupting on the ocean floor about 290,000 years ago and the volcano now covers an area above sea level of about 166km^2 and has a volume of some 80km^3. Água de Pau was more ex-plosive than its neighbours, giving off a large pro-portion of trachytic pumice and pyroclastic density currents in addition to its basaltic and trachytic flows. One section on the south coast near Ribeira Chà, for instance, shows that 65 explosive eruptions occurred over a period of 34,000 years. Trachytic pumice also now mantles the flanks of Água de Pau and hides much of the detail of its geological history. Most of the volcano seems to have grown between about 100,000 and 40,000 years ago, and the summit may have collapsed several times during this period.

The two most recent calderas have been identified. The larger, outer, caldera formed during a great trachytic eruption about 33,000 years ago and was 7km long and 4km across, and probably had a volume of 5km^3. The eruption inaugurated about 15,000 years of calm, which was disturbed only when three trachytic pumice rings erupted 18,600 years ago. Then, about 15,200 years

Figure 5.7 View of Lagoa do Fogo, Água de Pau. Its scalloped rim may indicate that the caldera was formed by intermittent collapse (photo by Perri.G – Wikimedia Commons).

ago, came another great trachytic outburst that created the caldera enclosing the Lagoa do Fogo, whose white trachytic beaches form an unusual scenic feature in the Azores. Like its predecessor, this eruption also heralded a dormant period, which lasted for 10,000 years. Fogo caldera probably collapsed intermittently because it has steep sides that are often coated with pumice fragments, its edges are scalloped into distinct lobes, and the lake occupies three intersecting hollows that mark the shifting focus of the eruptions.

Activity resumed within the caldera between about 4,675 and 4,435 years ago with the short Plinian outbursts, perhaps spread over decades or centuries, which expelled the Fogo A pumice. Pyroclastic density currents and ash falls distributed a thick blanket of buff-grey pumice and made a valuable indicator bed that is still 5m thick on the south coast and covers about 200km². Mudflows then also swept down to both coasts and covered about 24km². Four similar but less powerful eruptions followed. The Fogo B pumice came from a vent just north of the caldera some 3242 years ago. The Fogo C ochre pumice, which is notably rich in lithic fragments, was expelled from a vent in the south of the caldera. Fogo D erupted masses of pumice from the centre of the caldera. The latest Plinian eruption of Fogo took place in 1563. Severe earthquakes began on 24 June and lasted for five days. On 29 June, a huge Plinian column burst from the Fogo caldera. The wind spread most of the pale trachytic ash and pumice that

covered more than 200km² over the east of the island before the eruption ended on 3 July. This eruption may have induced the latest episode of collapse in the caldera.

Água de Pau also has more satellite cones than its companions, with the trachytic Cerrado Novo and the basaltic Monte Escuro among the most prominent. About 15 eruptions have occurred within the past 10,000 years on its western flanks; and some of the older cones and flows, for example, have been dated to 8700 and 6500 years ago. The basalt flow that forms a lava delta on the north coast at Ribeira Seca erupted 1790 years ago, the Pico das Mos about 1250 years ago, and the Mata das Feiticeiras only about 1000 years ago. Other unnamed cinder cones erupted near the south coast, including the cone rising to 153m, which emitted the fine delta of alkali olivine basalt on which the village of Caloura now stands. This cone has been dated to the period between the Fogo A eruption and the D eruption from Sete Cidades caldera, that is, between about 5000 and 3500 years ago. The vitality of Água de Pau volcano is also shown on its northern flanks, where there are hot springs of sufficient power to be harnessed by the geothermal station of Caldeiras, 4km southeast of Ribeira Grande.

The saddle of the Região dos Picos

The saddle of the Região dos Picos, stretching across São Miguel between Sete Cidades and Àgua de Pau, marks the greatest concentration of recent basaltic activity on

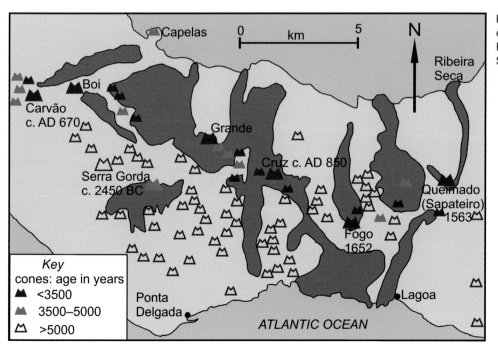

Figure 5.8 Recent eruptions in the Região dos Picos, São Miguel.

the island (Fig. 5.8). Although it began to form about 50,000 years ago, its surface features are much younger. It is dominated by about 250 cinder cones and their cultivated ash-covered lava flows, which originated along en-echelon fissures trending slightly obliquely to the main northwest–southeast grain of the western half of the island. Many vents were so closely spaced that the cinder cones are joined together, and also tend to be smaller than their more widely separated companions, which are often nearly 200m high. Although several lava flows reached either the north or south coasts, most of them are thin and less than 5km long.

The older cones, erupted before the Fogo A explosion 5000 years ago, occupy a broad band parallel to the south coast and lying 5–6km inland, which extends up to Sete Cidades volcano. Several cones erupted after the Fogo A eruption and before the Sete D eruption of about 3500 years ago. These include the Serra Gorda, which has, no doubt, dominated the saddle ever since it erupted about 4400 years ago. As its name suggests, this large cinder cone stands over 200m high and 1km across, and is easily recognized by its double-hump summit slashed by a fault-guided breach. The vent also expelled several basaltic lava flows that spread in an irregular apron around its southern base. At about the same time, Surtseyan eruptions formed the Capelas cone on the north coast and the Ilhéu Rosto de Cão off the south coast.

At least 18 eruptions have occurred from fissures in this area within the past 3000 years. The fissure some 2km long between the Pico do Cedro and the Pico do Enforçado gave rise to eight cinder cones and a broad basaltic flow spreading to the northeast. However, the backbone of the saddle is the fissure stretching 5km long that formed the Pico Grande, Mato do Leal and Pico Cruz, as well as seven other smaller cinder cones and two explosion pits. Pico Cruz erupted trachytic fragments some 1100 years ago. But perhaps the main contributions of this fissure were the large basaltic flows that spread to both the north and south coasts, where they form a delta, 4km wide, west of Lagoa.

Activity here since the settlement of the island has been on a smaller scale. In June 1563, a basaltic eruption took place at the same time as the great explosion in the Fogo caldera. It occurred when a short fissure cracked open the trachytic dome of Queimado (or Pico Sapateiro), formed several little craters on its crest, and gave off a thin lava flow that swamped part of the village of Ribeira Seca on the north coast. On 19 October 1652,

an eruption from a vent at the base of Fogo 2 cone formed a conelet, Fogo 1, and a lava flow that reached the sea 4km away.

Sete Cidades stratovolcano

Sete Cidades stratovolcano covers all the northwest of São Miguel (Fig. 5.9). It grew up from a base 2000m deep on the ocean floor and was a separate island until eruptions in the central saddle joined it to the rest of São Miguel. It is now 14km across at sea level, with a volume above sea level of some 70km³. Its smooth slopes rarely exceed 10° and are covered with farmland and plantations of the conifer cryptomeria. The Sete Cidades caldera (Fig. 5.10) crowns the stratovolcano and owes its name to the legend that seven cities were engulfed when it collapsed. Almost 5km in diameter and 6km³ in volume, it is home to rich vegetation, ranging from laurel and cryptomeria to hydrangea and the now rampant ginger lily (*Hedychium gardnerianum*). Several squat wooded cones and the village of Sete Cidades occupy its floor, but its most striking features are two lakes, the Lagoa Azul and the Lagoa Verde, which often reflect the blue and green of their names. This majestic caldera, whose extraordinary serenity belies its violent

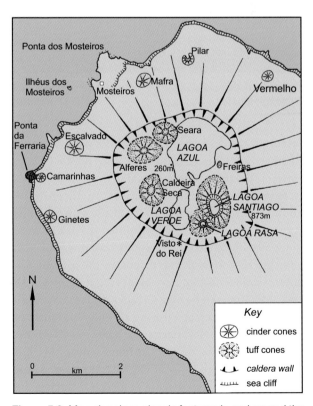

Figure 5.9 Map showing volcanic features in and around the Sete Cidates Caldera.

Figure 5.10 Panoramic views of Sete Cidades caldera. **A**) Highlights the Green Lake (left) and the Blue Lake (right) divided by bridge (photo by Maros (Wikimedia Commons). **B**) View looking over Lagoa de Santiago (Shutterstock/Henner Damke).

origins, is the deepest and perhaps the most beautiful in the Azores. The scene from the Visto do Rei ('king's view') is the most photographed in the islands.

Sete Cidades stratovolcano began to grow up about 290,000 years ago, at about the same time as Água de Pau. The early eruptions of basalts were gradually succeeded by trachytes, and it was only in later life that the volcano turned to the violent trachytic explosions that have blanketed its slopes with layers of pumice. The Sete Cidades caldera collapsed about 22,000 years ago during a Plinian outburst of trachytic pumice. Its appearance bears out its youth: the rim is undissected and the inner scarps have not weathered enough to form large basal screes. Many pyroclastic density currents surged out during the eruption, especially to the southeast, and massive deposits of pumice are still 60m thick on the coast,

13km from the caldera. After the caldera formed, a lava flow erupted near Feteiras on the southeastern flanks of the stratovolcano about 20,890 years ago, and domes and pyroclastic density currents erupted about 17,160 years ago near the Ponta do Escavaldo on the west coast. But the caldera itself stayed quiet for 17,000 years.

Just over 5000 years ago, a series of a dozen violent trachytic eruptions began, which sprang from vents along a ring fracture situated about 1km inside the caldera walls. They formed a ring of squat cones – Seara, the Caldeira do Alferes, the Caldeira Seca – and the unnamed steeper tuff cones containing the crater lakes of Lagoa Rasa and Lagoa de Santiago. In addition, the Lagoa Azul and Lagoa Verde are probably maars resulting from hydrovolcanic explosions. These seven features may have been at the origin of the Seven Cities legend.

These eruptions have been designated Sete A–L. About 5000 years ago, a Surtseyan eruption from a lake within the caldera built the Seara tuff cone and distributed the first (Sete A) layer of fine grey ash over the stratovolcano. An eruption on dry land formed the Alferes trachytic cinder cone perhaps about 4000 years ago. Most of the remaining eruptions came either from the craters of Lagoa Rasa or Lagoa de Santiago, or from both of them. They not only ejected thick layers of ash and pumice well beyond the caldera, but also built two prominent tuff cones within it. For all its eruptive vigour, however, the Lagoa Rasa crater now occupies a shallow basin, choked with thick pumice and ash. But the Lagoa de Santiago is a fine explosion crater, made sinister by the steep wooded slopes of pumice rising 200m around it. Some of the later eruptions from these craters have been dated. The Lagoa de Santiago expelled the Sete J layer 1860 years ago, and the Lagoa Rasa expelled the Sete K layer 1570 years ago. The last ash layer (Sete L) erupted in a lake from the vent that built the Caldeira Seca at the same time. Its slopes are scarred by many closely spaced gullies, which make it look much older than it is, because carbon-14 measurements indicate that it erupted either 660 or 500 years ago. These dates could correspond with an eruption that is said to have occurred in 1444. The small cinder cone of Freiras, beside the Lagoa Azul, has also been proposed as a source of this eruption. However, the description is a much better match to a more violent hydrovolcanic eruption. Indeed, an early chronicler, Gaspar Fructuoso (1591), recounted how the first Portuguese explorers had made a note from their boat of a distinctive peak at Sete Cidades to aid future navigation. When they returned with a group of colonists a year later, in 1439, they saw that the sea was cluttered with pumice and tree trunks, and that the distinctive peak had disappeared, because 'fire had arisen and burned it in the interim. And, for most of their first year of settlement, the colonists could hear, from their straw huts, the roars and bellows given out by the Earth, with great tremors from the subsidence and burning of the peak that had disappeared.' But the attribution of this eruption to the Caldeira Seca also poses problems. The first settlers could not have seen the Caldeira Seca from their ship because it would have been hidden within the Sete Cidades caldera: it could not have been 'the distinctive peak' that they noted. It has thus been suggested that the eruption might have come from the Lagoa do Canário, on the upper eastern flanks of the stratovolcano.

In view of all the commotion in the caldera during the past 5000 years, it is surprising that no volcanic activity has disturbed the caldera since those early days of Portuguese settlement. However, the earthquakes of low magnitude during the summer of 1998 could have been the first indications of a revival.

Activity on the Sete Cidades volcano during the past few millennia was not confined solely to the caldera. Even more basaltic emissions erupted from satellite vents all around the lower flanks of the volcano and in Maciço das Lagoas, which stretches southeastwards from the caldera. Eruptions have also been recorded offshore.

Maciço das Lagoas

Maciço das Lagoas (the Lakes Massif) was named thus because many craters are filled with lakes, and yet more lakes have formed where the drainage has been impounded between the cones. The eruptions occurred where three major fissures extend southeastwards from the caldera of Sete Cidades. In fact, it forms the highest upland area on the stratovolcano, culminating in the Pico das Éguas at 873m with its small lagoas (Fig. 5.11), and the vents were so closely spaced that the basaltic cinder cones often merge together. The three fissures were active at different times. The central fissure erupted about 3500 years ago, just before the Sete D layer exploded from the caldera; the Pico de Eguas erupted 2700 years ago on the southwestern fissure; and the 200m-high cinder cone of the Pico do Carvão erupted on the northeastern fissure 1280 years ago. The cinder cones are up to 200m high and their craters are unusually deep, which suggests some hydrovolcanic influence in their construction.

Eruptions and erosion on the coast around Sete Cidades volcano

The eruptions on the coastal fringe have formed cones and lava flows. Surtseyan activity on the northeastern coast formed the Capelas volcano, which is now reduced to a half-eroded promontory. However, the remaining eruptions usually took place, on radial fissures, on top of a 100m cliff. Many of these cones, such as Várzea, Lagoa do Pilar, and Mafra, are accompanied by lava flows that cascaded down the cliffs and formed lava deltas along the shore. These eruptions have taken place throughout the past 5000 years, but their blunt outlines suggest that they are not very young. The most recent

Figure 5.11 Pico das Éguas at 873m with the Lagoa das Éguas Norte in the foreground (Shutterstock/PRILL).

eruption on the stratovolcano on land probably took place 840 years ago. It built up the cone of the Pico das Camarinhas and emitted a lava flow that now extends seawards as the Ponta da Ferraria, the westernmost point of São Miguel.

One of the most active fissures in the Azores during historical times extends northwestwards from the Ponta da Ferraria (Fig. 5.12). It has produced three recorded submarine eruptions. On 3 July 1638, after a week of earthquakes, 'a great fire hurled masses of black sand as high as three church steeples placed one upon the other, and the billowing vapours joined the clouds. The sand that fell back down formed an island that only disappeared with the advent of winter, and left behind a vast and dangerous shoal' (recorded account in Weston, 1964). Calm returned on 28 July. In December

Figure 5.12 View of the Ponta da Ferraria (photo by José Luís Ávila Silveira/Pedro Noronha e Costa).

1682, violent earthquakes rocked the whole island. 'The earthquake, while the morning preachers were preaching on 13 December, was so powerful that people thought that they were going to be cast from the Earth. But it pleased God to raise up fire in the sea during the following week, almost four leagues off Ferraria … roasting a quantity of fish that were thrown up onto the coast. So much pumice was floating on the sea that a big sailing ship en route from Angra [in Terceira] could not force a passage through it.' This brief but powerful eruption lasted two or three days and seems to have taken place at a depth of about 300m.

After seven months of earthquakes, there was another eruption on 31 January 1811, on the same fissure, but in shallower water, less than 2km offshore. It produced a Surtseyan display of explosions, and jets of ash and cinders that lasted for eight days and spread ash and sulphurous fumes as far as Ponta Delgada, 20km away. But it never formed an islet. This lapse was rectified when another Surtseyan eruption broke out just to the northwest, on the morning of 14 June 1811, after four days of vigorous earthquakes. Great clouds of ash, fumes and cinders thundered from the sea, and an islet started to form on 16 June. It was 1600m in circumference and about 70m high when the eruption waned on 22 June and began to emit nothing but fumes. The British frigate *Sabrina* was in São Miguel at the time and, on 4 July, Captain Tillard landed on the island, planted the British

Don João de Castro Bank

Don João de Castro Bank marks the crest of a growing stratovolcano and could be the site of one of the next major islands to form in the Azores. In 1941, a shoal called the Don João de Castro Bank was discovered 64km northwest of São Miguel and 51km southeast of Terceira. A spectacular submarine Surtseyan eruption had begun there on 9 December 1720 and could be seen from both islands. In four days, it gave birth to an islet, which apparently reached nearly 1km across and 180m high during the ensuing two weeks. Towards the end of December, the activity calmed down enough to enable visitors to approach by boat. They observed that the island was composed of ash and fragments and that one vent was still operating. Activity waned during the following months and the winter waves had destroyed the island by March 1721, leaving only the submerged shoal behind. This shoal was the location of several earthquake spasms in 1988 and 1989. A magnitude 5.5 earthquake occurred on 27 June 1997 and was followed by 45 earthquakes exceeding magnitude 4.0 and some 2000 weaker tremors during the following two months. Although no fragments reached the open air, such seismic activity suggests that another eruption could have occurred at that time. A map of the islands highlighting some of the known offshore events is given in Fig. 5.13.

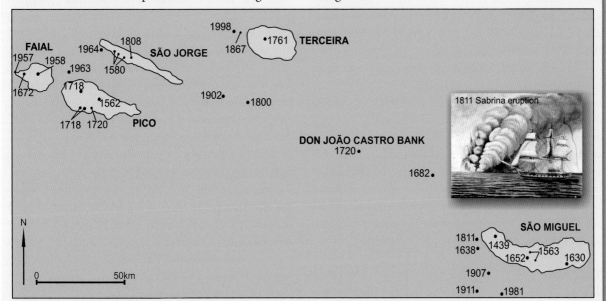

Figure 5.13 Locations and dates of eruptions during historical times in and around the Azores (note: many offshore eruptions, e.g. off Terceira, São Miguel and the Don João de Castro Bank).

flag upon it, claimed it for King George III, and named it after his ship. However, his imperial enterprise was not rewarded, for Sabrina's fragile tuffs succumbed to the Atlantic waves before October was out.

The relationships between coastal volcanic activity and marine erosion are beautifully displayed at Mosteiros in the far northwest of São Miguel. Off shore, the four islets of Ilhéus dos Mosteiros are all that remains of the eastern half of a Surtseyan tuff cone. Off the northern end of Mosteiros village lies an older and much more eroded cinder cone, which must have erupted on dry land when sea level was lower than at present. When sea level rose, the waves planed off the cone to such an extent that only its concentric black layers of spatter jutted out from the sea. Then, Mafra and a smaller vent nearby erupted, and their lavas formed a rugged delta that covered the southern part of the eroded cinder cone on the shore. In turn, the Atlantic waves planed off the surface of the lavas, when the sea stood some 15m higher than today. The village of Mosteiros stands on this surface. Now, the present waves are starting their work all over again and attacking the exposed fringes of the lava delta.

Santa Maria

Santa Maria is a small island, 16km long and 8km wide and rising only to 590m. It has shown no sign of activity since it became the first island in the Azores to be settled in 1439. Basaltic eruptions built up the island from a depth of about 2000m on the floor of the Atlantic Ocean. The oldest basalts yet dated, near Vila do Porto, are 5.14 million years old, and similar lavas from Pico Alto in the centre of the island are 5.11 million years old. Some flows contain layers of marine sandstones and limestones, which mark a period of subsidence in the history of Santa Maria, which has apparently no parallel in other islands. About 4.2 million years ago, renewed activity formed the Pico do Facho in the west and much of the undulating zone in the centre and east. All these volcanic masses have been deeply weathered and eroded by both fluvial and marine erosion, and São Lourenço volcano, for example, has been almost removed by the sea. Santa Maria has witnessed neither earthquakes nor thermal-spring activity during historical times.

Terceira

Terceira is so named because it was the third island in the Azores to be discovered and settled around 1450. It forms an ellipse, 30km long from east to

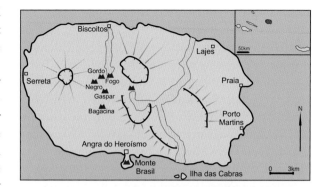

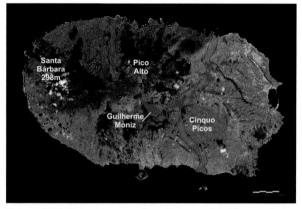

Figure 5.14 Recent cones, lava flows and caldera walls on Terceira (satellite map – USGS).

west, and 15km wide from north to south, with an area of 406km² (Fig. 5.14). On the fertile coastal plain, the volcanic rocks have weathered enough to provide viable farmland and to support an almost unbroken necklace of settlements around the island. It is often separated from the sea by fretted lava cliffs about 20m high, but the beautiful capital of Terceira, Angra do Heroísmo, grew up where the Surtseyan cone of Monte Brasil protects the most sheltered bay on the south coast. The eastern third of the island is low and gently undulating, but its higher and hillier central and western parts in the interior form grassy moorlands that rise to 1021m in the Serra de Santa Bárbara. It is here that the lavas are younger, where the latest eruption took place on the island in 1761, and where fumaroles, such as the Furnas do Enxofre, are still active. Terceira has also suffered several earthquakes since the Portuguese settlement; the one on 1 January 1980 caused much damage, especially in Angra do Heroísmo.

Terceira rises from a depth of over 1500m on the floor of the Atlantic Ocean, and only just over a tenth of the volcanic pile rises above sea level. The submarine part forms a basal basalt shield that was expelled during the past million years. However, most of the exposed rocks of Terceira are much younger, and the oldest dated lavas are 300,000 years old. The rocks also evolved to a number of compositions from basalts (e.g. olivine basalts, alkali basalt) through trachyandesites to trachytes and even to pantellerites (a peralkaline rock), which outcrop nowhere else in the Azores. In general, the peralkaline eruptions give rise to lava flows and domes rather than to fragments. For most of the history of the island, basalts and more alkaline lavas erupted in almost equal proportions; however, during the past 23,000 years, eruptions of alkaline lavas were notably more voluminous than basalts, although they were less frequent.

The Terceira Rift curves across the island from northwest to southeast. At present, it forms an active fissure zone about 2km wide, with over two dozen recent cinder cones, basaltic flows, spatter cones, intruded dykes, and elongated clefts. But it is a rather enigmatic feature, because it hardly ever forms a distinct morphological unit, especially where it has been recently less active in the east. However, the four stratovolcanoes that dominate the scenery of Terceira all lie on, or close to, the rift zone. They formed in succession from east to west; caldera-forming eruptions have decapitated their summits, and the western pair is still active.

Cinquo Picos stratovolcano

Cinquo Picos stratovolcano occupies much of southeastern Terceira, and its caldera, 7km across, is one of the largest in the Azores. It makes little visual impact because it is the oldest, lowest, smoothest, most gently sloping and most eroded on the island. Only two sectors survive in the landscape: the smooth Serra do Cume in the northeast, and the even smaller Serra da Ribeirinha is all that remains of the opposite rim. The stratovolcano is composed chiefly of basalt lavas, with some flows of olivine basalt, trachybasalt and trachyte on its flanks. Rocks near the summit of the Serra do Cume have been dated to 300,000 years ago. Sometime afterwards, much pumice erupted and the caldera formed. The floor of the caldera is covered by much more recent ash and flows, and makes a vast chequerboard of meadows, which is decorated by the five much younger basaltic cinder cones – the 'cinquo picos' – that gave the caldera its name.

Guilherme Moniz stratovolcano

Guilherme Moniz stratovolcano and the rim of its caldera are represented by the crescent of the Serra do Morião, which reaches 632m and forms the northern backcloth to Angra do Heroísmo. The outlines of Guilherme Moniz caldera are altogether sharper than those of its neighbour. The inward-facing scarps of the caldera are often vertical and more than 150m high. They are composed of thick flows and domes of trachyte, with few layers of fragments. Thus, effusive activity must have dominated the life of the stratovolcano until the violent eruptions that created the caldera. This event has not been accurately dated, but on morphological grounds it could well have taken place less than 100,000 years ago. The fringes of Pico Alto volcano have covered the northern sector, and recent basalts from satellite vents on its lower flanks invaded the Guilherme Moniz caldera, which therefore became a rare example of a caldera floored by lavas that erupted from another volcano.

Pico Alto stratovolcano

Pico Alto has perhaps the most complicated history and structure of any volcano in Terceira, or, indeed, in the Azores. It occupies much of north-central Terceira, culminates at 808m, and is composed almost entirely of trachytes and pantellerites. It is probably less than 60,000 years old. It developed a caldera, but subsequent eruptions from over a dozen vents filled it so quickly and completely that the initial hollow has been replaced by an isolated upland wilderness of confused relief around Pico Alto itself. Several explosions of pumice occurred, and one of the largest formed the narrow train of the Angra ignimbrite about 23,000 years ago. The more widespread Lajes ignimbrite has been dated to 18,600 years ago at São Mateus on the south coast and to 19,680 years ago at Lajes, which could indicate that there were two separate eruptions. Subsequent eruptions emitted trachytic flows reaching 50m thick, and extruded domes, some of which sagged down to form additional thick flows. The north flanks of the volcano reveal lava flows, now wooded, that made it down to the coast at the Ponta do Mistério (Fig. 5.15). The southernmost flanks of Pico Alto impinge upon the Terceira Rift, where some vents are still active; the Furnas do Enxofre (Fig. 5.16), for instance, now constitutes the major area of solfataras and fumaroles on the island.

Figure 5.15 B/W contrast image highlighting the wooded lava flow forming the Ponta do Mistério and the flanks of the Pico Alto Volcano.

Figure 5.16 Hot springs at the Furnas do Enxofre prove a popular tourist spot (photo by José Luís Ávila Silveira/Pedro Noronha e Costa).

Santa Bárbara stratovolcano

As Pico Alto grew to maturity, Santa Bárbara stratovolcano began to develop 10km to the west. Santa Bárbara occupies the western part of Terceira and its diameter at sea level is about 12km. It is the most clearly defined, highest, steepest and youngest stratovolcano in Terceira. Flows of basaltic-trachyandesite and trachybasalt form a shield up to about 700m that is surmounted by a steeper central cone that reaches 1021m, where the trachyandesites found near the summit are less than 29,000 years old. A violent eruption destroyed the crest of Santa Bárbara and formed a small caldera about 25,000 years ago. Much of it was then filled with trachytic domes and pitons, but about 15,000 years ago a small caldera formed a narrow hollow within its predecessor. Several trachytic domes have already extruded within it, and it now also contains a small lake.

During the past few thousand years, several cinder cones and domes have erupted on the flanks of Santa Bárbara. In the south, for example, the Cerrado das Sete and the Pico dos Enes probably lie on one of its radial fissures, and the Pico da Catarina Vieira and the Pico dos Padres arose on another. The prominent trachytic dome and lava flow of the Pico Rachado, which stands 200m high on the upper northern flanks of Santa Bárbara, could also belong to a radial fissure. However, the radial fissures on both the eastern and western flanks of the volcano merge with those developed on the Terceira Rift zone. These eruptions are probably among the youngest on the stratovolcano and form both trachytic domes, such as Serreta and Negrão in the west, and cinder cones such as the Pico da Candela in the east.

The Terceira Rift

The Terceira Rift has its finest expression – and its latest eruptions – in the upland saddle between the Pico Alto and Santa Bárbara volcanoes. They gave rise to basaltic cinder cones that are typified by Bagacina, whose very name is the local word for volcanic cinders. It is a fresh cone, 120m high, that was apparently constructed within the past few thousand years at most, and it is breached by a trachybasalt flow that reaches the south coast 5km away near Angra do Heroísmo. Pico Gordo, its counterpart some 2km to the northwest, expelled an even younger tongue of olivine basalt to the north coast, where it now juts out to sea in the spectacularly jagged reefs of the Ponta dos Biscoitos. Between Pico Gordo and Bagacina is a fine and unusually large spatter cone, the Pico do Gaspar (Fig. 5.17), where a wall of spatter encircles a crater about 100m deep. Its horned profile, forming the background to the Lagoa do Negro, is one of the most striking in Terceira. Further east, the Algar do Carvão erupted 2115 years ago on the lower flanks of the Pico Alto volcano. It still retains both the gaping abyss of the main vent and the magnificently rugged pinnacles on its lava flow. The basalts invaded the Guilherme Moniz caldera, where they were ponded back in a lake of lava. At length, part of the flow spilled from the caldera, swept eastwards around the slightly older cones of the Pico da Cruz, Areeiro and Gualpanar, and escaped in two fine rugged tongues, one extending to the north coast and the

Figure 5.17 Pico do Gaspar in the Terceira Rift (photo by José Luís Ávila Silveira/Pedro Noronha e Costa).

other to the south coast. Even the calmer easternmost part of the rift zone has also experienced some recent basaltic eruptions, and the latest flows jut out to sea in a lobe at Porto Martins.

All this activity probably occurred within a few millennia of the Portuguese settlement, but the only historically recorded volcanic activity in Terceira took place in April 1761. It occurred in two distinct areas lying just to the north of the rift zone. After three days of earthquakes, the Mistério Negro extruded in the west on 17 April and oozed down the lowest fringes of Santa Bárbara volcano. It now forms a squat, rounded hump of black pantelleritic trachyte made up of domes and short lava flows about 50m thick. On 21 April, an entirely different eruption began some 2km away to the east. It produced three small basaltic cinder cones, of which

the appropriately named Pico do Fogo is the largest. It emitted a short lava flow to the west and a flow some 2km long to the north-northeast. But the longest flow of all, a tongue of trachybasalt, travelled 5km as far as the church at Biscoitos and partly covered the slightly older alkali olivine basalts that had previously been expelled from the Pico Gordo. These eruptions ended on 28 April 1761.

Lave tubes and caves

Terceira Island is home to a myriad of tubes and caves. It is second only to Pico in the amount of volcanic caves, and boasts some 28 lava tubes. Some are quite off the beaten track but some are becoming more popular and accessible. Many of the tubes are located along the Terceira rift and along the lava flows that flow down to the coast from there. Some of the more touristic examples include the Algar do Carvão and the Gruta do Natal (Christmas Cave – see Fig. 5.18A). Another great example is the Gruta dos Balcões (cave of branches), which is a branching set of tubes that is found in the historical lava flow of Pau Velho (1761) in Biscoitos. Here a network exists that totals some 4.6km of tubes, preserving many lava tube features including very well preserved polymorph stalactites (lava drops) on the ceilings (Fig. 5.18B). There is still much to explore in these underground networks, and they are home to a number of flora and fauna including mosses and cave-dwelling arthropods.

Figure 5.18 Examples of lava tubes on Terceira. **A)** Gruta do Natal. **B)** Gruta dos Balcões with inset photo showing lava drops (photos by José Luís Ávila Silveira/Pedro Noronha e Costa).

Submarine eruptions (1800, 1867, 1902, 1998–2000)

Four submarine eruptions have been recorded off Terceira. A small eruption began on 24 June 1800, when the sea glowed, gave off fumes, and rumbled like thunder for three days. In 1867, five months of earthquakes culminated in a strong shock on 1 June that damaged 200 houses in and around the coastal village of Serreta. That night, an eruption boomed out some 6km from the Ponta da Serreta (see map in Fig. 5.19) and, during the next week, great fountains of water, steam and occasional masses of cinders gushed skywards from six or seven vents. The third eruption, which took place nearby, was even more discreet. It was identified only when the telegraph cable between Pico and Terceira was broken, burned and buried in tuffs at depths varying between 450m and 1400m. It caused little comment, because it occurred during the night of 7–8 May 1902 between the notorious eruptions of the Soufrière of St Vincent and Montagne Pelée in Martinique. The latest activity has been called the Serreta eruption because it occurred out to sea, 9–14km northwest of that village. It began with small earthquakes in November 1998. During the following four months, white fumes rose intermittently from vents at a depth of about 500m, and orange lights were observed on 23 December. On 8 January 1999, incandescent lava blocks, up to 3m across, kept on rising to the surface in half a dozen places, and each floated for several minutes (see Fig. 5.19). They were composed of alkali basalts that were riddled with gas holes, and it seems that they were emitted as hot lava balloons, rather like detached pillow lavas, that lost their enclosed gas and sank when the congealed skin cracked on contact with the air. This latest eruption episode lasted until March 2000, and the location of the eruptions suggests activity along NE–SW oriented faults (Fig. 5.19). The volcano has been termed the 'oceanic volcano' Serreta.

These eruptions formed no islets, presumably because they occur in waters that are still deep enough to stifle any Surtseyan activity. However, the southern coast of Terceira has experienced several Surtseyan eruptions in the recent past. The remains of two tuff cones lie off Angra do Heroísmo; and the Ilhéus das Cabras, 6km away to the east, mark the remnants of an older and more exposed Surtseyan cone. It has already been cut into two by the waves, although its original conical structures are still evident.

In terms of settlement, the formation of Monte Brasil was by far the most important of all the recent Surtseyan eruptions in Terceira, because it created the sheltered bay (or angra) that led to the foundation of the capital, Angra do Heroísmo (Fig. 5.20). A basaltic eruption in shallow water formed the bulky Surtseyan tuff cone of Monte Brasil, which rises 205m above the waves and is over 1km in diameter. Its crater is 170m deep, and powerful explosions kept open the southern sector of the cone. The whole mass is composed of thin beds of fine yellow and brown tuffs, and a high isthmus of windblown fragments joins the cone to the mainland. Although the sea has cut cliffs around much of Monte Brasil, it has also been partly protected by a skirt of black basaltic flows that emerged during its later Strombolian phase. As a result, Monte Brasil still forms a worthy bastion for Angra do Heroísmo, in spite of centuries of marine attack.

Graciosa

Graciosa has a greater proportion of cultivated land than any other island in the Azores, which is a reflection of its lower altitude, drier climate, deeper soils

Figure 5.19 Eruptions of Serreta volcano offshore from Terceira: **A)** map outlining approximate positions of offshore eruptions, **B)** during the 1867 eruption and **C)** during the main 1999 eruptions.

Figure 5.20 Monte Brasil sheltering the harbour of Angra do Heroísmo. **A**) Modern-day view (photo by Franzfoto, Wikimedia Commons). **B**) The bay of Angra do Heroísmo, Terceira, with boats and the Fortress of St. John the Baptist; Lebreton engraving, 1850 (Wikimedia Commons).

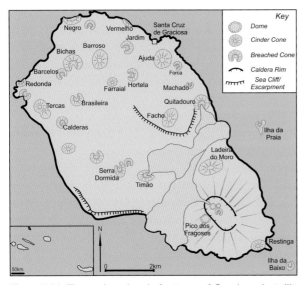
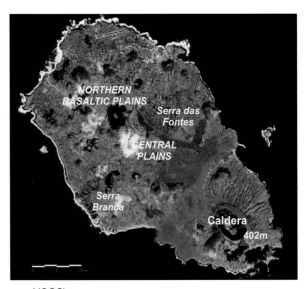

Figure 5.21 The main volcanic features of Graciosa (satellite map – USGS).

and gentler slopes. The centre, and oldest part, of the island is composed of two upland blocks that face each other across a broad axial valley (Fig. 5.21). To the north lies an extensive plain scattered with fresh cinder cones, but to the south the scenery is dominated by the stratovolcano, Caldeira. Basaltic flows and fragments often predominate throughout the island, but notable amounts of trachybasalt, basaltic-trachyandesite and trachyte also erupted, especially as the stratovolcano developed. No volcanic eruptions have been recorded since Graciosa was first settled in about 1480 and, until the repercussions of the Terceira earthquake were felt in 1980, the only previous major seismological event had occurred when Praia and Luz were damaged in 1730.

Eruptions began at a depth of 1500m or more on the ocean floor, but the oldest emerged rocks occur in the piles of horizontal flows comprising the central upland blocks. Trachybasalts forming the base of the Serra das Fontes have been dated to 620,000 years ago, trachytes at the base of the Serra Dormida are 350,000 years old, and those near the summit of the adjacent Serra Branca date from 270,000 years ago. These uplands are demarcated by steep straight scarps, often over 100m high, which seem to have formed by faulting. Thus, the two older upland areas of Graciosa seem to form up-thrown fault blocks that are now separated by a central fault trough.

Caldeira is now the most striking natural feature of Graciosa. It probably first emerged as a separate island to the south of the up-thrown lava blocks. Rising to only 402m and stretching only 5km across, the stratovolcano is relatively young, and erosion has been limited to valleys etched only two or three metres deep near the summit, and to rugged sea cliffs that are usually less than 20m high, except around Restinga in the far southeast.

The stratovolcano was built up by eruptions of basaltic flows and layers of cinders, lapilli, ash and tuffs 20cm to 8m thick. They enclose only two thin fossil soils, indicating that the volcano grew up rapidly. The inner walls of the caldera are composed almost entirely of lava flows, which suggests that effusive eruptions predominated for much of the life of the volcano. The transition to violent eruptions was therefore sudden and belated. The caldera is elliptical, stretching 1600m from northwest to southeast, but it is only 875m broad, and its rim is notched in the northwest. It is certainly less than 200,000 years old, and the sharp rim, steep walls and paucity of weathered material all suggest that it would be surprising if the caldera were, in fact, more than 20,000 years old.

The original amphitheatre might have been more than 400m deep, but later eruptions have already covered its floor. In the northwest, two basaltic domes rise 100m high and a flank of the western dome has collapsed to reveal its typical onion-skin layering. In the centre of the caldera, three small cinder cones form grassy mounds between 50m and 100m high. A basaltic flow also erupted and escaped through the notch on the northwestern rim of the caldera, and then spread around the flanks of the stratovolcano. The smaller branch flowed eastwards to the sea near Praia, and the larger western branch stretched out into Folga Bay. It is possible, indeed, that this flow first joined Caldeira to central Graciosa. However, the most original, and probably the latest, feature in the caldera lies in its lowest southeastern part in the Furna do Enxofre ('sulphur oven'). This is a dark vent, 47m deep, with two main fumaroles and an underground lake with a floor below sea level.

Meanwhile, other eruptions had taken place on the flanks of the stratovolcano. The Pico da Ladeira do Moro formed an impressive double cinder cone, some 150m high and nearly 1km across, that now dominates Praia (Fig. 5.22). Close to the southwestern rim of the caldera, twin domes form the Pico dos Fragosos, which have bristling, rounded summits of trachybasalt and basaltic-trachyandesite. The eastern dome collapsed and formed a viscous lava flow that congealed in a thick, rugged, steep-sided tongue stretching 1km down slope. Other satellite vents of the stratovolcano erupted the cones, the much-eroded remains of which now form the islets of Ilhéu de Praia and the Ilhéu de Baixo. Although they now lie in the sea, they are composed of ash and cinders that erupted on land, probably when sea level was lower several thousand years ago.

The northwestern third of Graciosa is a low-lying plain, covering about 25km². It is composed of many basaltic flows that are often weathered deeply enough to support lava-walled vineyards, meadows and even arable land. The flows reach the coast in intricate and very rugged minor cliffs. Dotted along fissures generally trending from northwest to southeast are 20 or so breached cones 50–100m high. Most of these are cinder cones, with sharply defined outlines that are probably only a

Figure 5.22 The town of Praia, Ladeira do Moro and the flanks of Caldeira (photo by Angrense).

few thousand, or possibly even only a few hundred, years old. Although the majority are Strombolian cones, one of the most prominent, Ajuda, is a typical Surtseyan cone, which dominates the skyline of the capital, Santa Cruz. Within several other coastal cones, such as Redondo, Negro and Vermelho in the northwest, layers of yellow tuffs between the cinders reveal Surtseyan episodes in their formation. These cones have already lost two-thirds of their volume to the sea, but yet have scarcely been eroded more than Capelinhos in Faial, which erupted as recently in 1957–8.

About 15 breached cinder cones also erupted, at much the same time, on the plateaux of the Serra das Fontes, the Serra Branca and the Serra Dormida. Here, relief has played an important role, for several cones formed on the crests of the bounding scarps. Quitadouro, for example, erupted at the summit of the southern edge of the Serra das Fontes. The southern sector could not form because lavas cascaded down the scarp towards Praia. The lava flow and the slumped cinders now form a chaotic tongue issuing from the reddish-brown armchair of Quitadouro. Its counterpart, the Pico Timão, is a fresh cone, about 100m high and with a clearly marked crater rim, which occupies a commanding position at the top of the edge of the Serra Dormida. Timão forms a crescent of bare, reddish cinders that is breached where a lava flow escaped down the scarp and prevented the northeastern sector of the cone from forming. The basalts spread southeastwards and entered the sea, 4km from the vent, in a 1km-wide lobe north of Praia, and they have maintained their rugged aa surface, which plantations of eucalyptus, cryptomeria and laurel cannot disguise. In their progress, they also covered parts of the flow that had issued from the caldera, as well as the snout of the flow from Quitadouro. The morphology suggests that Timão cannot have erupted long before the first colonization of Graciosa in about 1480.

São Jorge

Seen from Graciosa, the island of São Jorge looks like a defensive wall protecting the conical fortress of Pico beyond. Its crest rises about 800m above sea level and culminates at 1053m in the Pico da Esperança. São Jorge forms a spine, rarely more than 6km wide, running in a straight line for 54km from northwest to southeast (Fig. 5.23), and it is the only island in the Azores where fissures have played such an overwhelming role in the scenery. The straight and steep cliffs demarcating the

island further emphasize the rectilinear theme in the landscape. São Jorge bears no trace of a stratovolcano or a caldera.

Activity began at about 1000m down on the floor of the Atlantic Ocean. The eruptions above sea level produced fluid flows of alkali basalt, with some trachybasalts and a few basaltic-trachyandesites, whereas rather more explosive eruptions formed the many cinder cones, about 100m high, that dominate the backbone of the island. The older parts of São Jorge lie in the east and the most recent parts in the west. There have been two terrestrial and perhaps two submarine eruptions in São Jorge during historical times.

São Jorge is divided into two unequal parts by a band of faults trending from north-northwest to south-southeast, which cut slightly obliquely across the predominant tectonic trends on the island. The Ribeira Seca stream has cut deeply into this band of fractured rock. The eastern third of São Jorge forms the Serra do Topo, which is composed of piles of thin flows of alkali basalt and trachybasalt. They are sometimes interspersed with layers of weathered cinders, exhibiting various phases of repose, and yellowish tuffs, which suggest hydrovolcanic eruptions. The dissected relief of the plateau lavas and the smooth outlines of the cinder cones, which have almost invariably lost trace of their craters, indicate that the Serra do Topo is the oldest part of São Jorge. The morphological inference is corroborated by dates ranging from 550,000 to 110,000 years in age. Both the northern and southern coasts are apparently faulted down towards the sea, and the Serra do Topo formations are also brought to an abrupt end where the Ribeira Seca faults cross the island.

The remaining two-thirds of São Jorge lying to the west of the Ribeira Seca faults are clearly younger than the area to the east. However, the same eruptive style continued. The Rosais complex, about 30,000 years old, and the younger Manadas complex, about 24,000 years old, make up most of the broad ridge forming the centre and west of São Jorge. Thin alkali basalt and trachybasalt flows once again emerged repeatedly from fissures and piled up in a total thickness of over 500m. Older cinder cones were often buried beneath newer accumulations, but the fresh outlines and sharp craters of the youngest cones dominate the scenery along the spinal crest of the island. The Manadas cones, for example, near the northwestern end of São Jorge, were probably formed only during the past few thousand years.

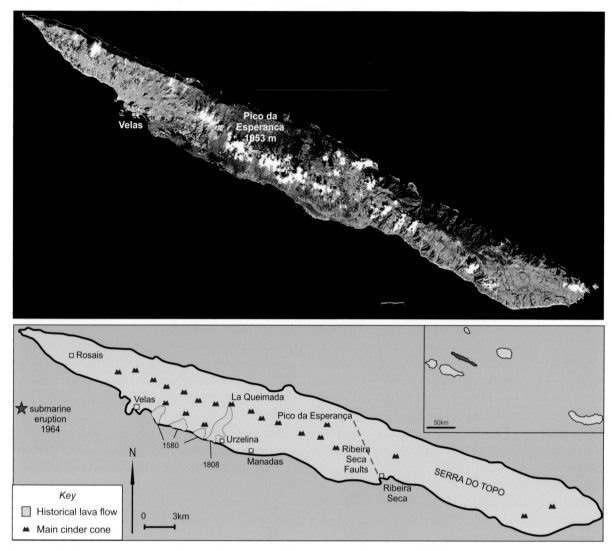

Figure 5.23 Recent features and historical lava flows of São Jorge (satellite map – USGS).

The cones are larger and more closely spaced in the centre of the island, but they become smaller and more scattered in the northwest. As in the eastern zone, most of the lava piles end abruptly in majestic straight cliffs that were probably created by faulting. These cliffs are so sheer that the few coastal settlements cling to a succession of fajãs – lava deltas or lobes of rubble that have fallen from cliffs.

These eruptions have continued into historical times. After over a hundred earthquakes had rocked the island during the previous two days, the first historic eruption on São Jorge began just east of Velas on 1 May 1580. The most remarkable event of this eruption was the appearance of pyroclastic density currents. Not only are they unusual in such basaltic fissure eruptions, but this was in fact the first time that they were named. The

local eye witnesses described them as *nubes ardentes*, or scorching incandescent clouds, but the term was largely forgotten until the famous French geologist Lacroix described their more lethal counterparts that destroyed Saint-Pierre in Martinique in 1902 (he used the term 'nuée ardente', which is now described as pyroclastic density currents). The basaltic eruption seems to have pursued a habitual course at first: two cinder cones quickly formed at Queimada, and lava flows made their way quickly to the shore. At this juncture, about a dozen men took a boat to rescue valuables from a house on the coast. Some of them were still searching the house when it was suddenly enshrouded by a pyroclastic density current. One of the men ran out towards the boat, and he was so badly burned that his skin peeled off. Those left in the house perished. Another pyroclastic density current then caught the five men who had stayed behind on the boat, and their burns were so extensive that they only just escaped with their lives. The first two vents were active only for a few days, but the third went on erupting until the end of August. It is said that 4000 head of cattle were lost in the eruption. Other pyroclastic density currents killed more people as they tried to rescue their possessions from the far less dangerous threat of the advancing basaltic flows. When the islanders realized that the volcano was unleashing still more of these unpredictable, inescapable and lethal clouds, they fled in terror from São Jorge. The lavas formed the Mistério da Queimada.

Some believe that 18 volcanic islets formed off São Jorge during an eruption in 1757. However, it seems more likely that the large earthquake on 9 July 1757 provoked major landslides that were mistaken for volcanic eruptions when they crashed into the sea. The major eruption on São Jorge was in 1808.

The latest volcanic event on São Jorge began on 15 February 1964, when more than 500 tremors shook the island during the next nine days. At the same time, a strong smell of sulphur fumes permeated the air on the west coast and the sea became discoloured. Although no cinders were thrown into the air from the ocean, it is most likely that a submarine eruption occurred about 100m deep off Rosais.

The eruption at Urzelina on São Jorge in 1808

The chief historical eruption on São Jorge began, exactly 228 years after the first, on 1 May 1808, when three spatter cones and lava flows gushed out northwest of Urzelina. But the main activity took place a little afterwards from the vent of La Queimada, to the northeast of the village. There was a thunderous rumbling of the ground, and no fewer than eight large, and five smaller, cinder cones exploded forth for several days. The *mistério* of alkali basalt poured down the scarp bordering the plateau and eventually reached the sea. It is said that some men took the poles that were used to hold the Holy Sacrament during processions and dipped their ends into the molten lava in a propitiatory gesture. Others took the lava-coated sticks to use as divining rods.

However, the greatest danger was again to come from pyroclastic density currents. By the middle of May the craters had gone silent, and it seemed that the lavas had done their worst. But on 17 May the first pyroclastic density current suddenly burst out from La Queimada,

and surprised and killed some goat-herders and their flock up on the ridge. Some of those who tried to run away were burned when their breadbaskets caught fire. With prodigious force and a terrifying noise, the pyroclastic density current surged over the ground, igniting everything in its wake, and descended in a huge black cloud onto Urzelina. In the church, the village priest thought that his last hour had come, but the building protected him from the worst of the hot blast. When he emerged from the church, he wandered, in shock, through the devastation until he met two priests and a group of victims; all had been badly burned. A few were fit enough to return home, but many others sought shelter in the church and in the houses nearby. Many had lost the skin from their exposed limbs; some people were so swollen and blackened that nobody could recognize them; and some had even had their legs broken by the blast and flying masonry. But all those who had breathed the fiery cloud had died at once. In all, more than 30 people were killed, including an elderly couple who were blasted away along with their home, and one victim that the pyroclastic density current threw into the sea. A little later, a lava flow swept down from the new volcano and half buried the church at Urzelina. The dire combination of flows and pyroclastic density currents plagued the area until 10 June. The survivors claimed that they had seen a vision of hell. A memorial sign of the eruption was erected 200 years after the event (Fig. 5.24) at Urzelina, and the remnants of the flows that reached the coast can be seen in exposed shallow cliffs and inlets (Fig. 5.25).

Figure 5.24 Commemorative sign at Uzell, erected 200 years after the 1808 eruption (photo courtesy of Breno Waichel).

Figure 5.25 Lava flow structure exposed at the coast in Urzella from the flows that made it to the sea (photo courtesy of Breno Waichel).

Pico

Pico is the second-largest island in the Azores (Fig. 5.26), and its name is proof enough that its outstanding landmark is a majestic volcano, rising to 2351m – the highest summit in Portugal, and twice as tall as any other peak in the Azores. Pico is the only stratovolcano in the archipelago that has not been decapitated by a caldera-forming eruption. The island of Pico is almost 50km long, and the stratovolcano forms its broadest part in the west, where it reaches a width of 16km, but its eastern tail is mostly less than 8km across. These distinctive parts reflect different eruptive styles: a cluster of vents has built up the stratovolcano, but fissure eruptions have predominated in the east, where aligned cones decorate a high basaltic plateau, from which lava flows have cascaded down the steep cliffs to the sea. The eastern parts of Pico contain the oldest lavas, but they are apparently less than 37,000 years old. The whole island is young: Pico is the newest addition to the Azores, and it is higher, rockier and more obviously volcanic than any of its neighbours. It is also by far the stoniest place in the archipelago, largely because it has witnessed some of its most recent eruptions. These eruptions have made Pico the main home of the *mistérios*, the black lava flows that were so awe-inspiring and inexplicable to the early settlers, and which still seem to emerge menacingly from the recesses of the hills.

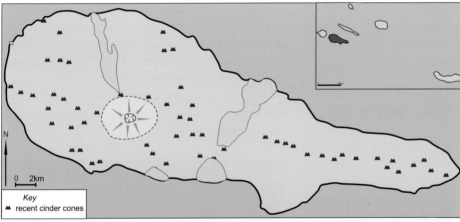

Figure 5.26 Pico stratovolcano, recent cinder cones, and *mistérios* on Pico Island (satellite map – USGS).

Key
▲ recent cinder cones

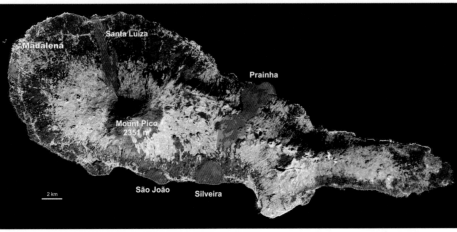

The Pico do Pico

The stratovolcano of Pico was built up from the ocean floor at a depth of 2000m and it has grown into the bulkiest volcanic pile in the Azores. Except for a few short *barrancos* on its southern flanks, vast tracts of Pico have retained the slopes inherited from the latest eruptions. The Pico do Pico has two distinct parts: a gently sloping basal shield, decked with many cinder cones, forms its plinth, and a sharp break of slope at about 1100m demarcates the steeper summit cone. The whole volcano is apparently composed mainly of cindery flows of alkali basalt, which are blanketed at times by basaltic ash, lapilli, cinders and scree. These give the bare upper parts of Pico their typical blue-grey colour and offer a precarious home for grass, myrtle and heather. Further down, the flanks of Pico are wooded and scattered with laurel, tree heaths and vineyards with high walls. There is no hint of a caldera.

The basal shield or plinth of Pico

The basal shield of Pico is 16km wide at sea level and its slopes range from 10° to 20° and rise to about 1100m. It contains most of the 125 or so cinder cones that have sprouted on Pico in recent millennia. Most of them rise on the lower reaches of the shield, near the present coast, but it is possible that later eruptions buried any older cinder cones in the upper areas of the shield. The cones commonly rise about 100m and they are often breached by lava flows on their downslope sides. They are often aligned on either radial or regional fissures. In the southeast, for instance, one major radial fissure clearly gave rise to the cinder cones of Forçado, Bois, Cabeço de Cima and Cabeço de Baixa near São João. In the west, one of the longest fissures gave rise to Tamusgo, Manuel João and das Casas. Although there have been no Surtseyan eruptions around the fringes of Pico in recent times, the two islets off Madalena (the Ilhéus Deitado and EmPé), are the eroded remains of an older Surtseyan cone.

The basal shield is broader in the north than in the south, because its southern flanks have been faulted down along a straight scarp, 500m high and 5km long, that runs westwards from São João towards São Caetano. The fault scarp is so well preserved that it could have formed within the past few millennia, although the most recent lavas have covered both its western and eastern ends. The eastern edges of the basal shield lie on top of, and are thus younger than, the lavas of the eastern plateaux of Pico.

The summit cone

The growth of the upper cone of the Pico do Pico marked the concentration of eruptions into a central cluster of vents, and its slopes approach 30°. However, similar alkali basalts built both the basal shield and the upper cone, and lava flows mantle the surface, along with coarse lava screes and cinders. Thus, if the surface exposures are typical, the concentration of eruptions into a central focus was not accompanied by an increase in their explosive power.

The summit of Pico is instantly recognizable (Fig. 5.27). The smooth rim of the crater encloses a hollow that is 500m wide, but it is now only about 30m deep, because pahoehoe lavas have filled it almost to the brim (Fig. 5.27B). The cinder cone of Piquinho rises 60m above these lavas to decorate the crest of the volcano, and a basaltic cornice wraps around its northeastern rim (Fig. 5.27C). Piquinho, in turn, has a small crater

Figure 5.27 Pico stratovolcano. **A)** Moon over the classic outline of Pico as seen from the island of São Jorge (photo courtesy of Breno Waichel). **B)** View at Pico summit (photo by Unukorno – Wikimedia). **C)** Aerial view of summit area (photo courtesy of Deep in the Forest ©).

The *Mistérios*

The *mistérios*, the grim, black, rough and rugged aa lava flows, reach great sizes on Pico. Even from a distance, their presence is betrayed by planted eucalyptus, cryptomeria or laurel, or by stone-walled Verdelho vineyards, by the *biscoitos* of clinkery lava, or by rugged low promontories jutting out to sea. These lava flows inspired the awe of the early settlers, who called them *mistérios* long before the first eruptions were recorded on Pico in 1562. This suggests that several such flows retained their original and awesome appearance, and therefore had probably erupted shortly before the island was first settled in 1466. One of these eruptions took place between the Pico da Urze and the Cabeço da Fajã, and caused a *mistério* that displaced a stream to the west. Similarly, the three fresh unnamed cinder cones, rising to 137m, 182m and 139m above sea level between São Mateus and São Caetano on the south coast, clearly cannot have erupted long before the settlement of the island, and their wide *mistério* bulges 1km out into the sea. At about the same time, the Cabeço Grande and the Cabeço Pequeno erupted the lavas around the well-named hamlet of Biscoitos, near Madalena.

The four eruptions recorded during historical times on Pico emitted fluid flows of alkali basalt that are mostly less than 3m thick. Each covered more than 10km², and each created a large *mistério*. The first eruption to be reliably recorded in the Azores formed the broad Mistério da Praínha. After three weeks of earthquakes that shook the whole island, explosions of ash and fumes started the eruption early on 21 September 1562 and built the Cabeços do Mistério cinder cones in about a week. On 24 September, fluid basaltic lavas, in a mass 3km wide, set fire to the dry scrub and woodland, cascaded down the 500m-high scarp bordering the plateau, and entered the sea in the wide lobe that now forms the Ponta do Mistério. At times, 40 glowing streams of lava were lighting up the whole island at night. This eruption is reputed to have lasted for two years.

On 1 February 1718, a fissure trending almost from north to south split across the basal shield of Pico, but passed underneath the summit cone apparently without disturbing it. The eruptions in the northern sector built up seven small cinder cones, of which Laurenço Nuñes is the largest. The vents emitted the Mistério da Santa Luzia in about two weeks. Its black alkali basalts spread 5km as far as Santa Luzia and entered the sea at Cachorro in a magnificent array of black pinnacles (Fig. 5.28). At the same time, the southern sector of the fissure opened, and four vents erupted the cinder cones of Cabeço de Cima and Cabeço de Baixo and emitted the Mistério de São João, which destroyed the original village of that name. It now forms a black basaltic headland at the Ponta de São João. Two people are said to have died in this eruption, which apparently lasted for ten months. On 11 February 1718, an undersea eruption also began along the same fissure, 100m off São João. It formed an islet that was joined to the coast before activity stopped ten months later.

On 10 July 1720, another vent opened at Soldão, on the edge of the eastern plateaux, only 3km northeast of São João. It formed the cinder cone of the Cabeço do Fogo and gave off noxious gases that poisoned much livestock in the vicinity. It also emitted the Mistério da Silveira – or Soldão – which spread sideways so easily that it formed a lobe 3km across, which scarcely makes an impression on the coastline as it enters the sea. This was the most recent eruption to occur on Pico.

Figure 5.28 The *mistérios*. **A)** Black lava pinnacles at Santa Luzia. **B)** Pahoehoe structures on the flows (photos by José Luís Ávila Silveira/Pedro Noronha e Costa).

that gives off occasional fumes. Although the crest of Pico may not have erupted in historical times, the fresh lavas at the summit are probably less than a thousand years old.

Few satellite cones interrupt the regularity of the stratovolcano. However, Cabras, Capitão, Lomba de São Mateus, Queiro, João Duarte and Torrinhas, for example, erupted at the base of the summit cone, some 2km from the main vent. Between them rise lines of three or four deeply cratered basaltic spatter cones and their accompanying lava flows, which were all emitted from small fissures radiating from the main cone. They all have the sharp outlines of recent formations.

Eastern Pico

Eastern Pico stretches out eastwards like a comet's tail for 25km from the stratovolcano, and it makes one of the emptiest and most isolated areas in the Azores, where settlement is restricted to the narrow coastal belt. Many grassy basaltic cinder cones rise from the plateau, and they are surrounded by weathered lava flows that blend imperceptibly into its surface and are thus clearly older than their counterparts on the flanks of Pico. The eruptions took place along two sets of fissures trending from northwest to southeast, slightly oblique to the overall trend of the island. Where the island narrows in the east, the fissures are short and close together, and they produced breached cinder cones less than 100m high. Explosions also formed small maars that are so well preserved that they seem to be among the youngest volcanic features in the area. In the broader western and central parts of the plateau, the fissures are fewer and wider apart, and have given birth to larger cinder cones, such as Grotoes and Caveiro, which are over 100m high. They are surrounded by lava flows that have impounded many lakes.

The backbone of eastern Pico ends abruptly on both sides along a series of fault scarps, with remarkably straight sections, 500m or more in height. They rise direct from the coast at Terra Alta in the north and between Santa Cruz and Calheta in the south. Valley exposures in the scarps reveal the piles of horizontal trachybasalt and basaltic-trachyandesite lava flows that make up the whole plateau. These have been dated to less than 25,000 years ago at Terra Alta and to less than 37,000 years ago at Arrife in the south. Further west, from Santo Amaro to São Roque on the north coast, and from Santa Cruz to Arrife in the south, the scarps

rise between one and two kilometres inland, but often approach 700m in height. Here, lava flows have tumbled down the scarps and then spread out in extensive aprons at their base. On the south coast, for instance, a few thousand years ago at most, a narrow lava flow filled an old valley cut into the scarp and spread its rugged black basalts into the sea to form the Ponta dos Biscoitos at Santa Cruz (Fig. 5.29). On the north coast, the towns of Praínha, São Roque and Cais do Pico are built on similar recently erupted lobes. Sometimes, copious lava emissions, like the Mistério da Praínha, the Mistério da Silveira and the steep accumulations near Cais do Pico, have buried the original scarps under fans of lava that splayed seawards in wide promontories. All these lava lobes can be more than 20m thick, and they form extremely rugged shores, but they have undergone little marine erosion, and there is no evidence of raised beaches upon them. It is most likely, then, that they came from the most recent eruptions during the past few thousand years.

Figure 5.29 The lava delta of Ponta dos Biscoitos at Santa Cruz das Ribeiras, Pico (photo by Mike Norton).

Faial

Faial is a compact island, 21km long and 12km wide (Fig.5.30), whose scenery is dominated by the strato-volcano called Caldeira, rising to 1043m. A string of villages forms an almost complete ring on the gentle slopes that form a balcony above the rugged lava cliffs, and only Horta, in the shelter of Monte Guia, really touches the sea. Faial epitomizes the predomi-nant traits of the Azores. It has lush meadowland, thick hedges of hydrangea; coppices of cryptomeria, laurel and eucalyptus; a large stratovolcano with a

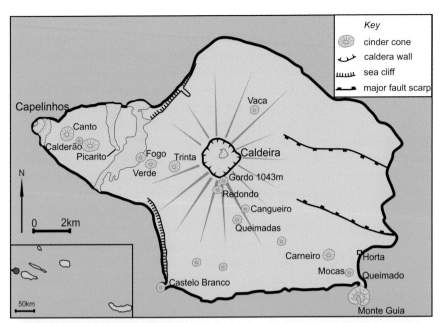

Figure 5.30 Volcanic features of Faial, with zones of most recent activity concentrated in the Capelo Peninsular in the west (satellite map – USGS).

summit caldera; a well-defined rift zone; long fissures, with lines of recent cinder cones; and – best of all – the birth of Capelinhos.

Faial grew up from a depth of 1500m on the flanks of the Mid-Atlantic Ridge. It sprang from the same set of fissures, trending northwest to southeast, which also gave rise to Pico. In fact, the Canal do Faial, between Faial and Pico, is only 100m deep and 7km across, and the two islands could be the emerged parts of a single volcanic complex. Of all the central group of the Azores, Faial lies closest to the Mid-Atlantic Ridge,

which passes 100km to the west. Probably as a result, Faial has registered many earthquakes that have usually been the mild shocks typical of the mid-ocean ridges. Nevertheless, the tremors in May 1958, during the eruption of Capelinhos, destroyed the nearby village of Praia do Norte, and on 9 July 1998 an earthquake of magnitude 5.8 on the Richter scale killed 8 people and wounded 154 others.

Faulting has apparently played a major role in delimiting the coasts of Faial. The interplay of fault blocks defines the promontories and bays of the east coast alongside the Canal do Faial. On the west coast, faults have thrown the fringes of the stratovolcano down below sea level and they now form majestic straight cliffs often over 200m high. On the other hand, although Faial lies close to the Mid-Atlantic Ridge, it is surprising that it has witnessed only two eruptions since the island was first settled in 1466. Both occurred on the western peninsula. The Cabeço do Fogo and Picarito erupted in 1672–3, and Capelinhos produced its display in 1957–8 at the western end of the self-same fissure system. A small submarine eruption was also noticed in 1963 in the Canal do Faial, but no lavas reached the surface.

Olivine-rich alkali basalts predominated in the areas of most recent activity in the west and southeast. Trachybasalts and basaltic-trachyandesites occur especially in the central rift zone and on the stratovolcano. Trachytes and trachyandesites outcrop on the lower flanks of the stratovolcano and in its caldera.

The growth of Faial

The basalts that floor the Mid-Atlantic Ridge around Faial are about 5.5 million years old. Faial began to grow up from this basement much later, and the oldest lavas, exposed along parts of the Canal do Faial, are about 730,000 years old. These are the oldest rocks exposed in the central Azores, and they prove that the individual islands were not formed in succession towards the Mid-Atlantic Ridge. The basal lavas accumulated over a long period, for one of their youngest flows is only about 30,000 years old. The fault trough, forming the central rift zone, foundered after these lavas erupted; and its clarity in eastern Faial also suggests a relatively recent origin.

It is about 7km wide and has three, and sometimes four, fault scarps, each rising between 60m and 130m high. The rift zone is made up of three segments, each offset by about 1km, and it can be traced right across the island from the Canal do Faial to Capelinhos.

Caldeira (Fig. 5.31), the stratovolcano, grew up within the rift zone after it had foundered. It is composed of flows and fragments of trachybasalt, basaltic-trachyandesite, trachyandesite and trachyte, which spill beyond the rift zone and reach the sea. Its flanks are blanketed with trachytic pumice from the Plinian eruption that formed the central caldera. It is 1850m in diameter and 470m deep, and the undissected inner walls are almost vertical, with few screes of weathered debris – suggesting that the caldera probably collapsed much less than 30,000 years ago. But the caldera is old enough for small eruptions to have already resumed within it, for its floor contains a cinder cone and a dome of trachyte that are both less than 30m high. Meanwhile, several basaltic cinder cones, each about 100m high, erupted on the upper flanks of the stratovolcano. They include Cabeço Gordo (1043m), which now forms the highest point of the caldera rim, as well as Cabeço Redondo to the south, Rinquim to the northwest and Cangueiro and Queimadas in the southeast. Castelo Branco, the Surtseyan cone of white trachyte fragments resembling an offshore fortress, probably also belongs to this period, for it has been much eroded by the sea, and only a small blunted crater survives on its summit.

About 11,000 years ago, the focus of activity shifted to the southeastern corner of Faial, when about a dozen basalt and trachybasalt cinder cones and lava flows erupted. Most of the flows are weathered and covered by Horta and its suburbs, but their

Figure 5.31 View into Caldeira, in the centre of Faial (Wikimedia, anon).

original ruggedness is revealed where they form the intensely fretted black cliffs on the southern coast. The eruptions took place along three fissures that trend from northwest to southeast – and the longest continues, in fact, from the Canal do Faial right to the western peninsula. It gave rise to a line of cones including Monte das Mocas and Carneiro, which dominates Horta.

Where one of these fissures reaches the Canal do Faial, basaltic eruptions formed Monte Queimado and Monte Guia (Fig. 5.32), which stand side by side, making the promontory that protects Horta harbour. They are a dissimilar couple. Monte Guia is a typical Surtseyan cone, 1km in diameter, which erupted in shallow water and is composed of innumerable thin layers of fine, yellowish, and often welded tuffs. The southern sector is missing – and probably never formed – and its wide double crater is deep enough to be invaded by the sea. Monte Guia probably erupted during the past few thousand – perhaps even few hundred – years. It is, in fact, better preserved than Capelinhos, which was formed in 1957–8, albeit in a more exposed position.

Its companion, Monte Queimado, is quite different. Its burnt appearance comes from the black and reddish cinders of a cone that erupted on land. But it could not have formed under present conditions, because its base now lies below sea level. Most probably, therefore, Monte Queimado erupted before Monte Guia, when the sea was lower, and could not reach the vent. When the sea level rose, Monte Guia erupted, and at the same time, the waves trimmed the edges of Monte Queimado, and redistributed the fragments to form the beach that now joins both cones to Faial.

The Capelo Peninsula

Both the northwestern and southwestern fringes of the stratovolcano foundered below sea level along two faults that intersect where they meet the line of fissures on the axis of the western, or Capelo, peninsula. This juts out 7km westwards into the Atlantic Ocean and it ended, before 1957, in a smattering of islands called the Ilhéus dos Capelinhos. Its eruptions have now replaced some of the land lost when the flanks of the stratovolcano foundered below sea level. The eruptions formed a broad prong

Figure 5.32 Monte Queimado and Monte Guia at Horta. **A**) View of Horta and Monte Queimado from Monte Guia. **B**) View across Horta to the two cones. **C**) Flooded by the sea – the double crater of Monte Guia (**A**, **B** and **C** photos by José Luís Ávila Silveira/Pedro Noronha e Costa).

of 16 cinder cones, usually called *cabeços* ('heads'), which arose on five en-echelon fissures, each about 2km long and 500m apart, that trend obliquely across the spine of the peninsula from northwest to southeast.

Most of the peninsula grew out westwards as a result of eruptions on land, where water could not usually invade the erupting vents. Its base consists of a thick cake of black basalt and trachybasalt flows that forms a plinth tapering from a height of 600m near the stratovolcano to 94m at the old Capelinhos lighthouse. Higher sea levels formed cliffs on some of the older flows, like those due west of Capelo, but others, such as those near the Ponta da Varadouro in the south, still form rugged, scarcely eroded lava deltas. All of these flows have formed an effective buffer against marine attack, and they illustrate a net gain of volcanic materials at the expense of the ocean. Only Capelinhos and the Costa da Nau volcano, at the very tip of the peninsula, were created primarily by Surtseyan activity, although, before Capelinhos erupted, the Ilhéus dos Capelinhos also marked the wreck of another Surtseyan cone that had lost its battle with the waves.

The cones crowning the Capelo peninsula were all built by lapilli and cinder eruptions, and their bases stand at least 100m above present sea level. The oldest eruptions, whose products are still visible, formed the Cabeço do Pacheco, Caldeirão, Furno Ruim and, most notably, the Cabeço do Capelo, which is over 200m high, and is young enough for its crater still to be apparent. At the same time, a Surtseyan eruption formed the Costa da Nau tuff cone, off what was then the westernmost point of the peninsula. The Cabeço da Trinta erupted later on the flanks of the stratovolcano, and the Cabeço do Canto and Caldeirina probably followed soon afterwards. Their trachybasalt lava flows partly covered the Costa da Nau and formed most of the western snout of the peninsula before 1957. In turn, the flows from the Cabeço do Canto were covered by tuffs from the Surtseyan eruption of the cone, the remains of which still survive as the Ilhéus dos Capelinhos. The next eruptions occurred inland, where they formed the Cabeço Verde and Pingarotes, but the true sequence of the remaining eruptions on the peninsula has yet to be established.

Historical eruptions

The eruptions recorded during historical times illustrate the same theme. In the east, the Cabeço do Fogo and Picarito erupted between 24 April 1672 and 28 February 1673. Seven months of intermittent earthquakes preceded the arrival of the magma at the surface 'with such a din that it seemed as if the end of the world had come' (recorded account in Canto, 1881). Ash spread all over Faial, and flows escaped from the base of the cone. Cabeçodo Fogo stopped erupting at the end of April 1672, but the baton was at once taken up by Picarito. Its flows buried 307 houses in the villages of Praia do Norte and Ribeira Brava (which was rebuilt as Ribeira do Cabo), and made 1200 people both homeless and landless. Three men were surrounded and incinerated by the flows when they went to inspect them at close quarters. Two new *mistérios* were expelled, one to the north coast, and the other to the south coast. They spread, some 4km across and 5km long, over an area already well armoured with lava flows. The place name Biscoitos, on the south coast, reveals the rugged nature of their surfaces and, although they are now wooded, they still make the most impressive *mistérios* outside Pico. The lavas completely cut off some 70 families who had unwisely escaped westwards along the peninsula, and they had to be supplied with food by boat until a new track could be made, long after the eruption ended on 28 February 1673. The eruption added some 360 million m³ of volcanic material to the peninsula, and little has been eroded away. The island was to wait another 284 years (until 1957) before the next eruption, and one that would further build out the island westwards and evict people from their homes (Fig. 5.33).

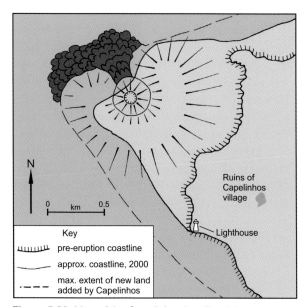

Figure 5.33 Map of the Capelinhos headland, indicating the original shoreline, the maximum extent of the new land by the end of the eruption, and the eroded remnants today.

The eruption of Capelhinos in 1957–8

The second eruption in Faial during historical times started in waters 80m deep near the Ilhéus dos Capelinhos. Thus, the new eruption was only regaining ground previously lost to the Atlantic Ocean. Capelinhos arose between 27 September 1957 and 24 October 1958, during which time Surtseyan activity gradually gave way to Strombolian activity.

There were over 200 tremors between 16 and 27 September 1957. At 08.00 on 27 September, the sea began to boil about 400m west of the Ilhéus dos Capelinhos. On 29 September, black basaltic ash was fired 1km into the air in typical Surtseyan pointed jets, shaped like cypress fronds or cocks' tails, and billowing white clouds of steam often rose 4km high around them. Each explosive interaction of magma and water happened at two-hourly intervals and lasted about half an hour. When this activity stopped on 29 October, a tuff cone 100m high had been built; but by 1 November, all that remained of it was a shoal around the Ilhéus dos Capelinhos.

After a week of total calm, eruptions resumed on 7 November 1957, from a new vent about 500m east of the first. A new tuff cone quickly grew up, and windblown ash built an isthmus that joined the new volcano to Faial for the first time on 12 November 1957. Then, as the eruptions began to wane in early December, this second cone began to slump into the sea, and its days, too, seemed numbered.

However, at 22.30 on 16 December 1957, molten lavas emerged and indicated that sea water had failed to penetrate the vent for the first time. But the vent was blocked only intermittently, and Surtseyan explosions continued to dominate during the spring of 1958. However, the effusive phases gave off enough lava to form an apron around the tuff cone that both retarded erosion and kept the sea out of the vent. By the end of March, the tuff cone had grown to a height of 150m and a diameter of 1km, and had covered the Ilhéus dos Capelinhos in the process. The Surtseyan jets were still firing 1800m into the air, and lava bombs were whistling down over the promontory. The villages had to be abandoned. A thick and noxious blanket of ash, and the sodium chloride that had evaporated from the ocean, destroyed crops and buried houses. Soon, only the now useless and badly damaged Capelinhos lighthouse remained defiant (Fig. 5.34).

Figure 5.34 The Capelinhos eruption and deposit. **A)** Commemorative stamps show the eruption offshore and around the lighthouse (stamp images sourced from UPU – Universal Postal Union). **B)** Old photo showing the ruins of Capelinhos village and the lighthouse with the new volcano on the left horizon. **C)** Close-up of the famous Capelinhos lighthouse (photo courtesy of Breno Waichel). **D)** The Capelinhos volcano with the Strombolian cone in the centre, surrounded by the Surtseyan tuff cone, now much eroded by the sea.

The last episode of the eruption began when over 450 earthquakes were recorded between 12 and 14 May 1958. They were centred on the caldera, not on Capelinhos, but the effect on the new volcano was radical. The Surtseyan activity virtually ceased. The most spectacular phase of the eruption began when huge lava fountains spurted from within the crater of the tuff cone and constructed a basaltic spatter cone that had attained 75m before the end of May 1958. This phase continued until the eruption suddenly stopped on 24 October 1958. By then, about 2.4km² had been added to the area of Faial. The core of Capelinhos was a steep spatter cone of jagged wine-red basalt 160m high. It was encircled by a smooth and more gently sloping tuff cone, 150m high and 1km across, which was composed of fine tuffs like buff-coloured sand. Its surface was strewn with remarkable basaltic bombs the size and shape of tortoises, where the 'legs' had burst from the molten interior as they hit the ground. The tuff cone was also almost completely surrounded by lava flows.

However, Capelinhos did not produce enough lava to protect itself properly. Fifty years later, only one square kilometre of the new land remained. Almost half the spatter cone had been removed and its jagged crater stood on the brink of towering cliffs. The sea had swept away part of the tuff cone, revealing the main islet of the old Ilhéus dos Capelinhos once again. Capelinhos needs another eruption to save it from destruction.

Flores and Corvo

Flores and Corvo lie west of the Mid-Atlantic Ridge and belong to the North American plate. The main trends of both islands, emphasized by dykes running from north-northeast to south-southwest, are thus quite different from those of their neighbours to the east. Flores is a compact elliptical island, 17km from north to south and 12km from east to west. It is unique in the Azores because it is dominated neither by long fissures nor by a large caldera. The Funda Caldera is less than 1km across, yet is adorned with beautiful *lagoas* (Fig. 5.35A). Its fissures are subsumed within a complex volcanic pile that issued from an array of eruptive centres of similar size; waterfalls cascading over innumerable colonnades of ancient lava flows attract the eye as much as the volcanic hills. Flores forms a cluster of small stratovolcanoes that crown a large accumulation rising more than 1000m from its base on the ocean floor. There was a broad succession of basalt, basaltic-trachyandesite and trachyte eruptions, with a return to basaltic emissions in recent times. The oldest rocks of Flores tend to be

Figure 5.35 Flores and Corvo. **A)** Panorama of the Funda caldera and surrounds, Flores (photo by José Luís Ávila Silveira/ Pedro Noronha e Costa). **B)** Panorama of Corvo Caldera at the summit of Corvo island (photo by Angrense).

exposed around the coast and are dated to over 600,000 years old. The central zone is younger, especially where a blanket of fine exploded fragments surrounds the Morro Grande, the highest summit on the island at 942m. No eruptions have occurred on Flores in historical times, and the advanced age of the dated lavas, coupled with their deeply weathered and eroded surfaces, suggests that the volcanoes of Flores may now be extinct.

Corvo is the smallest island in the Azores, only 7km from north to south and 4km from east to west. It belongs to the North American plate, but its structure is apparently much simpler than its companion, Flores. The island is the emerged summit of a basaltic stratovolcano, rising to 718m, which culminates in a caldera 300m deep and 2km across (Fig. 5.35B). Subsequent eruptions have begun filling the caldera with small cinder cones that are now interspersed with small lakes. The smooth, gently sloping, inner slopes of the caldera indicate that it could be among the oldest in the Azores, and no eruptions have been recorded on the island during historical times. Thus, Corvo may not be active.

Chapter 6

Iceland

Introduction

Arguably the biggest playground a geologist could wish for, Iceland represents a dynamic wilderness of fire and ice. Iceland itself covers an area of 102,846km^2 and it is almost completely volcanic in origin. It constitutes the longest emerged segment along any of the world's mid-ocean ridges, and it is the largest volcanic area in Europe. With limited vegetation, the volcanic features are very well exposed throughout the island, interrupted only by the permanent cover of ice from its ice caps and glaciers. The landscape of Iceland offers a glimpse of how mid-ocean ridges develop, especially in the central axial zone of contemporary activity, which curves across the island from north to southwest in a swathe 60–80km wide (Fig. 6.1). Here, the North American plate on the west is diverging from the Eurasian plate on the east. Crustal accretion occurs at their edges through the injection of great swarms of sheeted dykes, often accompanied by eruptions of basalts onto the land surface. Broad expanses of plateau basalts stretch from both sides of the band of contemporary activity and cover three-quarters of Iceland. These were themselves erupted on older axial zones during the past 16 million years, but have been carried to extinction away from the axis, as crustal divergence, rifting and accretion have continued. Thus, in northwest Iceland, the exposed basalts are from about 15 million years old, whereas those in eastern Iceland date from 13 million years ago. They are bordered on their inner edges, alongside the band of present activity, by younger basalts 3.1–0.7 million years old. In the zone of contemporary activity, all the eruptions occurred less than 700,000 years ago. All of these basalts erupted in a broadly similar fashion and, therefore, the activity of the contemporary axial zone clearly indicates how most of Iceland was formed. Iceland has widened by some 400km from east to west since the divergence began. The axial zone is widening at an average rate of 1.5–2.0cm per year.

Iceland has been built up on a segment of the Mid-Atlantic Ridge, which lies between the Kolbeinsey Ridge to the north and the Reykjanes Ridge to the southwest. This segment displays many of the characteristics of rift systems, which can be seen as expressions of fissures, aligned volcanoes and faults as tears in the crust (Fig. 6.2). The 400km-long Icelandic segment is offset by two transform-fracture zones where most of the major earthquakes have occurred in historical times: the Tjörnes fracture zone off the north coast, and the south Iceland seismic zone crossing southwestern Iceland.

These tectonic aspects of Iceland are fundamentally similar to those displayed on submarine spreading ridges. Nevertheless, Iceland represents an unusual volume of volcanic materials – more than twice the thickness of the average ridge – built above sea level by the presence of a hotspot beneath the spreading axis (the Iceland Plume), which has been active since the formation of the North Atlantic Igneous province (~60Ma), where large volumes of predominantly basalts were erupted on both sides of the Atlantic from Greenland to the UK. This also represented the start of the separation of Europe from America. Volcanic rocks about 1500m thick are exposed in Iceland, and they lie on a volcanic basement three times thicker. The Icelandic hotspot seems to be a cylindrical zone about 300–400km in diameter and more than 400km deep. It is probably now centred just near to Vatnajökull.

Of the exposed volcanic sequences, basalts represent 80–85 per cent, silicic and intermediate materials about 10 per cent, and the remainder comprises fluvial or glacial sediments that are themselves derived from volcanic rocks. The basalts form distinct petrological and morphological units. Porphyritic basalts arise most often in large, rapid eruptions from fissures, and they form characteristically massive lava flows. Olivine-rich tholeiitic basalts are emitted as thin individual pahoehoe flows (called *helluhraun* in Iceland), which are often piled up into lava shields by repeated eruptions. They seem to come from magmas lying more than 10km deep. On the other hand, olivine-poor tholeiitic basalts

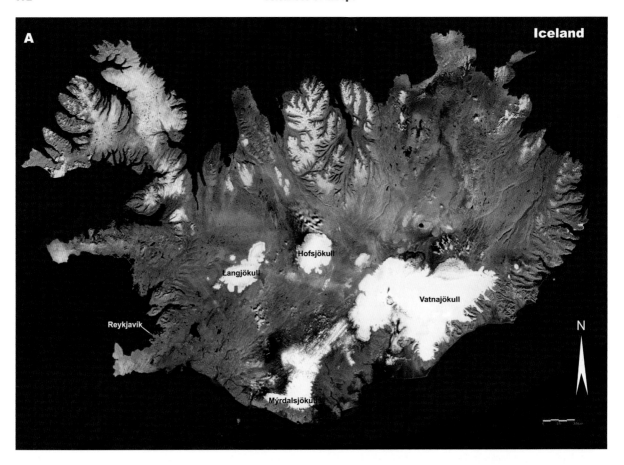

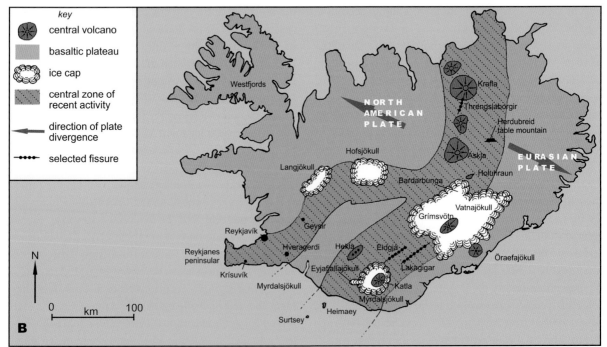

Figure 6.1 Iceland. **A**) Location satellite map of Iceland. Major ice caps are clearly visible as well as the relatively barren and exposed nature of the terrain (USGS). **B**) The main volcanic features of Iceland, with the central zone of recent volcanic activity between the diverging Eurasian and North American plates.

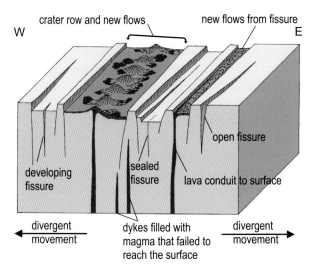

W

crater row and new flows   new flows from fissure   E

open fissure

developing
fissure

sealed
fissure

lava conduit to surface

divergent
movement

dykes filled with
magma that failed to
reach the surface

divergent
movement

Figure 6.2 Schematic of a typical section through an Icelandic rift and fissure swarm (after Imsland, 1989).

emerge chiefly from fissures and the central volcanoes, after having spent time in shallow reservoirs. They form as aa lava flows (called *apalhraun* in Iceland) Alkali olivine basalts erupt from reservoirs about 3km deep outside the main zones of rifting, and tended to develop after, and cover the products of, the tholeiitic activity. The greatest differentiation, probably derived after the longest periods in shallow reservoirs, is represented by the silicic eruptions, which are often associated with violent explosions, which have reached Plinian proportions at Hekla, Askja and Öraefajökull, amongst others. Among the more silicic rocks, about two-thirds form lava flows or intrusions, but one-third comprises layers of welded tuffs and ash that are often of a rhyolitic or rhyodacitic nature. Hekla, Öraefajökull and Askja have been their main sources in historical times.

Many of the characteristic genetic, tectonic and petrological features of the mid-ocean ridges imparted to Iceland can be studied in the open-air landscape, without recourse to the expensive submarine equipment required where the ridges remain submerged. However, high precipitation and proximity to the Arctic Circle have helped maintain four main ice caps (see Fig. 6.1). Vatnajökull, with an area of 8400km^2, is the largest ice sheet in Europe. The smaller ice caps of Langjökull, Mýrdalsjökull and Hofsjökull together are about 2500km^2 in area. They create additional problems when eruptions take place beneath them, for they often generate destructive jökulhlaups or glacier bursts (meltwater floods). Much of the rest of the country lies in a periglacial subpolar climatic environment, where the

general absence of vegetation and thick soils reveals the eruptive features with a stark clarity that makes Iceland one of the finest volcanic exhibitions on Earth. Moreover, for a mid-ocean ridge environment, the volcanic landforms display a surprising variety and, although basaltic effusions have been predominant, almost every kind of volcano can be found. Volcanic systems composed of an elongated fissure swarm, with a central volcano in its midst that might even have developed a caldera, are some of the most striking elements of contemporary activity. They are accompanied by broad basaltic shields, much steeper table mountains formed beneath the old ice sheets, occasional domes of silicic lava, as well as maars and tuff rings, and a whole gamut of hydrothermal features.

The Norse settlement of Iceland began in AD874, and thus historical records of volcanic activity, however vague or conflicting in the early centuries, are available for more than a thousand years. On average, a magmatic eruption is under way in Iceland every fifth year. However, most Icelandic eruptions are among the least dangerous in the volcanic repertoire, although they are often of great scientific interest. The formation of Surtsey in 1963 not only created a new island in the Vestmannaeyar but also did much to clarify ideas about these particular Surtseyan eruptions. The eruption on the neighbouring island of Heimaey ten years later marked an important step in the development of procedures for civil protection and damage limitation. The studies of the rifting and eruptive episode that began at Krafla volcano in 1975 gave greater insights into the way that active mid-ocean ridges operate. Finally, the recent eruptive episodes at Eyjafjallajökull (2010) and Holuhraun (2014–15), occurred with sophisticated monitoring systems in place, providing a wealth of new direct and remote data for volcanologists to digest for the years to come.

The plateau basalts and old volcanoes

The plateau basalts cover about three-quarters of Iceland, forming open and rather stark uplands of rounded hills and ridges, between 750m and 1250m high, each scarred by the outcrops of layered lava flows (Fig. 6.3). They have been scraped bare by repeated glaciations, and eroded by streams in the non-glacial periods. The lava flows now dip gently down towards the axial zone from both west and east. Most of the plateaux developed from eruptions of tholeiitic basalts that emerged from deep fissures as

Figure 6.3 Sheets of the plateau basalt flows forming the headland north of Suðureyri, in the Westfjords (photo courtesy of Morgan Jones).

wide-spreading fluid lava flows, 5–15m thick. Central vent volcanoes may also amount to as much as half of the volume of the plateaux, although they have often been buried by later lava flows. There are perhaps as many as 50 or more such central volcanoes. They mainly form shields and occasional stratovolcanoes composed chiefly of basalts, but also have some andesites or even rhyolites in their makeup. They have gentle slopes and modest heights. Breiddalur, in eastern Iceland, for instance, had a volume of 400km^3 and gentle slopes of 9°, but it probably never stood more than 600m above the surrounding lava plains. As they grew, the central volcanoes were periodically swamped by thicker and more fluid flood basalts disgorged from the fissures around them, so that they now form insignificant features in the landscape. With

Meet the Scientist – Steffi Burchardt

How did you first start getting into research on volcanoes?
I've always found volcanoes fascinating, even as a kid when I learned that Snaefell (Snaefellsjökull) volcano is the entrance to the centre of the Earth. However, when I participated in a field trip to the Canary Islands as an undergraduate student, I completely fell in love with volcanoes and have studied them ever since.

How have you used new techniques to map out and visualize some of Iceland's volcanic rocks?
I have used a lot of structural field work to study the architecture of eroded volcanoes and the features related to magma transport and storage. I have used the data collected in the field to build 3D structural models of magma reservoirs and channels, which is a useful technique to, for instance, visualize subsurface features and reconstruct the volume of magma reservoirs.

What has been your most exciting discovery about it?
From my studies in Iceland, I have learned a lot about how magma reservoirs are born and how they look like in 3D, e.g., that they are not just spheres of magma in the crust, but actually complex and dynamic systems that evolve during the lifetime of a volcano. Apart from that, I think my most exciting discovery on Icelandic volcanoes is that I found a collapse caldera that was previously unknown, geological *terra incognita* so to say.

How important do you think the volcanoes of Iceland have been in our understanding of volcanology?
Iceland is a central puzzle piece in our understanding of how volcanism works at mid-ocean ridges. It is here we can directly witness the spreading of the ocean floor. Another important aspect is that the deep structure of the rift zone and its central volcanoes is exposed in the fjords along Iceland's coast. A lot of what we understand today of the inner workings of rift zones and volcanoes goes back to George Walker's pioneering work in these deeply eroded volcanoes.

Do you have a favourite Iceland Volcano?
I like many of the Icelandic volcanoes, as they all have their individual character. If I have to select one though, then I'd choose Katla, one of Iceland's most hazardous volcanoes and named after a short-tempered witch.

Figure 6.4 A) Exposures of the Slaufrudalur Pluton, Southeast Iceland. **B)** Inset showing 3D model of the exposed intrusion (photo and image courtesy of Steffi Burchardt).

the extensive erosion of these older landscapes, the intrusive centres that cut up through the older crust can be visualized. The granitic Slaufrudalur intrusion (approx. 6Ma), is one such example, and its extensive exposures allow the reconstruction of its complex 3D geometry to be realized (e.g. Fig. 6.4).

An essentially similar pattern of eruptions continued within the more recent series, initiated 3.1 million years ago, that covers a quarter of Iceland along each side of the active zone and is more than 500m thick. Glaciations introduced greater variety, not only by depositing glacial debris between the lavas, but more importantly, by provoking the formation of pillow lavas, pillow breccias and palagonites (altered glass) during sub-glacial eruptions. These palagonites are known as *móberg* in Iceland.

The active volcanic zone

The axial zone of contemporary activity is demarcated geographically by the inner borders of the basaltic plateau lavas and defined geologically by eruptions in the past 700,000 years. During the glacial periods, eruptions occurred under the ice sheets, which led to the formation of subglacial table mountains and thick layers of palagonitic basalts and tuffs. Apart from activity beneath

present ice caps, post-glacial eruptions have also taken place in the open air.

The same general pattern of eruptions has continued into modern times and, as in the past, basalts have composed the vast bulk of the output. However, the active zone is more complex than it at first appears. It curves south-westwards in a belt 60–80km wide from the north coast and then branches into two arms: a western rift zone, running across the Reykjanes Peninsula to join the Reykjanes segment of the Mid-Atlantic Ridge; and an eastern rift zone, which is propagating towards the Vestmannaeyar Islands. The eastern arm may be the site of contemporary rift jumping at the expense of the western arm (see the map in Fig. 6.5 for a detailed overview).

The rifting zones cover about 18,000km^2 (60%) of the present active area, and they account for 77 per cent of its volcanic production. They are dominated by fissure swarms arranged en-echelon and running parallel to the trend of the active zone. The swarms of fissures have often developed into more complex volcanic systems, where a low and broad elliptical central volcano has grown up in their midst. Some 29 volcanic systems have been active, and they have produced between 400km^3 and 500km^3 of lava, covering about 12,000km^2.

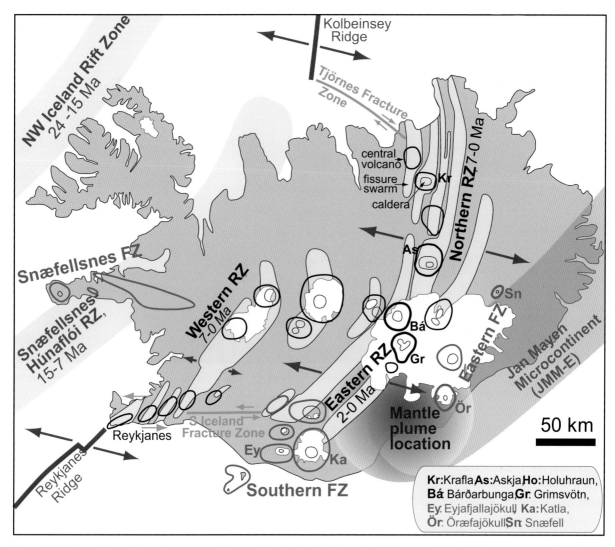

Figure 6.5 Map of the major volcano-tectonic trends in Iceland with marked rift zones (RZ) and fracture zones (FZ) (image courtesy of Reidar Trønnes).

Southern Iceland has been the most productive region. These volcanic systems are 40–100km long and 2–10km wide. In the active zone as a whole, plate divergence first causes continual crustal stretching for 100–150 years. It is then brought to an end by a short episode of active rifting, commonly limited to one volcanic system at a time, which lasts for less than a decade. The rifting opens fissures and faults that usually develop parallel to the trend of the active zone, at right angles to the directions of plate divergence. From these fissures the basalts often emerge during the eruptions that distinguish Icelandic activity.

In their initial stages, the fissure swarms produce voluminous hot effusions of fluid tholeiitic basalts. In time, slightly more evolved lavas are emitted, which are rather more viscous and explosive in character, and

which tend to be expelled from vents concentrated in the centre of the fissure swarms or volcanic systems. This concentration is associated with the development of a shallow reservoir in the crust, which facilitates the evolution of the magma and builds up a broad central volcano. Although eruptions apparently increase in frequency, the volume expelled during each active episode decreases. However, the central volcano may develop a caldera, as a result of either an explosive eruption of dacitic or rhyolitic fragments or of the lateral migration of magma from the reservoir into fissures. The Askja caldera was apparently formed by lateral magma migration, but the Öskjuvatn caldera collapsed within it in 1875 after a large rhyolitic explosion. The fissure eruptions of Krafla between 1975 and 1984 perhaps represent an intermediate stage of small eruptions accompanying marked fissure formation on the flanks of a central volcano that has already developed a caldera.

Not all of the rifting causes eruptions at the surface, although magma almost certainly invades the fissures at depth. Near Reykjavík at Thingvellir, rift fissures are particularly well marked. They first developed about 9000 years ago. Now, often bordered by recently formed cliffs, fissures continue for 2–10km across country. This area is significant historically in Iceland as the site of the original Icelandic Parliament (established AD930), and is probably the most striking place where you can see the separation of the two continental plates. You can stand, or indeed dive, and feel as if you were touching two sides of a rifting continent (see Fig. 6.6), but in truth the rift itself is a complex arrangement of rift faults across a rift zone, as opposed to a single obvious fault. Small earthquakes sometimes form gaping clefts at the surface, and it is said that a larger earthquake in 1789 lowered the whole area by 67cm. At all events, the extent of fault-trough subsidence in this area may be

judged by the fact that Lake Thingvallavatn lies below sea level.

Fissure eruptions in Iceland are beautiful, often spectacular, usually brief, and rarely dangerous. With the notable exception of outbursts on the scale of the Lakagígar (Laki) eruption in 1783, they soon reach a brief crescendo, followed by several weeks of waning emissions. Their activity usually begins with lava fountaining, often in continuous curtains 1km or more in length. The lavas are hot and fluid, reaching the surface at temperatures of about 1200°C. After several hours, emissions become concentrated in vents, usually 250–750m apart, where moderate explosions form a row of individual cones of spatter and cinders. At the same time, many cones are breached by quiet effusions of lava that quickly form widespread and relatively thin flows. The fissure is most active, and the greatest volumes of material are expelled, during the early stages of the eruption, which commonly last for only a few hours or a few days. Effusion rates can be high during the first few hours and

Figure 6.6 Thingvellir and Silfra Lake. **A)** View over Thingvellir with rift faults visible. **B)** Onland expression of the rifting apart of the continents. **C)** Silfra Lake, the gateway to the underwater rift. **D)** Touching two continents. Diving in the extremely clear cold waters of Silfra reveals deep rift faults where you can touch the sides of the rifting continental plates.

The Skaftár Fires: the Laki eruption in 1783–4

Most famous of Icelandic fissure eruptions was the Skaftár Fires or Laki eruption of 1783–4, also notorious for its far-reaching effects. It sprang from Lakagígar, the fissure extending northeastwards across the Sída Highlands, through the 818m-high mountain of Laki, which had itself formed along older fissures extending 100km from the subglacial volcano of Grímsvötn (Fig. 6.7). The eruption lasted eight months, during which its tholeiitic basalts never varied in composition, which suggests that they originated at depth, formed a reservoir beneath the solid crust, and then rose quickly to the surface. The events were described by the Revd Jón Steingrímsson, the Minister at Prestbakki, in the Sída Lowlands.

After earthquakes had rumbled for a week, the fissure opened on the southwestern flanks of Laki on Whit Sunday morning, 8 June 1783, and hot, fluid tholeiitic basalts gushed forth. The initial discharge was extremely high: two-thirds of the total volume were emitted – at an average discharge of 5000m³ per second – during the first 50 days. The lava quickly invaded the gorge of the River Skaftár and, on 12 June, emerged onto the lowlands more than 35km from the fissure. Two days later, the lava had filled the gorge to the brim, the ground was shaking, and an odd foul-smelling blue fog was veiling the sun. On 25 June, the fissure lengthened, and explosions and lava fountains extended the line of cones across the mountain. More eruptions followed extensions along the fissure, until 25 July. Four days later, an extension of the fissure opened up on the opposite, northeastern, flank of Laki. This section also emitted great volumes of lava, but this time they swamped the Hverfisfljót gorge to the east. The vigour of the eruption began to decline in mid-August and more markedly from the end of October, so that the emissions had weakened considerably before they stopped altogether on 7 February 1784. By then, two vast lava flows had formed prominent deltas in the Sída lowlands, and more explosive activity had also formed no fewer than 140 cones, ranging from 30m to 70m high, on a fissure 27km long (Fig. 6.8). The lava flows, reaching a volume of 14.7km³, overwhelmed an area of 565km², and a further 8000km² was covered with a thin mantle of fine ash. A few churches, farms and homesteads were damaged, but no one was killed directly.

The factor that gave the Skaftár Fires their notoriety was the emission of poisonous gases. About 8 million tonnes of fluorine contaminated the farm animals and their pastures, and, as a result, more than three-quarters of the sheep and horses, and half the cattle, died. The dreadful famine that ensued killed 20 per cent of the population. About 122 million tonnes of sulphur dioxide was released in columns that rose 6–13km high and generated some 250 million tonnes of sulphuric acid aerosols that remained aloft for about two years. The aerosol spread as a curious dry blue fog and caused consternation all over Europe during the summer and autumn of 1783. Even the experts did not know where it came from or what it portended, and only one, the French naturalist Mourgue de Montredon, correctly linked it with Iceland on 8 August 1783, a year before Benjamin Franklin gained undue credit for some speculations about the event. This sulphuric aerosol certainly

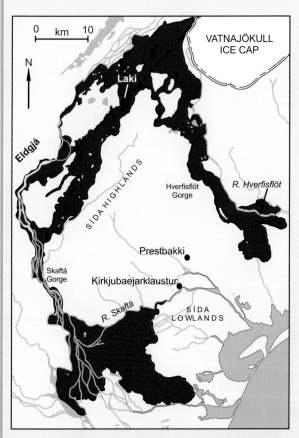

Figure 6.7 Map of the distribution of the main lava flows during the Laki eruption of 1783–84 (courtesy of Thor Thordarson).

seems to have been responsible for the three long and abnormally cold winters in the northern hemisphere from autumn 1783 to spring 1786, although it may not have caused the abnormally hot and stormy July in western and central Europe for which it has also been blamed.

Figure 6.8 The Laki cone row, showing cones of various sizes formed along the fissure in the 1783–84 eruption.

sometimes exceed 1000m^3 per second. Emissions then become weaker and are separated by lulls that increase in length until they stop almost unnoticed. The fissure then usually becomes extinct. The lava solidifies, seals the cleft and welds together the older lavas alongside it, forming a subterranean dyke that may be 4–6km deep and more than 10km long. The old fissure becomes a band of strength, 1–2m wide, which is added to the increasing bulk of Iceland. The stress of rifting is taken elsewhere until another fissure develops alongside it. Thus, a multitude of lava-clogged dykes border each new fissure. Typical fissure formation with the emission of lava flows and the growth of a cone row is well illustrated by the Threngslaborgir and Ludentsborgir cone rows east of Lake Mývatn.

Since settlement in AD874, the southeastern branch of the axial zone has been the most prolific source of basaltic fissure eruptions in Iceland. The fissures here are mainly related to two volcanic systems, both of which are partly masked by contemporary ice caps: Grímsvötn lies beneath Vatnajökull, and Katla lies below Mýrdalsjökull. The Eldgjá eruption of about AD935 came from a fissure that extended northeastwards from the Katla volcanic system; the Lakagígar 'Laki' eruption of 1783 occurred on a fissure stretching southwestwards from the Grímsvötn volcanic system. Eldgjá, like Katla, erupted transitional alkali basalts, whereas Lakagígar emitted tholeiitic lavas, as did Grímsvötn. The Eldgjá and Lakagígar fissures run parallel to each other and are only 5km apart, and they produced two most voluminous outpourings of lava in historical times.

The eruption along the Eldgjá 'fire fissure' in about AD935 was the largest emission of flood basalts on Earth during the past millennium. It is one of the largest fissures in Iceland, made up of four slightly offset sections, with an overall length of 57km, that are 140m deep and sometimes as much as 600m wide (see Fig. 6.9). The eruption featured at least eight distinct episodes

Figure 6.9 View along the Eldgjá fissure that opened around **AD**935.

and may have lasted from three to eight years. The cleft opened, lava fountaining followed, and activity became concentrated in individual vents that formed cinder cones and spatter cones, and gave off much sulphur dioxide. At the same time and immediately afterwards, $19.6km^3$ of fluid flows were discharged that covered $780km^2$. The fissures released about 184 million tonnes of sulphur dioxide into the atmosphere, troposphere and lower stratosphere, and the lava flows also released about 35 million tonnes into the lower troposphere. Thus, Eldgjá produced the greatest volcanic pollution during historical times, and about 1.8 times as much as its only rival, Laki, in 1783. However, the climatic effects of the Eldgjá eruption were not as intense, perhaps because emissions were distributed over several years. Nevertheless, the date of about AD935 coincides with an acidity peak in cores from the Greenland ice sheet.

The northern active volcanic zone

The active volcanic zone finds perhaps its clearest expression in northern Iceland between Vatnajökull and the Atlantic Ocean. Several active volcanic systems range en-echelon along the axial zone. Askja–Dingjufjöll

erupted in 1875, sporadically between 1921 and 1926, and again in 1961. The Krafla system was active between 1724 and 1729, and again from 1975 to 1984. On the other hand, the Theistareykir system underwent rifting without eruption in 1618; the Fremri–Namur system has not yet been active in historical times. Thus, the results of plate divergence have not been uniform, at least within the human timescale. The rifting at Krafla, between 1975 and 1984 in particular, threw light on problems associated with the mechanisms of rifting, the formation of magma reservoirs and the lateral transfer of magma to fissure swarms. This was somewhat reinforced by the the recent 2014–2015 fissure eruption at Holuhraun, which was the largest eruption since Laki, in a fissure extending northwards from the Bárdarbunga volcano.

Askja–Dyngjufjöll

Dyngjufjöll lies in the Ódádahraun ('the desert of crimes'), a vast and stark region covering $3000km^2$ north of Vatnajökull. It is a broad massif, with an area of $250km^2$, that rises about 700m from its base. It is the eroded remains of a stratovolcano, and Askja caldera crowns its summit (see Fig. 6.10). Dyngjufjöll is the

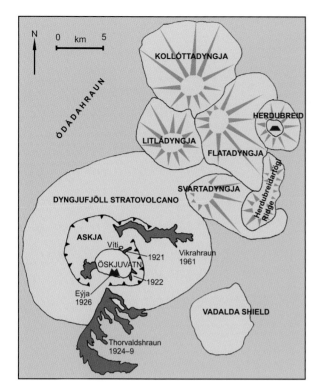

Figure 6.10 Askja, Dyngjufjöll, with the most recent lava flows, neighbouring shield volcanoes, and the table mountain of Herdubreid.

centre of a volcanic system that is 100km long and some 20km wide, with two major fissure swarms, one trending north-northeastwards through the present western edge of Askja caldera, the other running northeastwards through the eastern parts of the caldera. The eruptions became concentrated in the centre of the fissure swarm when the glacial ice sheets still covered the land. They built up a thick and widespread series of basaltic palagonites, including pillow lavas, pillow breccias and tuffs, formed in a subglacial water body. When the volcano rose above the level of the ice and meltwater, basaltic lava flows erupted, with occasional layers of breccias, tuffs and rhyolitic fragments. When the ice retreated from this area about 7000 years ago, eruptions from the western fissure chiefly gave rise to extensive basaltic lava flows, and small explosion craters formed on its eastern counterpart. It was probably between 4500 and 6000 years ago that the Askja caldera formed, because it took place before the deposition of ash erupted from Hekla about 3800 years ago. On the whole, lava production was about 30 times higher 8000–4500 years ago, when it expelled 5.98km^3 of lava per thousand years, than during the period from 2900 years ago, when it emitted lava at a rate of only 0.32 km^3 per thousand

years. This period of rapid volcanic accumulation could be explained by unloading of the pressure of the ice on the crust, which therefore reduced the pressures upon the crustal magmas below.

Askja ('the box') is one of the most impressive sites in the whole of Iceland. It is a circular caldera with an area of almost 50km^2 and a volume of 8.3km^3, and its walls are between 100m and 400m high. The collapse of Askja was apparently not associated with an explosive eruption, but rather with migration of magma from the reservoir out into the fissure swarm. After a rest of several thousand years, rifting and basaltic eruptions resumed within the Askja caldera and sometimes flowed out via the down-faulted Öskjuop col onto the flanks of Dyngjufjöll. However, these eruptions were neither frequent nor voluminous, but they continued until medieval times. They were then succeeded by an interlude of almost total quiescence lasting 400 years.

The Plinian eruption of Askja in 1875 was the first eruption recorded with certainty in the area. It was preceded by vigorous fumarole activity, rifting and earthquakes, and tholeiitic basalts entered the magma reservoir during the latter part of 1874 and early 1875. Basaltic eruptions occurred on 18 February, 40km north of Askja, where they formed the Sveinagjá lava flow, and 25km south of Askja, where they formed the Holuhraun lava flow (not to be confused with the recent activity and lava flow in 2014–15).

Early on Easter Monday, 29 March 1875, a Plinian eruption expelled a large column of ash and pumice. It expelled 2km^3 of rhyolitic pumice in less than 12 hours, which eventually spread as far as Scandinavia and even formed ephemeral pumice islets in the Atlantic Ocean. Much of the surface of Askja caldera is strewn with black ash and white pumice from this outburst. Towards the close of the eruption came the formation of the small hydrovolcanic Víti ('hell') crater, which is still filled with warm waters.

After this great expulsion of material, the north-eastern part of Askja caldera collapsed at intervals until 1907. This new caldera, nested within the larger Askja caldera, covered an area of 3km by 4km and was by then largely filled by Lake Öskjuvatn (Fig. 6.11). Small basaltic flows erupted alongside it in 1921 and 1922, and between 1924 and 1929 rifting formed a 6km-long fissure trending north-northeastwards on the flanks of Dyngjufjöll outside the Askja caldera. It emitted a large lava flow covering 16 km^2. The fissure also extended into both calderas and in 1926 erupted a

Figure 6.11 View looking south over Viti crater towards Öskjuvatn lake (photo courtesy of Morgan Jones).

small basaltic cinder cone that now forms the Eyja islet in Lake Öskjuvatn.

The eruptions of 1961 within Askja caldera were heralded by tremors from 6 October. Solfataras started up just north of the Víti crater on 10 October, and powerful geysers and mudpots developed, all of which are rare at the inception of an Icelandic eruption. On 26 October an 800m-long fissure opened nearby, producing lava fountains that rose 500m, and then the Vikrahraun lava flow, which eventually grew to a length of 7.5km. The lava had an aa surface at first, but its later emissions developed fine pahoehoe forms. In addition, three craters were formed on the fissure, which together expelled 3 million cubic metres of fragments. When the eruptions stopped in early December 1961, both the Askja caldera and the smaller Öskjuvatn caldera within it had foundered a little. No subsequent eruptions have been as violent as that in 1875, but they are slowly filling Askja caldera. However, they are being counterbalanced to some extent by continued foundering, punctuated by periods of uplift, which are both centred in the midst of the caldera. It has been suggested that these deformations could be related to pressure fluctuations in the upper part of the

Iceland mantle plume. Moreover, prolonged activity at Krafla between 1975 and 1984 apparently also caused continuous subsidence at Askja.

Krafla

Krafla is the core of a major volcanic system in northern Iceland, which rises to 818m and has a diameter of some 20km. Its centre is marked by a caldera about 10km across. Its fissure swarm, about 80km long and up to 10km wide, stretches from Axarfjordur on the north coast and extends to the area east of Lake Mývatn (see Fig. 6.12).

The eruptions of the Krafla volcanic system began under the ice, and its basement is thus composed of subglacial tuffs and palagonitic basalts, as well as of voluminous tholeiitic basalt flows. They are interspersed by layers of welded dacitic fragments that were expelled when the broad summit caldera collapsed. The caldera probably formed during an interglacial period and has since been almost filled by repeated emissions of basaltic flows. About 18 eruptions have occurred in and around the caldera and a further 15 in the south of the volcanic

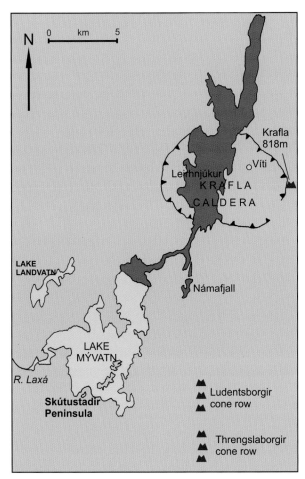

Figure 6.13 Rootless cones, formed about 2100 years ago when the Younger Laxá flows entered the shallow eastern shore of Lake Mývatn.

Figure 6.12 Lake Mývatn and Krafla caldera: lava flows of 1725–9 and 1975–84 (in grey).

system. The first main burst of eruptions occurred just after the ice sheets melted. A quiescent period then lasted at least 4000 years. The second main burst of activity has lasted until the present time. It began about 2500 years ago when a hydrovolcanic eruption formed the maar and tuff ring of Hverfjall. About 2100 years ago, eruptions along two major fissures formed the great pair of cone rows of Ludentsborgir and Threngslaborgir, both of which run from north to south and are only 250m apart. Threngslaborgir is 4km long, and its extremities are offset some 250m to the west of its main section. Ludentsborgir curves slightly north-northeastwards and reaches a length of 2.5km. Both fissures apparently operated simultaneously and formed two rows of about 30 often-intersecting cones less than 30m high. Most of the cones were breached as they formed when the vigorous effusions of the Younger Laxá basalts gushed from the fissures. They had, for Krafla, an unusually large volume of 2–3km³ and they covered an area of

220km², ponded back Lake Mývatn, flowed northwards for 63km, and almost reached the north coast. The lavas entered Lake Mývatn and formed a whole series of rootless cones that are scattered across the islets in the lake and form most of the arc-like Skútustadir Peninsula on its southern shores (Fig. 6.13).

A similar pattern was repeated between 1724 and 1729 during the Mývatn Fires, which sprang from the Leirhnjúkur fissure on Krafla. The episode was inaugurated on 17 May 1724 with a brief hydrovolcanic explosion that was possibly caused when snowmelt penetrated the fissure and encountered the rising magma. First, rhyolitic ash and then basaltic cinders and lapilli exploded from the vent and piled up in a tuff ring about 100m high, and the spectacular Víti maar, 300m across and 33m deep, formed in the hollow during the summer (Fig. 6.14). The rifting caused many earthquakes, which developed fault troughs and up-thrown fault blocks and lowered Lake Mývatn by about 2m. Six smaller hydrovolcanic eruptions formed aligned craters in the immediate neighbourhood at the same time.

The real Mývatn Fires started on 11 January 1725, when very minor eruptions of mud and lava began on the Leirhnjúkur fissure in the western part of Krafla caldera. This fissure was the main source of activity when the Mývatn display began in earnest in August 1727. The lava fountaining, in both curtains and individual jets, was accompanied by the emission of copious *helluhraun* (pahoehoe) basaltic lava flows that covered 35km² northeast of Lake Mývatn. From time to time, other parallel fissures nearby joined the display, including the Hrossaldur cone row in 1728 and the Bjarnarflag cone row in 1729. After the Mývatn Fires burnt out in

Figure 6.14 The spectacular Víti maar, Krafla, formed by the hydrovolcanic explosion on 17 May 1724 that initialled the Mývatn Fires (Shutterstock/Daria Medvedeva).

August 1729, their sole magmatic aftermath was the small eruption north of Leirhnjúkur in 1746. However, mudpots and solfataras continued to issue from many of the vents for the rest of the eighteenth century; and some have maintained extensive fumarole activity until the present day. Two centuries of repose then followed.

The new outbreak of rifting at Krafla was betrayed by a marked increase in earthquake activity and uplift in the latter part of 1975. The rifting episode of Krafla from 1975 to 1984 was characterized by repeated slow uplift or inflation, followed by shorter, quicker periods of subsidence or deflation. At the same time, repeated rifting and widening took place in a narrow, highly fissured band within the Krafla fissure swarm. Generally speaking, the uplifts and subsidence were less marked after the 1980 eruptions.

The episode began on 20 December 1975 with rapid subsidence and a small basaltic eruption some 2km along the fissure stretching northwards from Leirhnjúkur. The accompanying earthquakes migrated along the fissure as far as the north coast, 30–70km from Krafla. It was here that the greatest crustal widening took place, and formed gaping fissures and faults up to 2m high. This general pattern was to repeat itself about 20 times during the whole rifting episode. Slow inflation (or uplift) was followed by rapid deflation

(or subsidence), rifting, and sometimes by eruptions. However, in some cases, especially during 1977 and 1978, deflation occurred without lava emissions onto the surface, although earthquakes, fissures and subterranean dyke formation were always registered. The periods of inflations, amounting to about 2m or less in the central area of Krafla, usually lasted from one to seven months, whereas the deflation periods lasted only from one to twenty days. In general, the eruptions seem to have been linked with the fastest deflation rates.

The eruptions occurred on 20 December 1975, 28 April 1977, 8 September 1977, 16 March 1980, 10 July 1980, 18 October 1980, 30 January 1981, 18 November 1981 and 4 September 1984 (see examples in Figure 6.15). When emissions were most frequent, from 10 July 1980 to 18 November 1981, each was also more voluminous than usual. The eruption that started on 8 September 1977 had an unusual consequence about 9km south of Leirhnjúkur at the centre of the Krafla caldera. Here, in the Námafjall geothermal field, five boreholes had been drilled to extract steam for the power station. Borehole 4 was 1138m deep, and, at 23.45 on 8 September, it began to erupt. A roaring eruptive column of steam and ash 15–25m high lasted for about one minute. Calm ensued for the next 10–20 minutes, broken by occasional red flashes. The episode then concluded with

Figure 6.15 Examples from the 1984 eruptions at Krafla. **A** & **B**) Views of the fissure eruptions. **C**) Running and photographing an emerging fissure as it tiers up the hillside. **D**) Resultant lava flow from the newly opened fissure (photos courtesy of Torgeir Andersen).

a minute of explosions of glowing cinders. The magma had apparently been injected from a fault intersecting the borehole at a depth of 1038m. The little eruption gave off a thin cover of ash and cinders some 50m long and 25m wide, which had a volume of about 26m³. In contrast, the largest eruption of the series was the last, on 4 September 1984. Deflation started at 20.10 and the eruption ensued at 23.49. Lava fountains expelled some 24km² of lava flows from a fissure stretching 9km north of Leirhnjúkur, which was fed by a dyke 1m wide and 2.5km deep. Most of the eruption was over in 12 hours, but weaker activity persisted from one vent for two more weeks.

During the rifting episode, magma rose into a shallow reservoir about 2.6km or 3km that may have been supplied by a magma layer lying 8–30km below. Alternatively, the shallow reservoir could have been supplied by three individual reservoirs relaying magma towards the surface and lying at depths of about 5km, 20–25km and 30km. The rifting fissures opened below the surface and magma was quickly inserted, blade-like, into them at a rate approaching 500m³ per second, to

the accompaniment of earthquakes. The fissures were 4km or more deep and often over 50km long in a major swarm, and were arranged more or less vertically. The eruptions usually reached the surface 10–30km from the reservoir.

The reduction of volume and magma pressure entailed by the lateral migration of the magma from the shallow reservoir immediately caused a rapid deflation or subsidence at the surface. Magma nevertheless continued to rise into the shallow reservoir so that it was replenished in about three to five months. Thus, it filled more quickly than it emptied and, consequently, periods of inflation were much longer than those of deflation. Each filling, of course, slowly inflated the central zone of Krafla and the whole process was repeated. In all, these episodes of magma intrusion widened the fissure swarm by about 5m, with a maximum of some 9m in the centre. The Krafla fissure swarm, as a whole, has apparently widened by an average of about 10–15m during every millennium over the past 10,000 years. The whole series of eruptions covered some 36km², but their volume amounted to less than 0.3km³.

The Holuhraun fissure eruption 2014–15 at Bárdarbunga volcano

The largest eruption on Iceland since the Laki flow in 1783 occurred just north of the Vatnajökull glacier, between 31 August 2014 and 27 February 2015. The first signs of activity were recorded as earthquake tremors from magma movement, which highlighted that magma was intruding its way progressively northwards from the Bárdarbunga volcanic centre. On the 29 August 2014 there was a short-lived minor eruption lasting just a few hours, but the main eruption phase started on the morning of 31 August. This marked a period of six months of continuous activity. The eruption was characterized by lava fountaining from fissure eruptions (see Fig. 6.16), and an extensive basaltic lava flow that measured some 1.6 km³ in volume when it finished

Figure 6.16 Examples from the Holuhraun fissure eruption. **A**) Fire fountaining from main fissure. **B**) Lava flow in foreground with fissure eruption behind. **C**) Fountaining and sulphur gas plume from the volcano. **D**) Lava breakout flow from the side of 'Heiturpottur' the 'hot pot', officially named Suðri (southern) crater after the eruption. **E**) Lava flowing over a glacial outwash river; inset photo showing detail of lava breakouts. **F**) Sampling the fresh lava and the water at front of the flow. (Photos taken on 10 September 2014 by a team from University of Oslo, VBPR, DougalEARTH and the University of Iceland – Dougal Jerram, Sverre Planke, Morgan Jones, John Millett and Sigurður Gíslason.)

Watching the birth of a volcano

The Holuhraun fissure eruption was remarkable in many ways. There was significant seismic activity from magma movement leading up to the eruption from around 16 August up to and during the eruption itself. As the tremors from magma movement underground continued, the monitoring at and around Bárdarbunga allowed the 3D location of the earthquakes to be visualized. Therefore, as the magma migrated through the crust, it was observable virtually by its epicentre trail. Even after the fissure had started to erupt, the dynamics of the system were constantly under surveillance with earthquakes and ground movements being recorded.

Fig. 6.17 shows an example of the epicentre locations for tremors associated with the event illustrated in real time in 3D.

Although we now know the eruption as the Holuhraun fissure, there was debate as to how the volcanic event should be named, given that it did not occur at Bárdarbunga. It was announced later in 2015 that the eruption would have an official name, and the local council in Skútustaðahreppur voted to name it the Holuhraun volcano. The council had four proposed names for the site, which were: Flæðahraun, Holuhraun, Nornahraun and Urðarbruni.

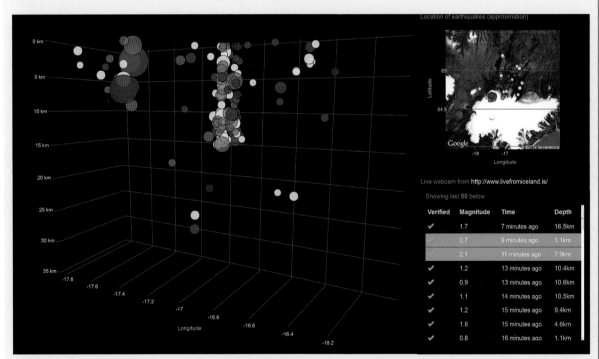

Figure 6.17 Monitoring the activity around Bárdarbunga and Holuhraun. Seismic activity with a magnitude greater than 2 over 36 hours from 29/08/2014, 13.25. (Larger circles are magnitute greater than 4, blue colour over a day ago; red most recent.) (Images courtesy of the web resource '3dBulge' created by Bæring Steinþórsson using data provided by the Iceland Met Office).

in February 2015. At its height the maximum fissure length was about 1.5km and the final flow covered an area of some 84km².

Significant ash and gas emissions were associated with the eruption, but due to its remote location and weather conditions at the time, this caused little impact on humans and the environment. It is estimated that

some 11 megatonnes of SO_2 were associated with the eruption, which is about ten times less than the Laki eruption (120Mt). Much of this was dispersed by winds, though locally villages in Iceland had air quality warnings about high sulphur values. Another aspect of the eruption that reduced environmental impacts was that the final eruption site was not under the ice. Had the

Meet the Scientist – Sigurður Reynir Gíslason

Sigurður Reynir Gíslason is a research professor at Institute of Earth Sciences at the University of Iceland, with many years of experience working in Iceland. His main areas of interest are the geochemistry of water, looking at aspects of precipitation, surface water, ground water and geothermal water. He is particularly interested in how magma affects the chemistry of water. After a trip into the site of the Holuhraun eruption in September 2014 to sample water in front of the active lava flow, Sigurður explains why!

Is this the first time you managed to get right up close to a lava flow?
No, no, I was born and raised on volcanoes; my grandfather galloped a horse in front of the Katla 1918 eruption in SE Iceland. I graduated in geology in 1980 and then the first job as a geologist I worked with eruptions on Krafla and another in Hekla.

You were on an active volcano and you were just sampling the water, not the lava: why is that?
If the volcanic eruption starts under the ice, then the magma will melt the ice. The ice can collect at the eruption site and eventually be drained into the local rivers, causing some hazards.

So how would measuring the water help?
When magma gases dissolve in the rivers, this changes their connectivity. The Met office in Iceland has conductivity meters in all the rivers that can be affected by glacial outburst floods associated with eruption under a glacier. We can look at these values through our network on the internet and check the connectivity. This can be a sign of meltwater associated with volcanic eruption under the glaciers being drained into the rivers, and we try and look at the details of what is happening. Today we had a unique opportunity: we saw the lava flow flowing into the river, and we can see how the connectivity changes before and after the contact with the water.

With Bárdarbunga volcano, if the magma activity was to move under the ice into the caldera, what are the main issues?
There are three main issues: if the magma makes it through to the ice it will result in volcanic ash production no matter what the composition; it could also be a more evolved magma, which could also mean more explosive eruptions with fine ash in the air, and effect airspace like the 2010 eruption. Finally, a reservoir can collect a significant amount of water, which can drain quickly as a major flood.

Is there any major infrastructure down stream?
Depending on where it breaches, there is a tipping point where it could go southwest, north or northeast. If it goes southwest it can enter a river system where there are five hydro powerplants, and that is a major energy production for Iceland and could be a very delicate issue.

So this is just a normal day at the office for you?
Definitely not! You become like a child every time you get close up to a lava flow. And you should let the child inside you get out and enjoy, and then you start to be objective. We are all excited when we get up close to the eruption, and we are doing our best to try and understand it.

Figure 6.18 Sigurður Gíslason and Morgan Jones returning with water samples from the edge of the Holuhraun lava flow where it enters a river; the fountaining eruption can be seen in the distance.

eruption been located at the Bárdarbunga volcano, there could have been major floods (*jökulhlaups*) resulting from melted glacial ice. The volcanic system is still being extensively monitored and clearly has the potential to erupt again in the near future.

Subglacial volcanoes

Volcanoes formed beneath ice sheets are among the most distinctive features of Iceland. Some developed under the extensive glacial ice sheets and were exposed when the ice melted, and others are still active beneath the four remaining ice caps. Subglacial volcanoes owe their special characteristics to the interactions between the ice cover, the molten materials expelled, and the meltwaters that were thereby generated. Morphologically, the now-exposed subglacial volcanoes are distinguished by their relatively flat summits and steep, sharply defined sides. They form table mountains (*stapar*) and ridges (*hryggir*), which contrast markedly with the conical crests of most volcanoes formed in the open air. Many older shields or stratovolcanoes, such as Hekla, Öraefajökull and Dyngjufjöll, for instance, probably spent much of their early lives erupting beneath ice sheets, although the typical forms have since been masked by the copious products of more recent eruptions in the open air. Grímsvötn (under Vatnajökull) and Katla (under Mýrdalsjökull) provide the finest examples of contemporary subglacial activity.

The simplest subglacial features are the flat-topped palagonitic (*móberg*) ridges formed by subglacial basaltic fissure eruptions, which would have given rise to cone rows and lava-flow vents if they had occurred in the open air. They are exemplified by the Herdubreidartögl ridge in the Ódádahraun desert and the Namáfjall ridge lying east of Lake Mývatn.

Table mountains are more complex than palagonitic ridges, chiefly because they erupted for several decades or more. Had they erupted in the open air, they would probably have developed as lava shields. More than 30 table mountains have been mapped in the central zone of activity. Most of them have a basal diameter of between two and five kilometres, and they rise by steep concave slopes to a regular, almost flat, lava-capped plateau. They owe their sharp outlines and steep sides to the confinement of the enclosing ice and the meltwater vault, generated by their eruptions, which stopped the fragments from spreading further outwards. Their flattish summits were formed when lava flows erupted in the open air after melting had pierced the ice sheet. Below the cap of lava, the bulk of each table mountain is composed of palagonitic tuff, breccia, and pillow lavas erupted in a mass of meltwater that was confined beneath the ice sheet.

Herdubreid, dominating the plains of the Ódádahraun, is one of the most majestic of Icelandic table mountains (Fig. 6.19). Its summit plateau is 2km across and rises about 1000m from its base. The lower 250m of

Figure 6.19 View of the table mountain Herdubreid rising up from the surrounding lava plains (Shutterstock/ Wolfgang Berroth).

the mountain is largely composed of pillow lavas piled up close to the vent in the meltwaters confined beneath the ice. They are succeeded by layers of tuffs about 200m thick, which were erupted by Surtseyan explosions when the ice sheet had melted to form a shallow open-air lake. In turn, these tuffs are succeeded by a 50m-thick layer of angular pillow-lava lumps, forming a breccia that was generated as lavas spread out across the lake. It is overlain by a band, 150m thick, of thin alternating layers of lava flows, broken pillow-lava layers, and cinders that were expelled in response to variations in the lake level as eruptions continued. A thicker lava flow forms the summit plateau at 1400m, and its eruption marked the final emergence of the volcano from its glacio-lacustrine constraints. This freedom was emphasized when explosions built a cinder-cone that now marks the summit of Herdubreid at 1650m.

Several Icelandic volcanoes, or volcanic systems, still erupt from time to time beneath the modern ice caps. Grímsvötn, Kverkfjöll, Bárdarbunga and Gjálp lie beneath Vatnajökull; Katla lies under Mýrdalsjökull; and Eyjafjallajökull lies under the ice cap of the same name. Kverkfjöll, which last erupted in 1875, is now marked by strong fumarole activity. Bárdarbunga has experienced many small historical eruptions. Eyjafjallajökull may have produced as many as 17 intermediate lava flows during the past 10,000 years, although it has erupted only three times during historical times: in 1613, 1821–3 and 2010. These were relatively small eruptions from the summit crater fuelled by explosive magma, and made more explosive by water from the melting ice cap. The fine, dense and dark-grey fragments expelled, for example, in 1821–3 attained a volume of only 0.004km^3. The spectacular eruptions of 2010, which wreaked havoc across the European air travel system, resulted from an estimated eruptive volume of ~0.25km^3.

Grímsvötn

Grímsvötn has an ice-covered caldera about 40km^2 in area that lies beneath Vatnajökull. It has been the source of at least two major prehistoric ash eruptions, as well as about 50 episodes of activity since AD1200. These eruptions have been basaltic, with insignificant quantities of silicic materials. Grímsvötn has given off between three and five cubic kilometres of material during historical times; its eruption in 1938, for instance, was probably the third largest in Iceland in the twentieth century. Many of these eruptions have been accompanied by powerful *jökulhlaups*, which commonly have discharges as vigorous as the world's major rivers. Grímsvötn is also the site of the largest field of active fumaroles in Iceland. Their heat maintains a lake that is trapped in the caldera beneath about 200m of ice. When the volcano erupts, the waters increase greatly in volume, lift up part of the ice cap, and then drain rapidly over the rim of the caldera in the form of *jökulhlaups*, which then rush seawards for several days.

At 10.48 on Sunday 29 September 1996 there was a strong earthquake, followed by 30 hours of smaller tremors. On 30 September, basaltic-andesite lava began erupting at Gjálp, under the 500m-thick ice cap between the calderas of Grímsvötn and Bárdarbunga. The ice cap quickly began to melt from below, and ashy fumes rose from its surface. Soon, all of the ice lying directly above the erupting vents had melted to form lakes. The fragments piled up closer and closer to the surface of the lakes, where the weaker water pressure could no longer stifle the explosions. At 05.18 on 2 October, Surtseyan eruptions were already blasting fine tuffs and black ash into the air. Meanwhile, meltwater was pouring into the Grímsvötn caldera and accumulating under the ice cap there. On 4 October, the water beneath the ice exceeded any levels previously recorded. On 13 October, the eruption stopped. It had built a tuff cone that rose 40m above the meltwater lake. But the ice continued to melt at Grímsvötn. On 2 November, the waters were 60m higher than the level at which *jökulhlaups* were usually unleashed. Then, at about 08.30 on 5 November, the waters finally lifted the dam of ice from the caldera rim, and the expected and overdue *jökulhlaup* finally cascaded down the Skeidar valley. Over 3km^3 of meltwater, with a discharge of almost 45,000m^3 per second, charged across the coastal plain. It was the largest *jökulhlaup* since 1938. No lives were lost in this sparsely populated area of the southeastern coastlands, but electricity cables, a main road bridge and 10km of the coastal ringroad were ripped away. Information displays about the 1996 event can be found along the south road, as well as the remnants of some of the distrusted structural foundations of the bridge (see Fig. 6.20).

Grímsvötn erupted again at 09.20 on 18 December 1998 and an eruption column soon extended 10km above the southern edge of the caldera. Slightly less vigorous activity continued the next day, when the column was reduced to a height of about 7km and ash was blown southeastwards. On 20 December, more ash was expelled and the eruptive column collapsed

Figure 6.20 Aftermath of the Grímsvötn Gjálp eruption in 1996. **A)** Photograph of the large *jökulhlaup* that spread out across Skeiðarársandur (glacial outwash plain) from the terminus of Skeiðarárjökull, an outlet glacier of Vatnajökull, during the eruption (photo by Oddur Sigurðsson, source USGS Multimedia Gallery). **B&C)** The information boards and remains of one of the washed-out bridges along the southern road route.

at least once and formed a pyroclastic density current. Thereafter, the eruption's power diminished, the column developed only intermittently, a lake grew up northeast of the vent, and a tuff ring was formed partly on the ice and partly on the adjacent exposed bedrock. Activity ceased on 28 December 1998. Two further eruptions occurred in 2004 and 2011, which both breached the

ice and resulted in ash clouds being ejected into the sky. The smaller eruption in 2004 caused issues with local airspace, but the 2011 eruption had much further effects. A total of 900 flights were cancelled as a result of the eruption in the period 23–25 May 2011 during the height of the eruption, which occurred between 21 and 25 May. The eruption column was recorded from

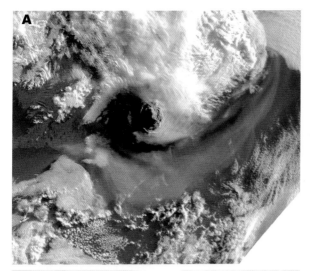

Figure 6.21 The 2011 eruption of Grímsvötn. **A**) NASA MODIS satellite image acquired at 05.15 UTC on May 22, 2011 shows the volcanic ash plume casting shadow to the west (source NASA/GSFC/Jeff Schmaltz/MODIS Land Rapid Response Team NASA Goddard Space Flight Center). **B**) Grímsvötn with surrounding ice covered in ash three months after the eruption –photo by Henrik Thorburn (Wikimedia Creative Commons).

space by the NASA MODIS satellite, and the aftermath of the eruption revealed a crater lake, and the ice cap draped with a dark cover of ash (see Fig. 6.21).

Katla

The Katla volcanic system lies beneath the Mýrdalsjökull ice cap in the eastern branch of the active volcanic zone, and the caldera in its centre is 700m deep and covers an area of about 110km². One of the most active volcanoes in Iceland, Katla has probably had more than 150 prehistoric eruptions, and about 20 have been recorded since the settlement in AD874. The active phases are commonly preceded by a few hours of strong earthquakes, but

eruptions normally last only two or three weeks. The main active episodes have usually expelled about 0.5km³ of fine tuffs, all of which have a similar alkali basaltic composition. The most important of these, the great Eldgjá eruption of AD935, occurred on the northeastern arm of the fissure swarm.

A shallow magma reservoir existed at least between 6600 and 1700 years ago when 12 silicic, largely dacitic, eruptions occurred, each concluding with basaltic emissions. The last five of these occurred 3600, 3139, 2975, 2660 and 1676 years ago respectively. However, they were not nearly so large as the main eruptions of Hekla or Öraefajökull, although pumice fragments that floated from Katla's explosions have been identified in Scandinavia and Scotland. During historical times, basaltic eruptions have often occurred at intervals of fewer than 70 years, and the most important took place in about AD935, 1485, 1625, 1660, 1721 and 1755–6. The eruption between 17 October 1755 and 13 February 1756 was not only the longest but also the most vigorous in historical times, and it gave off some 1.5km³ of fragments. The latest eruption took place in 1918, when close to 1km³ of basaltic fragments were expelled, and a 3km-long peninsula formed at the coast.

The *jökulhlaups* of Katla are its most dangerous features. They have periodically devastated the nearby coastal plain, which is in fact essentially composed of *jökulhlaup* debris. Indeed, they have extended seawards by some 4km during the past thousand years. Their discharge during the last major eruption on 12 October 1918 approached 200,000m³ per second for two days, which is a thousand times that of the River Thjorsa, Iceland's largest river, and comparable to the discharge of the River Amazon. These *jökulhlaups* make Grímsvötn and Katla the most dangerous volcanoes in Iceland. Fortunately, most of the regions invaded by *jökulhlaups* are sparsely populated, otherwise they would be notable killers.

Eyjafjallajökull – the 'Icelandic volcanic crisis' of 2010

In 2010 a global crisis occurred, closing airspace over most of Europe for six days and stranding thousands of people. This was caused by the eruption of Eyjafjallajökull, the 'Billion Dollar Volcano', and its associated ash cloud. Leading up to the eruption there were plenty of signs that the volcano was becoming restless. Iceland has a very good network of remote sensing equipment that monitors its many volcanoes. The Nordic Volcanological

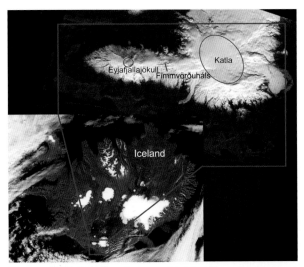

Centre (at the Institute of Earth Sciences, University of Iceland) and Iceland's meteorological service, reported the activity from eathquake tiltmeters and GPS stations that are sited around the volcanoes of Eyjafjallajökull and Katla. Seismic activity was on the increase from December 2009, with highly increased activity occurring from around 26 February. Also the ground was moving, evidenced by GPS stations around the volcano. The first eruption is thought to have begun on 20 March 2010 some 8 km east of Eyjafjallajökull's crater, on Fimmvörðuháls. This is situated on the saddle between Eyjafjallajökull and Katla volcanoes (see Fig. 6.22). During this early eruptive phase there were spectacular fire fountains, small explosions, and some lava flows (Fig. 6.23). Importantly, at Fimmvörðuháls there was no ice cover, and so the magma was erupting directly out of the ground with no ice to melt to cause flooding or to enhance the explosivity of the eruption.

Figure 6.22 Satellite images highlighting the location of Eyjafjallajökull and Katla volcanoes in South Iceland. Fimmvörðuháls is situated on the saddle in between (NASA/ GSFC and NASA/JPL).

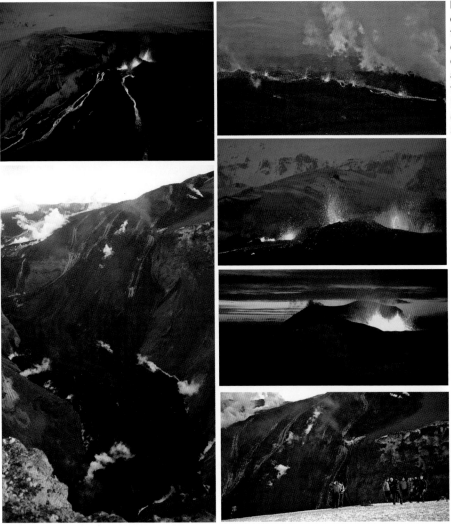

Figure 6.23 Montage of examples of the fissure feed eruptions and flows during the early stages of 2010 eruption located at Fimmvörðuháls, on the saddle between Eyjafjallajökull and Katla (photos courtesy of Jón Viðar Sigurðsson).

This situation was about to change. After a short pause, eruptions started on 14 April 2010 at the main Eyjafjallajökull volcano, under a small ice cap that is located over its summit. This change in the eruptive centre resulted in much more explosive volcanism, with a resultant ash cloud that billowed up into the sky. The new magma that had caused the eruption at Fimmvörðuháls now invaded the magma chamber beneath Eyjafjallajökull itself, where a larger batch of more evolved, explosive magma was residing. This sparked off the more explosive phase of the eruption. The more

evolved magma, and its dynamic interaction with ice, fuelled the plume of fine ash, which was erupted high into the atmosphere (Fig. 6.24). An estimated 250 million cubic metres of material (a quarter of a cubic kilometre) was ejected with a plume height of up to 9km (30,000ft). There were also *jökulhlaups* associated with the eruption (see Fig. 6.25), but with several hundred people being evacuated from the area, and with precautions in place to protect the bridges, effects were limited, with some damage to infrastructure and inundation of farmland.

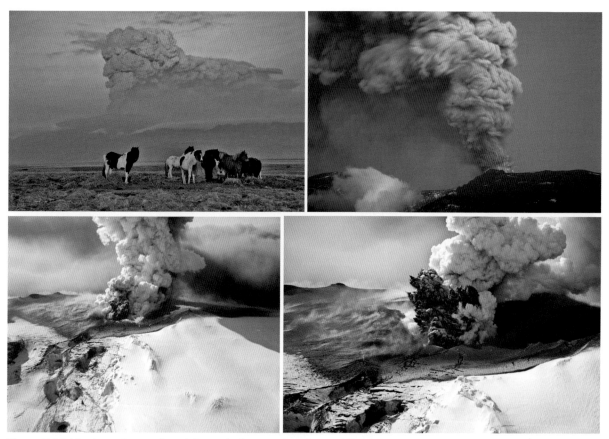

Figure 6.24 Montage of examples of the explosive eruption phase of Eyjafjallajökull itself as the eruption centre shifted to the main volcano under the ice cap (photos courtesy of Jón Viðar Sigurðsson).

Figure 6.25 Outwash into the sea from a *jökulhlaup* during the 2010 Eyjafjallajökull eruption (photo courtesy of Thórdís Högnadóttir).

The Billion Dollar Volcano

The so-called 'Icelandic volcanic crisis' of 2010 reputedly cost the global economy several billion dollars. Eyjafjallajökull; the 'Billion Dollar Volcano', was an event not witnessed before, and it has changed the way we deal with volcanoes in our modern lives. It created a situation unprecedented by other natural disasters by directly affecting the lifestyle of many millions of people, without causing death or directly threatening their lives. The volcano's site was remote, yet its direct influence on our lives was both very pronounced and also very rapid. The eruption sent a large ash plume into the high winds, and caused the airspace over most of Europe to be closed for six days, causing a global crisis with thousands of people stranded all over the world. The skies over cities such as London were quieter than people could remember for decades, and it was reported that residents heard birdsong in the skies around Heathrow Airport, London, for the first time. Satellite images from the time highlight the plume trail heading towards Europe (Fig. 6.26).

Before the eruption struck, civil aviation authority regulations regarding flights through volcanic ash were in place. A zero tolerance level for flying through ash meant that, although the ash was not in very high concentrations in all parts of the affected airspace, flying was not allowed. In the past, aircraft have literally dropped out of the sky as their engines have failed, clogged by volcanic ash. Luckily in these cases the planes have managed to restart their engines close to the ground, in time to avert disaster. A no-fly zone through most of European airspace was put in place, leading to massive disruption around Europe and felt all across the world. The eruption was lasting for days and there seemed no foreseeable end to the disruption. This had the potential to last for a long time; for example, a previous eruption at Eyjafjallajökull, in 1821, had lasted over a year. This called for more action with the flight ban in its sixth day, and following urgent tests and with mounting public pressure, the flight rules were changed and established a new tolerance limit of 2000 micrograms per cubic metre ($\mu g/m^3$), with anything more than 2000 $\mu g/m^3$ as a no-fly zone. At between 2000 and 200$\mu g/m^3$ planes are required to take extra precautions; below 200$\mu g/m^3$ no threat is recognized. For comparison, the EU recommendation for limits on particulates in air pollution is only 20$\mu g/m^3$ (yearly average). A tolerance of 4000$\mu g/m^3$ was also accepted for certain areas for limited periods in UK airspace, as this is one of the busiest on the planet, and the area that was being most affected by the plume. Had these limits been in place at the start of the crisis, very limited areas of airspace would have been closed, and for shorter time periods. The threat still remains that in favourable wind conditions certain types of eruption from Iceland and other volcanoes around the world have the capacity to shut down airspace, which has happened on a handful of occasions since the Iceland volcano crisis.

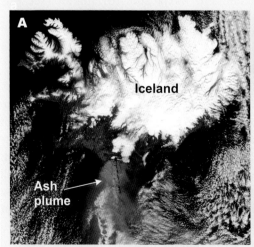

Figure 6.26 Trails of the ash plume. **A)** NASA's Aqua satellite – visible image of the ash plume from the Eyjafjallajökull volcano 17 April 2010 (NASA's MODIS Rapid Response Team). **B)** NASA's Aqua satellite image over Iceland's Eyjafjallajökull Volcano on 10 May 2010. (NASA Goddard/MODIS Rapid Response Team).

After the Eyjafjallajökull eruption attention turned to its sister, and much larger, volcano, Katla. There were concerns that Katla might erupt. In the past Katla has often erupted shortly after Eyjafjallajökull. Katla is some five to eight times the size of Eyjafjallajökull and has a much thicker and more extensive ice cap (see Fig. 6.22). Thus Katla has the potential to provide a much larger eruption, with even more profound effects on Europe and the world. Luckily in this instance Eyjafjallajökull was the only one of the two volcanoes to erupt.

Shield volcanoes

Although the volcanic systems composed of central volcanoes and fissure swarms comprise the major features of the active volcanic zone, smaller independent central volcanoes also occur, as well as independent fissure systems related to rifts. Shield volcanoes are probably the most widespread landforms created by concentrated vents; similar subglacial eruptions formed the table mountains; and apparently independent rift systems are best represented by Thingvellir. The shields and independent fissures commonly expel olivine tholeiite basalts, whereas the volcanic centres erupt quartz tholeiitic basalts.

Most of the shields formed 9000–7000 years ago, and they seem to have continued the eruptions that formed the subglacial table mountains. Shield-forming activity may possibly be waning, because the offshore eruption of Surtsey in 1963–4 was the only such episode during historical times. They were formed by accumulations of basaltic flows that were emitted at the low average rate of about $5m^3$ per second. The shields have gentle slopes, broad basal diameters, ranging from 5km to 10km across, and their heights rarely exceed 500m. They are almost entirely composed of copious outpourings of olivine tholeiitic basalts, which were erupted in a hot and fluid condition.

Skjaldbreidur is perhaps the most conspicuous shield volcano in Iceland. It forms a characteristic accumulation of fluid lava flows with gentle slopes radiating from a central hub rising to 1060m, marked by a crater 350m across. Ketildyngja and Kollóttadyngja, each of which has emitted copious lava flows, provide other examples in the Ódáðahraun, to the northwest of Askja. About 1900BC, Ketildyngja erupted the enormous Older Laxá lava flow, which covered 330km², reached the sea 82km away, and also formed the first lava dam that impounded

Lake Mývatn. These lavas, erupted in a hot fluid state, now have a distinctive *helluhraun* (pahoehoe) surface. Kollóttadyngja is a similar shield. It is 5km across and its gentle slopes rise to a height of 460m. It is probably about 7000 years old.

Plinian eruptions in Iceland

The Plinian eruptions in Iceland are generated when more evolved lavas develop after relatively long sojourns in magma reservoirs. They are richer in volatiles and more viscous in character, and are marked by a transition to alkali basalts, dacites, and even occasionally to rhyolites. Fragments and lava flows are expelled, but overall production rates are low. Their lava flows typically develop *apalhraun* (or aa) surfaces. This activity usually springs from vents clustered on very short fissures, which form slightly elongated, often rather small and compact, shields or stratovolcanoes. They frequently lie on top of an older volcanic basement that had been generated predominantly by eruptions from fissures.

These violent eruptions occur in four regions on the flanks alongside the main areas of fissure activity: the southernmost area of the eastern arm of the active volcanic zone, the Snaefellsnes Peninsula, the alignment of stratovolcanoes stretching northeastwards beneath the Vatnajökull ice cap from Öraefajökull to Snaefell, and the two volcanic areas under and around the Langjökull and Hofsjökull ice caps in central Iceland. Much of the explosive material originates from Hekla, Öraefajökull, Eyjafjallajökull, Askja and Torfajökull (see Fig. 6.27).

Torfajökull

Torfajökull is probably one of the most active silicic volcanoes in the world. It covers an area of about 450km² and forms the largest expanse of rhyolitic rocks in Iceland. A ring fracture some 30km across along its major axis may, in fact, represent the rim of an old caldera. Since the most recent glacial period, Torfajökull has erupted at least 11 times, and ash from one of these explosions was deposited in the Orkney Islands about 5560 years ago. It most probably experienced two episodes of activity during historical times: the eruption that took place in AD870, soon after the settlement of the country, and the other in 1477. The outbursts of

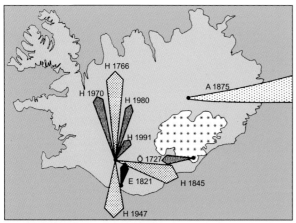

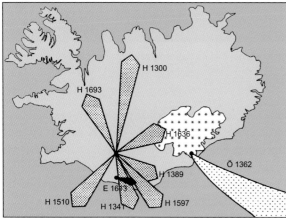

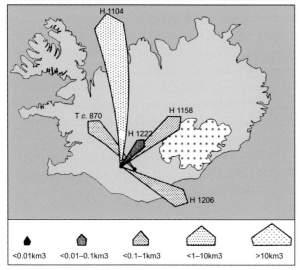

Figure 6.27 The main axes of silicic layers explode from Askja (**A**), Eyjafjallajökull (**E**), Hekla (**H**), Öraefajökull (**Ö**), and Torfajökull (**T**). Arrow indicates order of magnitude; length indicates relative volume (after Larsen et al., 1999).

Torfajökull seem to be set off by, and occur virtually at the same time as, eruptions from the nearby tholeiitic volcanic system of Veidivötn–Bárdarbunga.

Öraefajökull

Öraefajökull is the second largest active stratovolcano in Europe after Etna. It is also the highest mountain in Iceland, rising to 2119m on the southern edge of the Vatnajökull ice cap. Several distinct peaks surround an elliptical ice-filled caldera, stretching 5km from north to south and 3km from east to west. The volcano is older than the last glacial period, as its base is composed not only of basaltic lava flows but also of palagonitic breccias and pillow lavas, as well as fluvioglacial deposits. This mass was then transected by rhyolitic intrusions, one of which forms the Hvannadalshnukur, the highest point of both the volcano and Iceland. Most of the summit cone of Öraefajökull probably formed during the last glacial period, when basaltic fragments exploded into the open air above the ice cap. It seems that an eruption from this volcano was responsible for the white pumice found on the 9000-year-old Trandvikan shoreline in Norway. There have been several other explosive but poorly dated episodes, notably one between 2000 and 4000 years ago, and another between 1000 and 2000 years ago.

It was during the eruption in June 1362 that Öraefajökull earned its name, which is derived from Oraefi, 'the waste land'. This was the largest explosion in Iceland in historical times. It apparently produced 10km³ of distinctive white rhyolitic pumice, about 2km³ of which fell on land. Some 4300km² in the area to the northeast of the volcano was covered in ash 10cm deep, and ash also fell on Ireland and Scotland. The explosion of hot fragments through the summit glaciers formed vigorous glacier bursts (*jökulhlaups*) that combined with ashfalls to devastate the adjacent farmlands to such an extent that they were abandoned for over a century. The smaller eruption of andesitic fragments in 1727 produced further *jökulhlaups*, which once again destroyed much of the land that had been reoccupied.

Hekla

Hekla (Fig. 6.28) is one of the most famous volcanoes in Iceland. It is also one of the most scientifically informative, because its eruptions spread layers of pale and easily identifiable fragments that are valuable indicator beds in establishing the chronological sequence of many postglacial events in the country. Indeed, Hekla has expelled more than half the total volume of intermediate and silicic materials erupted in Iceland since the end of the most recent ice age. Hekla

Figure 6.28 View of Hekla from the southwest along the axis of the Heklugjá fissure (commemorative inset stamps courtesy of Josef Schalch mountainstamp.com).

rises near the edge of the eastern branch of the active volcanic zone in the centre of a volcanic system that is about 40km long and reaches a height of 1491m. It forms an elongated stratovolcano aligned on the 5.5km-long Heklugjá fissure, which carries three distinct craters that do not always erupt in unison.

Hekla has been growing since the most recent glacial period from a base of intermediate and basaltic palagonites and lava flows, ash and pumice, and fluvioglacial deposits. Five major eruptive episodes were identified, dated and named H5 to H1, and they have recently been re-investigated. The first (H5) occurred about 6100 years ago, H4 about 3800 years ago, and H3, which produced about 12km³ of fragments, took place about 2900 years ago. However, the Selsund pumice of the original H2 eruption has now been found below the H3 deposits, and has been dated to 3500 years ago. The old H2 has therefore been renamed the HS eruption. The H1 eruption occurred in AD1104. All these eruptions apparently began with a major explosion of rhyolite or rhyodacite, and concluded with intermediate or basaltic-andesite emissions. Most of the material erupted since 1104 has been andesitic. Fragments from many of these eruptions have been identified as far afield as Scotland, Ireland, Sweden and northern Germany.

Since 1104, Hekla has usually behaved with a certain regularity. Earthquakes lasting less than two hours commonly provide preliminary warnings. Activity would begin with a violent Plinian explosion of fine rhyolitic or rhyodacitic fragments, forming an impressive column

for several hours, which usually gave off more than two-thirds of the total fragmentary output. As this episode concluded, first dacite and then basaltic-andesite flows would pour forth, quickly at first, then more slowly after a few days, and then more rapidly again for several weeks. The eruptions ended with the emission of lava flows from the summit Heklugjá fissure for several months. Few eruptions have lasted as much as a year, and the more violent outbursts only a few hours. Generally speaking, the more violent eruptions with the higher silica contents follow the longest periods of quiescence. The rate of emission has averaged 10 million cubic metres per year. The volcano has produced about 8km³ of materials during historical times, probably from a compositionally zoned magma reservoir at about 8km depth.

To the surprise of Hekla's students, it erupted again in 1970, after an interval of a mere 23 years. After a 25 minute prelude of earthquakes, the eruption began on 5 May 1970 at 21.23, when three vents rapidly developed lava fountaining and lava flows along a fissure, later called Sudurgígar, about 1km southwest of Axlargígur crater. They were matched by three more vents that developed 7km northeast of Axlargígur crater and were later called Hlidargígar. The Sudurgígar vents continued to erupt until 10 May 1970, and the Hlidargígar vents continued until 20 May 1970. That same evening a new fissure, Oldugígar, took up the baton about 1km north of Hlidargígar and gave rise to many lava fountains. Activity on the Oldugígar fissure gradually reduced until only one vent remained to build a cone, 100m high, before

The historic eruptions of Hekla

The H1 eruption of Hekla began in the autumn 1104 after a rest of almost 700 years. It was a violent explosion of some 2.5km^3 of rhyolitic fragments, which has been exceeded in historical times only by the Öraefajökull eruption of 1362. Hekla gave off no lava flows, and the explosive climax was probably brief. But the winds carried fine rhyolitic pumice northwards over half the country, where many farms had to be abandoned, and eventually as far as the British Isles and Scandinavia. Three minor eruptions occurred in 1158, 1206 and 1222. The next eruption began in July 1300 with a Plinian explosion that blanketed the north of the island with fine fragments and dislocated the rural economy so much that 500 people died in the ensuing famine in the Skagafjordur area in the north. Activity lasted for a year and ended with emissions of lava flows. The weak eruption in May 1341 emitted fragments, but the main damage was to the animals asphyxiated by poison gases. The eruption of 1389 saw the development of a new fissure near Skard and a new crater, the Raudoldur, south-southwest of the summit fissure. After more than a century of quiescence, Hekla built up enough energy for a violent Plinian explosion on 25 July 1510. The ashfall caused widespread damage, and volcanic bombs were even thrown 40km from the volcano. The weak eruption of 1597 was chiefly notable because it lasted six months, and the even more feeble activity beginning on 8 May 1636 lasted a year.

The eruption that continued for seven months from 13 January 1693 was in many ways a repeat of the events of 1300. There was the usual Plinian opening act; then a widespread blanket of pumice combined with emission of poisonous gas to annihilate part of the animal population, and vast lava emissions closed the performance. The longest historical activity of Hekla took place between April 1766 and May 1768. This duration was matched by the exceptional amount of lava, covering 65km^2, emitted from the central fissure. The ashfalls from the initial Plinian phase caused great destruction on the farms to the north, and a warm *jökulhlaup* later overwhelmed the Ytri–Ranga Valley. The seven-month eruption that began on 2 September 1845 was very similar. There was much damage from the blanket of ash and pumice, another *jökulhlaup* invaded the Ytri–Ranga valley, and the events ended with a lava flow. Hekla then rested for a century.

After the second longest period of quiescence since 1104, Hekla produced one of its most powerful eruptions in 1947, lasting for 13 months. The initial blast on 29 March 1947 sent a white column of steam soaring 27km from the summit fissure within 20 minutes. Soon afterwards, the column settled around a height of 10km as brown and then black dacitic ash was expelled. The ash covered 70,000km^2 in Iceland, and within two days it was falling over Finland, 2750km away. In the early hours of the eruption, the melting ice and hot ash generated many *jökulhlaups* that flooded, as usual, into the Ytri–Ranga Valley. Much less usual was the expulsion of lavas on the very first day from the summit fissure, and they covered more than 18km^2 in 18 hours. Lava emissions became concentrated on Hraungígur, the new vent at the southwestern end of the summit fissure. The other two vents on the crest of Hekla, Axlargígur and Toppgígur produced frequent explosions for about a month, but then only intermittently for the rest of the active period. On the other hand, the Hraungígur vent displayed 13 months of continuous basaltic-andesite effusions and produced some 800 million cubic metres of lava. Cattle were poisoned by carbon dioxide gas emissions, but, most unusually in this sparsely populated area, the eruption caused a fatality when a lava block fell upon a volcanologist.

the eruption ended on 5 July 1970. All of these fissures produced basaltic andesites covering 18km^2 in area. The vents also gave off much fluorine that killed farm animals in northern Iceland.

The 1970 eruption had several unusual features. Lava flows emerged during the very first stages of the eruption. This was probably because not enough time had elapsed since the 1947 eruption for evolved magma to develop in the reservoir. The 1970 eruption also – and possibly for the same reason – gave off proportionately more lavas and fewer explosive fragments than usual. But the oddest aspect of all was that the summit fissure of Hekla, the Heklugjá, took virtually no part in the eruption, probably for the first time in the present cycle. However, the three newly formed fissures most probably represent lateral eruptions from the Hekla reservoir.

Hekla erupted again at 13.27 on 17 August 1980, without prior warning, after the shortest dormant period since 1104. The summit fissure, Heklugjá, released a Plinian column 15km high, which eventually covered 17,000km² with fluorine-rich ash. At 13.30, as the active summit fissure lengthened to 8km, six basaltic-andesite flows were expelled and eventually covered 24km². Activity stopped on 20 August 1980, but not for long. On 9 April 1981 a short-lived, modest ash column, 6km high, exploded from the summit, followed by basaltic-andesite flows that emerged from new flank fissures and soon covered 6km². This brief episode ceased on 16 April 1981. The 1980 and 1981 eruptions seem to have been part of the same episode because they emitted similar lavas, probably from a magma reservoir about 7–8km deep.

Hekla resumed its eruptions once again on 17 January 1991. The outburst was preceded, as in 1970, 1980 and 1981, by ground deformations of some 3–4cm and about 30 minutes of earthquakes. It produced mainly basaltic-andesite lavas, which covered about 23km², and a Plinian column that rose 11.5km within 10 minutes. But after a day had elapsed, activity was limited to the main fissure, where the main crater then formed. The eruptions came to an end on 11 March 1991. The fissure and flow directions for eruptions at Hekla between 1970 and 1991 are given in Figure 6.29.

The agitated phase is continuing. Earthquakes began beneath Hekla at 17.00 on 26 February 2000. Warnings of the impending eruption were announced on the national radio at 18.00, and the volcano duly burst into activity once again at 18.19. The eruption reached its greatest intensity during the next hour. Lava fountains gushed forth over a fissure nearly 7km long, a column of ash immediately soared 10km into the air, and a light scattering of ash covered central Iceland. Next day, lava flows advanced down the flanks of Hekla from several craters just south of the summit and, on 28 February, one flow with a snout more than 8m thick had reached Lambafell, 5km south of the vents, but it had advanced at a speed of less than 3m an hour. On 29 February, activity almost ceased, but it revived after noon at the southern end of the fissure. On 1 March, seven craters were in moderate Strombolian activity and the lavas had covered about 17km² with a volume of about 0.1km³. The activity declined appreciably thereafter.

Tubes, tunnels and chambers

Given the extent of volcanism in Iceland it has its fair share of lava tubes and hollows distributed around the island (examples given in Fig. 6.30). Many of these can be accessed easily, with stops on the side of the road, and many are within a short distance of Reykjavík, making them popular with locals and tourists – for instance, the Maríuhellar caves (Maria's caves) just 15 minutes from Reykjavík. Some of the bigger tubes can be accessed as part of excursions organized by the many Iceland excursion companies. The Leiðarendi lava tube is one great example of the tubes that Iceland has to offer, situated approximately 30 minutes from the capital. Several tubes are less accessible, and others yet to be discovered or fully explored. In places, the roof of a tube may have collapsed, creating windows of light that illuminate the interior (Fig. 6.30A), but to explore deeper into the systems you need torchlight, which can expose the beautiful patterns left in some tubes by receding lava levels, as well as the exposed tops of frozen lava flows (Fig. 6.30B). Hydrothermal fluids may also work their way into the cracks and fissures to form eerie hot underground lakes, some of which are too hot to swim in (Fig. 6.30C).

One of the more spectacular underground sights is found at the volcano Thrihnukagigur, which has been dormant for the last 4000 years. This is located on the south peninsula, south of Reykjavík. Here magma has drained out from a chamber, leaving behind a gigantic cavern. This natural citadel is full of vibrant colours (Fig.

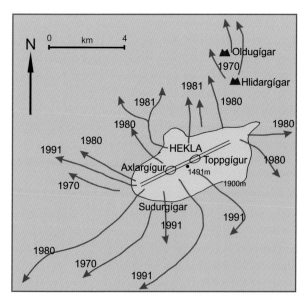

Figure 6.29 The Heklugjá fissure, with directions and dates of chief lava flows erupted from Hekla between 1970 and 1991.

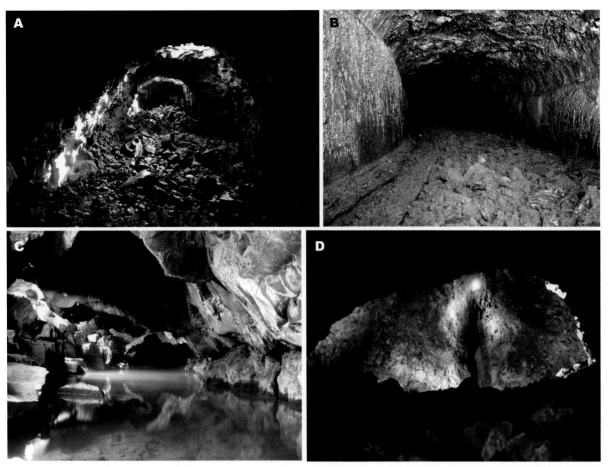

Figure 6.30 Iceland's tubes, tunnels and chambers. **A**) Skylights from roof collapse bring natural light into some caves. **B**) Frozen in time, lava flow tops within tubes and levels up the side of the tube indicating differing heights of the lava flow during the tube formation. **C**) Hydrothermal fluids can provide beautiful cave scenes, but in this case are too hot to bathe in. **D**) The drained magma chamber at Thrihnukagigur, the 'Inside a Volcano' experience (photos **B–D** courtesy of Morgan Jones).

6.30D) from hydrothermal alteration and oxidation. The 'Inside a Volcano' experience is only available, at some cost, through a guided tour.

Hydrothermal activity

Iceland is riddled with hydrothermal features, which have contributed much to the local place names, tourism and culture (Fig. 6.31). The basic cause of these features is the abundant precipitation that falls on fissured ground, which is heated by shallow magma. The lavas provide a maze of conduits that facilitate the downward penetration, lateral distribution and ascent of the waters. The varying geometry of the conduits also contributes to the different rates of water heating, mineral contamination and circulation. The drive for much of the hydrothermal system is the presence of high-level magma, which heats the adjacent rocks to high temperatures. The temperature of the rocks

increases rapidly with depth, and the average temperature rise in the hydrothermal zones is about 1°C for every 10m from the surface downwards, but in some areas it can exceed ten times that amount. The water circulation and the predominant temperatures also contribute to the chemical alteration of the enclosing lavas, and thus to the amount and type of material eventually carried by the waters towards the surface. Generally speaking, the areas with the higher temperatures give rise to solfatara fields, which have relatively low water discharges, much chemically altered rock, and a predominance of steam and sulphurous gases. The lower temperature areas, on the other hand, tend to produce geysers and hot springs (where the waters carry small proportions of altered rock), and mudpots (where the proportion of altered rock is higher).

The hydrothermal areas are concentrated in the axial zone and alongside its southwestern branch. The

Figure 6.31 Examples of Iceland's hydrothermal energy. **A** & **B**) The geothermal power station at Hellisheidi. **C**, **D**) The hydrothermal fields near lake Mývatn, Krafla. **E**) Natural underground hydrothermal cookers; inset Dougal about to try some volcano-cooked bread and eggs. **F**) Strokkur Geyser erupting, Geysir hydrothermal area.

largest high-temperature areas, for instance, are located at Torfajökull, covering 100km²; at Henzill, covering 50km²; and also beneath the Grímsvötn ice cap. The areas of lower temperatures are almost all confined to fissure zones in southwestern Iceland, where geysers and warm or hot springs at less than 100°C are generated. They are especially common, for example, in Hveravellir ('hot field') and Kelingarfjoll. The western end of the Reykjanes Peninsula has saline mudpots and hot springs resulting from the rather unusual infiltration of sea waters into the heated substratum.

Iceland is perhaps more famous for its geysers than for any of its magmatic eruptions and has, of course, given that word to the international vocabulary (Fig. 6.31F). Iceland now has only about a dozen geysers which, in fact, would operate only sporadically without artificial assistance. The much-fissured area around Geysir is the chief low-temperature thermal centre in Iceland, where the springs reach the surface at 80–100°C. The Grand Geyser itself is waning now after about 8000 years of activity. It erupted every half hour 200 years ago; a century ago it still managed once every three weeks; now it bursts out only after long irregular intervals unless it is stimulated by soap powder. Nevertheless, earthquakes in June 2000 seemed to give Grand Geyser a new lease of life. However, Strokkur, nearby, has been sustained artificially on a borehole life-support machine and still gushes 20m high every few minutes so as not to disappoint the customers. Like all hydrothermal systems, geysers are so fragile that they can be altered or destroyed by earth tremors or by natural or human-induced blockages. In compensation, other conduits may be opened by the same token, such as the Niyhver springs, which developed after an earthquake in the Reykjanes Peninsula.

Geysers form when water and steam are suddenly released to the surface, but most of the heated waters bubble out more calmly as hot springs that often form distinctly warm streams and rivers, or bubbling pools, such as Bolutler in Hveravellir. Many hot springs are naturally contaminated by dissolved chemical precipitates. They then give rise to mudpots, which bubble threateningly, such as, for example, Krísuvík in the Reykjanes Peninsula and those in Kelingarfjoll. Other hot springs precipitate highly coloured deposits of carbonate or sulphur compounds around their vents, which, for instance, add much to the beauty of Kelingarfjoll. In other cases, dissolved minerals or algae provide the colour, and Hveravellir, for example, has both a blue and a green basin. However, many minerals are often precipitated around the vents and conduits near the surface and, as a result, the systems can be blocked by natural means. This is especially noticeable where the geysers deposit mounds of silica, called geyserite, around their vents. In contrast, earthquakes could also widen fissures and rejuvenate the whole process.

So many of the Icelandic hot springs have been exploited that more than three-quarters of the population use geothermal heat. Boreholes up to 2000m deep have been sunk to bring up hot water to provide cheap, clean, ecologically sound heat for homes, industrial and commercial buildings, open-air warm swimming-pools, and greenhouses. Much of Reykjavík is heated in this way. There are currently five geothermal powerplants in Iceland that contribute to about a quarter of the country's energy needs. Krafla has a steam-powered electricity plant, and the locals still cook bread and eggs through shallow ovens that are dug into the hot ground (see Fig. 6.31E). At Hveragerdi, the geothermally heated greenhouses, covering 10ha, have become world famous because bananas, as well as less exotic fruits such as tomatoes, have been ripened almost within reach of the Arctic Circle.

The island volcanoes

Iceland originated from submarine eruptions on the Mid-Atlantic Ridge. It has also grown, since the most recent glacial period, along the southwestern branch of the active volcanic zone on the Reykjanes Peninsula. Nine reliably recorded eruptions, and probably at least three more, have also occurred off Reykjanes since the settlement of Iceland, and temporary islands were formed in 1211, 1422 and 1783. Where the active volcanic zone extends off northern Iceland, too, at least one eruption in 1867, and probably three more, have taken place in historical times. But no eruptions occurred on its main southeastern branch towards the Vestmannaeyar (the '-aeyar' suffix meaning 'islands') from the outbursts of Helgafell on Heimaey more than 5000 years ago until fires were reported southeast of Geirfuglasker in 1896. Thus, the eruptions of Surtsey in 1963 and on Heimaey in 1973 testify to the new vitality of activity here. At present the Vestmannaeyar (Fig. 6.32) rise from a platform that is often less than 130m deep. It could happen, within a short span of geological time, that these islands will be joined together by fissure eruptions to form a large peninsula projecting southwestwards into the Atlantic Ocean.

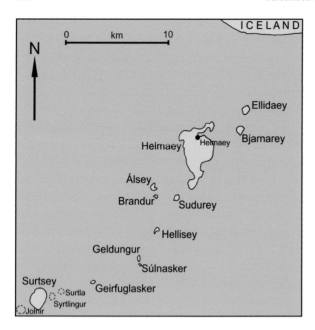

Figure 6.32 Surtsey, Heimaey and their companion islands in the Vestmanaeyar. The locations of Surtla, Syrtlingur and Jolnir, which existed for a brief time during the Surtsey eruptions, are also indicated.

The eruption of Surtsey 1963–7

The eruption of Surtsey set the standard all over the world for what became known as the Surtseyan style of activity. It was the first modern example of the emergence of a new island from a submarine volcano, and it has become one of the case studies on how new land gets colonized. The island of Surtsey was born in 1963 some 32km off the southern coast of Iceland and is now a UNESCO world heritage site (Fig. 6.33).

The first phase, which may have lasted a week, was the completely submarine eruption of basaltic pillow lavas and then fine tuffs, which built up a volcanic mound rising steeply from the submarine platform at about 130m depth. This phase passed largely unnoticed except that the sea was 2°C warmer in the area on 13 November, and there was a distinct smell of sulphur on Heimaey and at Vik on the mainland on 12 November.

The second, truly Surtseyan, phase then began when the eruption expelled its first column of steam and lava fragments into the air, which was first sighted about 33km from the mainland at 07.15 on 14 November 1963. Within four hours, the eruption had developed the full fury that it was to maintain with little respite until 4 April 1964. Every few minutes, sea water rushed into the vent, forming a slurry of ash, tuff and water, and met the hot lava and exploding gases rising to the surface. The gases and steam repeatedly shattered the lava into black jets and plumes of fine fragments that were expelled 500m into the air every second, as if they

had been fired from a gun, and the larger fragments formed bombs that whistled outwards 1km or more from the vent. A billowing column of white steam, often riddled with lightning and laden with fine dark ash, rose 8km high in calm weather, but gales often blew the column sideways so that it rarely exceeded 2km in height.

At first the eruptions occurred from two, three or four vents along a north-easterly fissure, but they gradually concentrated on one vent, called Surtur I. They formed a large horseshoe cone that reached 45m on 16 November, 100m on 23 November, 130m on 30 December 1963, and 174m on 5 February 1964. At the same time, the waves had cut a 100m-wide marine platform around the cone of unconsolidated tuffs. Yet the eruptions were intense enough to keep pace with such an onslaught. On 16 December, the first geologists to land on the island saw that Surtsey was producing hot olivine basalt, at a temperature of 1150°C, which was very similar to lavas erupted by Katla on the nearby mainland, and they deduced that the eruptions were being supplied by a deep-seated magma reservoir.

On 29 December 1963 a new eruption, called Surtla, broke from the sea 2.4km northeast of Surtsey. However, the expected new island never appeared, and it remained as a shoal 6m below the waves, because activity stopped on 6 January 1964. Activity on Surtsey

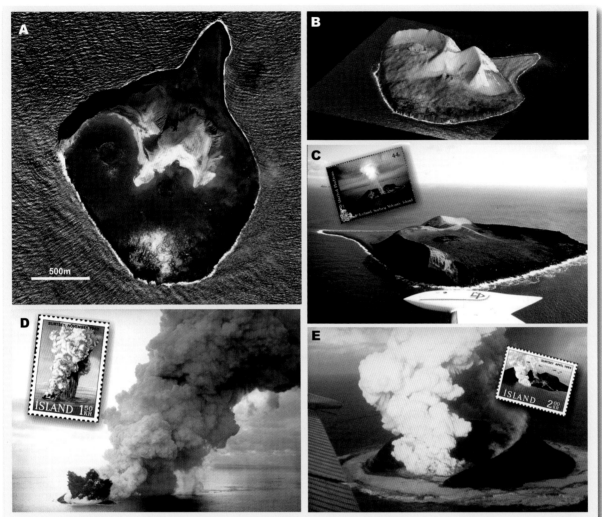

Figure 6.33 Examples from Surtsey. **A**) Satellite image of Surtsey (NASA Earth Observatory). **B**) Digital model of Surtsey topography made by scanning laser altimeter swaths collected on 22 July 1998 (NASA GSFC). **C**) Surtsey from the air. **D**) Emergence of the island of Surtsey with eruption plume (NOAA). **E**) Surtsey on November 30, 1963, 16 days after the beginning of the eruption (NOAA). C, D & E) Commemorative stamps depicting the eruption and building of Surtsey (inset stamps courtesy of Josef Schalch mountainstamp.com).

had waned during this episode, but it increased again when Surtla stopped, thus demonstrating that Surtla was only derived from a lateral vent of Surtsey.

Another, more permanent, change took place on 1 February 1964, when Surtsey developed a new crater called Surtur Junior on the northwestern flanks of the first cone. The original vent, Surtur I, never resumed activity, but this new vent maintained the Surtseyan effort unabated. By the end of March 1964, the island was 1750m long and 180m high. Surtur Junior had built a cone as high as its predecessor's and was erupting fragments so quickly that the sea could scarcely penetrate the vent. Thus, the typical Surtseyan explosions seemed doomed.

The third effusive and final phase duly began on 4 April 1964. A lava lake welled up the vent, lava fountains rose 50m high, and the first lava flows formed a barrier along the beach. The emissions discharged at such a rate that, by the end of April 1964, one-third of Surtsey was covered with lava, and the island had the armour-plating needed for survival. Surtsey gradually grew into a lava shield.

As Surtsey continued its less spectacular effusions, two further submarine eruptions occurred, most probably from lateral vents on the same fissure as Surtsey itself. On 5 June 1965, the first eruption gave rise to the islet of Syrtlingur, 500m northeast of Surtsey. But its explosions were never so continuous, nor were its accumulations so voluminous as on Surtsey. Thus, Syrtlingur intermittently grew by Surtseyan explosions and was destroyed by the ocean throughout the summer of 1965. It reached its maximum height of 65m on 15 September 1965, and was last seen on 17 October, at the start of a week of bad weather. On 24 October, Syrtlingur had stopped erupting and the waves had eroded away the 2 million cubic metres of fragments that it had expelled above sea level.

The second submarine eruption, which began on 28 December 1965, formed the islet called Jolnir, about 500m southwest of Surtsey. Again it had neither the explosive persistence nor the volume of fragments to survive for long. Like Syrtlingur, Jolnir appeared after every major emission of fragments and disappeared again when marine erosion gained the upper hand. Once, on 10 August 1966, Jolnir reached 60m high, but it disappeared altogether on 20 September 1966. In the meantime, on 19 August 1966, Surtsey had begun its last episode of effusive activity, which continued at a decreasing rate until the eruption stopped in July 1967.

By then, more than half of the island's area of 2.5km2 was covered with lava flows, with a total volume of almost 300 million cubic metres. However, the other eruptions formed shoals, which in time will form the supporting pillars for new eruptions on nearby fissures. Surtsey's eruption generated just over one cubic kilometre of volcanic materials, of which only 10 per cent formed the island above sea level. Surtsey demonstrated that shield volcanoes could be formed within a single eruptive episode, but it also confirmed that no clear-cut distinctions could be made between fissure eruptions and central eruptions, for Surtsey evolved from the first to the second, and eventually formed a circular shield. However, throughout the eruption the type of magma, its chemical composition and its gas content remained the same.

The eruption on Heimaey in 1973

Covering an area of only 16km2, Heimaey is the largest and also the oldest of the Vestmann Islands, or Vestmannaeyar, and Surtsey is now the second largest. But, in the shallow waters around them, there are many submarine volcanoes that either did not erupt enough material to reach sea level, or were too fragile to withstand marine attack. The northern part of Heimaey forms a long volcanic ridge, but the rest of the island is a low plain, where the cinder cones of Saefell and Helgafell erupted in the south about 5000 years ago. The population of 5300 is concentrated in the town of Heimaey, which is situated 1km due north of Helgafell (Fig. 6.34). In 1973 an eruption here was marked by a frantic and heroic effort to save the town of Heimaey. The eruption produced a dramatic new volcanic edifice, 'Eldfell', and lava flows added to the shoreline and nearly cut off the natural harbour (see Fig. 6.34)

The eruption in 1973 was preceded by some 30 hours of tremors of increasing intensity and frequency. At 01.55 on 23 January 1973, lava fountains burst more than 200m into the air from a fissure 1800m long on the eastern slopes of Helgafell. On 24 January, the lava fountains concentrated upon the northern end of the fissure, but they subsided when the first ash, lapilli and bombs were expelled and olivine basalts started to flow eastwards to the sea. Next day, explosions increased at the northern end of the fissure and rapidly built a cone and showered ash on the town. On 26 January, hot ash set fire to some houses and crushed others under its weight in its eastern parts. The fissure lengthened to 3km, enabling the lavas to flow in a more threatening northward direction towards the town. Both flows and fragments were discharged at high rates. On 30 January, the cone was already 185m high and the lava flows had added 1km2 to the island. By mid-February, 110 million cubic metres of lava covered an area of 3km2. On 19 February, the northern crater wall collapsed and the breach helped direct yet more lava towards the town and its harbour. It was saved only by human intervention and the end of the eruption in April 1973. In March, the discharge rates of the lava fell to only 10m3 per second, a tenth of the original rates of emission.

The disaster was well handled because the best use was made of contingency planning and the resources available for evacuation. This was the kind of eruption where human initiative was likely to be most successful, because the emissions were limited to a moderate

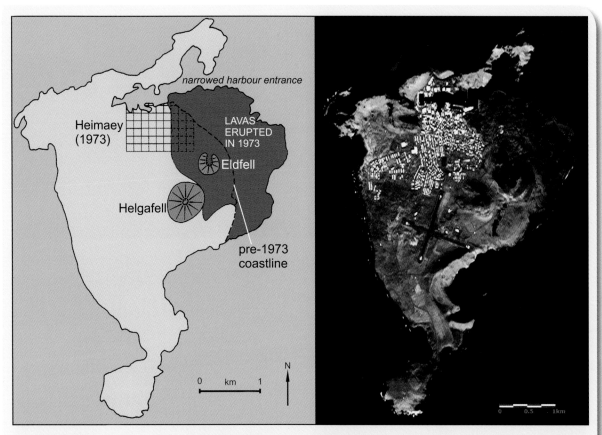

Figure 6.34 Map and satellite view of Heimaey. The extent of the new land developed during the eruptions, the original coastline, and the volcanoes of Helgfell and Eldfell are highlighted (Satellite Image – USGS).

spread of fragments and to rather slowly moving lavas. A seven-person committee was set up to direct operations. The whole population of the island was evacuated to temporary accommodation on the mainland within 7 hours, using the 77 fishing boats sheltering in the harbour from a storm that, luckily, had just abated. The sick and the old were moved by air. The cattle and the cars, US$1 million worth of deep-frozen fish that had been stored in the harbour, the money from the bank, and municipal documents were transferred mainly to Reykjavík. The national bank released cash for the evacuees, the government donated US$2 million and increased taxes to pay for the disaster.

From the onset of the eruption, only police and others with technical ability remained on the island, with suitable vehicles and equipment. They faced three dangers: poisonous gas, falling ash and lava flows. Invisible dense carbon dioxide and also hydrogen sulphide, which can at least be detected by its smell of rotten eggs, lurked in the cellars for days, but, once

their presence was known, they could be neutralized by gas masks. But it was carbon dioxide that caused the only fatality during the eruption.

Falling ash caused the greatest danger during the first weeks of the eruption, when the explosions were at their most powerful. Bombs damaged rooftops and broke windows, and hot ash often set fire to the wooden houses. The weight of the ash also caused roofs to collapse and was even thick enough to bury houses in the eastern part of the town, nearest the volcano. Boarding up windows with metal sheets, and shovelling ash from rooftops, could not keep pace with the falls, although volunteers worked well into March. In the end, the ash burned and buried over 80 houses, three-quarters of them being lost in the first week.

The lavas started to flow towards Heimaey after the fissure lengthened on 26 January and part of the crater wall collapsed on 19 February. They threatened the eastern part of the town and the entrance to its harbour, as well as the electric cables and the

Figure 6.35 Examples from Heimaey. **A**) Partly excavated house that was buried under ash from the 1973 eruption with Eldfell in the distance (inset photo shows detail of house). **B**) Pillars can be found all around town, which show the height to which that part of the town was buried under ash; this particular point was one and a half Dougal's height. **C**) Lava flows at the sea edge with Eldfell in the background (photo by Diego Delso, Wikimedia Commons, Licence CC-BY-SA 4.0). **D**) View to Heimaey from Eldfell volcano (Shuuterstock/David Varga. C & D) Inset commemorative stamps depicting the 1973 eruptions (inset stamps courtesy of Josef Schalch mountainstamp.com).

freshwater conduit from the mainland. The threat to the harbour was all the more important because the town accounted for over an eighth of Iceland's fish exports. There were, therefore, important financial and humanitarian reasons for trying to stop the advancing flow. In early February, volunteers began pouring sea water onto the lava's snout to solidify it and slow down its advance. When this was successful, in early March, the government sent out a ship whose pumps dowsed the lavas much more effectively. Powerful water pumps were then also despatched from the USA and installed on land. Soon, 20,000m3 of lava were being congealed every hour, stopping the advance of the flow. At the same time, the ash swept from the streets was compacted into a barrier around the perimeter of the flow to prevent its spreading sideways and westwards into the town. Moreover, the lavas that reached the harbour narrowed the entrance to 30m and thereby provided better protection for the fishing fleet than ever before.

But such methods could only be successful during a relatively docile eruption, where the water supply was plentiful and the lavas slow moving. The financial and human costs of seawater pumping and barrier building were high. In spite of all this endeavour, the electricity supply cable was cut, three of the five fish factories were destroyed, and the lava flow claimed over 300 houses. Furthermore, the lava flow was stopped only when the eruption was ending and the rates of discharge and advance had already waned considerably. Most of the town was rebuilt, and the population has reached its former total. Seismometers have been installed on the mainland nearby, and continuously recording tiltmeters have been set up on the island to warn of future crises. The island acts as a natural museum of the event, with columns in the streets depicting the thicknesses of ash that have been cleared from the town, and part-buried buildings still on show stand as testament to the eruption. The spectre of Eldfell still looms over the town, and the event is also marked on commemorative stamps (see Fig. 6.35).

Chapter 7

Norway

Introduction

Norway is not known for its volcanoes, and is a somewhat unlikely member of the Volcanoes of Europe team. However, there is volcanic activity in its offshore territories with subaerial eruptions as recent as 1985, and with ongoing hydrothermal activity in an underwater discovery, it warrants attention. The islands of Jan Mayen, Spitsbergen, and the underwater Norwegian waters between them (Fig. 7.1) are the focus for this short but intriguing chapter.

Jan Mayen

The lonely Norwegian island of Jan Mayen lies 650km northeast of Iceland in the North Atlantic Ocean at 71°N and 8°W (see Fig. 7.1). It is wholly volcanic in origin, covers an area of 320km², runs 54km from northeast to southwest, and is mostly less than 10km wide (Fig. 7.2). Historical records extend back about 400 years, but they are incomplete and scanty, for this grim, cold, isolated, inhospitable and often icebound island has supported settlement only rarely, and it now acts as

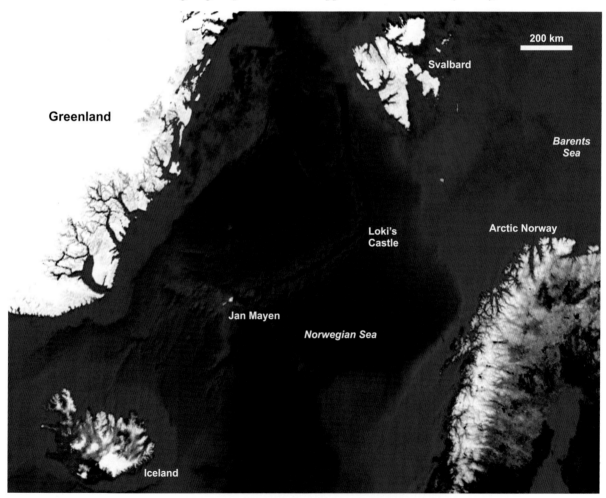

Figure 7.1 Satellite map of North Atlantic (Norwegian and Barents Sea area) highlighting location of Jan Mayen, Svalbard, and the submarine location of Loki's Castle (NASA WorldWind)

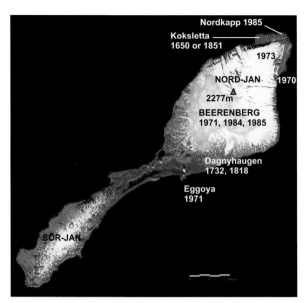

Figure 7.2 Satellite location map of Jan Mayen highlighting historic eruptions (map: USGS).

a weather and navigation station. The island has two distinct parts, each about 25km long, that are offset from each other by an isthmus. Both are dominated by fissures trending from northeast to southwest. The narrow southwestern area, Sör-Jan, is a hilly range rising to 750m above sea level, where fissure eruptions have formed a series of aligned trachytic domes and many cinder cones, from which recent lava flows have spread down to the coast. The northeastern part of the island, Nord-Jan, is wider, higher and more spectacular, for it is dominated by the basaltic stratovolcano, Beerenberg, the northernmost active volcano on land in the world. Beerenberg rises from 3000m below sea level to 2277m above sea level, its summit crowned by an ice cap that radiates glaciers down towards the coast (see Fig. 7.3). The central crater, Sentralkrateret, more than 1km across and 300m deep, feeds the most powerful of these glaciers, Weyprechtbreen. Beerenberg is often

Figure 7.3 A) The snow-capped Beerenburg volcano viewed from Egg-oeja, a peninsula on the west coast of Jan Mayen photo by Hannes Grobe; **B)** showing layers of volcanic ash (photo Alfred Wegener Institute – Creative Commons).

shrouded in fog or storm clouds, but in clear interludes the glistening ice and snow form a stunning contrast with the ice-free areas, which are often covered by fresh black and red cinder cones and lava flows erupted from fissures on its flanks. In spite of the presence of the great stratovolcano, it is the fissure eruptions that have dictated the relief and outline of Jan Mayen, and even Beerenberg itself is aligned in the predominant northeast–southwest direction. The same trends are also followed by the volcanic Stimen bank, 15km long and less than 100m deep, in the ocean southwest of Sör-Jan. The form of Jan Mayen is thus closely related to major regional tectonic features.

Jan Mayen lies near a major offset of the Mid-Atlantic Ridge. The Kolbeinsey Ridge forms its main mid-ocean spine stretching 650km north-northeastwards from Iceland. It is then offset 200km to the east before resuming its course towards Spitzbergen as the Mohns Ridge. The line of the offset is marked by the Jan Mayen fracture zone, generated by a sheer displacement that forms a continuous linear depression more than 2000m deep across the ocean floor. Jan Mayen rises where the fracture zone joins the Mohns Ridge, where enough lava could reach the surface to build up an island. However, the tectonic situation of Jan Mayen is more complex than this basic background to its volcanic activity would, perhaps, imply. It may also have grown up above a hotspot, probably now centred beneath the Eggvin bank, 150km west of Jan Mayen, near the northern end of the Kolbeinsey Ridge.

From the scarce evidence available, eruptions on Jan Mayen seem to be separated by about a century of rest. Although eruption rates could have been higher in the past, the rate of volcanic activity in historic time has been low in Jan Mayen, and the average eruption has produced only about 0.07km^3 of lava. Some 5.35km^3 of lava has erupted altogether during the past 10,000 years or so.

Many of the Jan Mayen lavas are alkali-olivine basalts of relatively high potash content. Ankaramites and magnesium-rich basalts are common on Beerenberg; more evolved basalts, associated with trachytes, are more characteristic of Sör-Jan. Sör-Jan is the oldest part of Jan Mayen, forming a range of aligned cinder cones and lava flows composed of ankaramitic basalts accompanied by several trachytic domes. Volcanic activity may have begun in the southwest and shifted northeastwards.

Beerenberg clearly represents the modern volcanic climax in Jan Mayen. The rocks are predominantly alkali-olivine basalts. Beerenberg grew up in four distinct phases, each of which is reflected in its morphological characteristics. Little is known about the first – submarine and longest – phase in its growth. It undoubtedly expelled the most lava, for the plinth on which the volcano stands extends southwestwards beneath the rest of Jan Mayen and probably includes some of the Jan Mayen Ridge as well. This submarine activity continued in a northeasterly direction until a phase of vigorous Surtseyan eruptions, now represented by basal tuffs, heralded the emergence of the island about 500,000 years ago.

The second phase of activity formed the broad basal shield of Beerenberg, which has a diameter at sea level of 15–24km. It is now represented by the various layers of the Kapp Muyen group, where the ankaramitic lavas lie between various glacial beds. This second phase had three unequal parts. At first, basaltic emissions from a central vent gradually built up the shield to a height of about 750m. This phase was interrupted by an explosive episode that expelled the Havhestberget formation, an ashflow of basaltic pumice that was directed mainly to the southwest, where it now covers glaciated older lavas in a wide plateau leading towards Sör-Jan. The last episode marked a return to effusive conditions, which built the rest of the basal shield. Innumerable fluid flows of ankaramitic basalts, usually less than 20m thick, radiated from the central vent, and together they compose the Nordvestkapp formation. This is the most voluminous formation in the whole of Beerenberg, and the hub of the shield eventually rose to a height of about 1500m. When these eruptions waned, erosion then carved valleys into the flanks of the shield.

The renewal of volcanic activity in the third phase was marked by a significant change. More viscous eruptions rapidly formed the steep-sided lava cone, about 750m high and 5km across, which lies like an enormous sand-castle on top of the basal shield. The formation of the main cone of the stratovolcano and its chief crater, Sentralkrateret, was probably completed about 6000–7000 years ago. There then followed a period of fluvial, glacial and marine erosion forming *barrancos*, small high-level glacial corries and steep cliffs. It was probably at the end of this third phase that displacements formed the great fault cliff that bounds the northeastern flanks of Beerenberg.

The fourth and latest phase of activity on Beerenberg during the past 6000 years or so has been dominated on its outer flanks by fissure eruptions that created several small cinder cones and widespread lava flows. These fissures trend along the axis of Jan Mayen and curve through the central cone of Beerenberg. The sequence of events has been established in relation to marine platforms around the perimeter of the stratovolcano. About 4000–5000 years ago, the chief fissure eruptions formed the Tromosryggen Ridge of cinder cones and flows, which is 300m high and 6km long. The lavas cascaded down the newly formed fault cliff on the northeastern end of Beerenberg and extended in lava deltas at its foot. These features were subsequently eroded by glaciers and cliffed by the waves. A similar episode of fissure eruptions later gave rise both to the Sarskrateret cone, in the far northeast of the island, and to the Koksletta lavas in the north. These lavas flowed down cliffs that had already been carved into the Tromosryggen basalts; they spread out into a broad platform, 4km long and 1km wide, at the northern tip of Jan Mayen. They may have erupted 2500–3500 years ago. On the other hand, they are so remarkably unweathered that they could have erupted even as recently as 1820–82. For example, the Koksletta lava platform was mapped in 1882, but it was not marked on Scoresby's quite accurate map in 1820, perhaps for the good reason that it did not then exist. It has been suggested that the Koksletta eruption could have taken place in 1851.

This fourth phase of eruptions has certainly continued into historical times, but, even then, isolation, frequent low cloud and the long spells without permanent settlement have combined to make observations of eruptions unreliable, intermittent and scanty. No activity appears to have taken place in Sör-Jan. On Beerenberg, an eruption may have occurred at Koksletta in 1650, but any resulting forms have not been identified. But the results of the more reliably dated eruptions in 1732 and 1818 can be seen in the fresh, unglaciated lavas and cones at Dagnyhaugen, low on the southwestern flanks of Beerenberg. The opposite northeastern flanks, near the Nordkapp, probably saw activity about 1850, and certainly in 1970, 1971, 1973 and 1985. All the eruptions developed near sea level on fissures trending northeast–southwest. However, the central crater of Beerenberg has been seen to erupt only twice, briefly emitting steam in March 1984 and April 1985, although bad weather might have hidden other activity. In 1970, for example, a severe storm hid the eruption for three days from the weather station situated only 30km away.

A fissure curving 6km from sea level to a height of 1000m on Beerenberg gave rise to the eruption in 1970. It opened on 18 September 1970 after a magnitude 5.1 earthquake occurred at a depth of 30km. Activity was first noticed on 20 September from a passing aircraft, when a high ash cloud had developed and the first lavas were probably emitted. Eruptions eventually concentrated on five vents, just over 1km apart, which produced tall lava fountains, spatter cones and lava flows cascading seawards. Almost incessant earthquakes of less than magnitude 4.0 (often 800 a day) were registered over the next month. Lava-flow emissions ended in late October 1970, but intermittent explosions of fragments continued until January 1971 or possibly until June 1971. Some ash and steam may also have been expelled after June 1971 from the summit crater of Beerenberg, and from the Eggoya satellite cinder cone low on its southwestern flanks, and fumaroles were observed in both places in 1974. In all, about 0.5km^3 of alkali-olivine basalt was emitted, adding about 4km^2 of new land to northeastern Jan Mayen.

The latest eruption was in January 1985 (Fig. 7.4) from a fissure prolonging the great fault cliff that dominates the relief of the northeastern arm of Jan

The eruption in January 1985

Several strong earthquakes approaching magnitude 5.0 on the Richter scale shook the island on 4–6 January 1985. The eruption began in the afternoon of 6 January, and lasted some 35–40 hours before it ended on 8 January. It occurred on a small fissure, 1km long, at the far northeastern corner of Jan Mayen, extending from sea level up to 200m on the north side of the Sarskrateret cone. Three main craters emitted gas and spatter, and the wind distributed fine ash from a dark-brown column that rose 1km above the vents. About 7 million cubic metres of volcanic material (about one-tenth of an average Jan Mayen eruption) was given out. Lavas also gushed vigorously out northwards into the sea and eventually added about 0.25km^2 to the land area of Jan Mayen.

Figure 7.4 **7** January 1985, the second day of an eruption on the northern flanks of Beerenberg, the eruption column turning towards the east (photo by Pall Imsland).

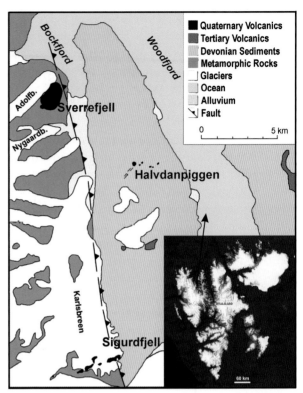

Figure 7.5 Location satellite map and geology of northern Spitsbergen (geology map courtesy of Allan Treiman; satellite map NASA WorldWind).

Mayen. It represents a reactivation of the fault that has already thrown down the lower northern flanks of Beerenberg. Also, the earthquakes accompanying the eruptions were unusually strong and deep, and seem to have been generated by regional tectonic displacements related to strike-slip shear motion in the Jan Mayen fracture zone. Thus, the magma probably rose up a line of weakness where stresses were released on the southern edge of the Jan Mayen fracture zone. Thus, the eruptions were caused by leaking tectonic fractures, and were not truly related to the main Jan Mayen system. In future, Beerenberg and Jan Mayen can therefore be expected to grow further to the northeast.

Svalbard archipelago – the volcanoes of Spitsbergen

In the Arctic Ocean north of Norway, about halfway to the North Pole, is a group of spectacular islands called Svalbard. They cover an area of some 61,000km^2, and are home to polar bears, scientists, hardy coalminers, and a landscape to die for. The main island is called Spitsbergen and makes up more than half of the archipelago (Fig. 7.5). It is renowned for its geology and wilderness, and many a field trip is taken to look at stunning outcrops, mainly of the Palaeozoic and Tertiary rocks there. There are three main aspects of its geology: Precambrian–Silurian basement, Late Palaeozoic–Cenozoic sediments, and relatively unconsolidated surficial deposits from the Quaternary period. Where not covered by glaciers the outcrops can be up to 100 per cent of the surface, and some complete sections through Earth's history can be realized.

The area is not specifically known for the presence of young volcanoes, but rather for lavas and sills dating from way back in the Cretaceous. However, in the northern reaches of Spitsbergen, a number of young volcanic remnants make up broadly three areas: Sverrefjellet, Halvdanpiggen, and Sigurdfjellet (all named after old Norse kings) (Fig. 7.5). These volcanic centres have erupted between the last glacial periods some 1Ma to 100,000 years ago, and are seen as fresh, dark volcanic rocks amongst the glacially scarred landscape around Bockfjorden, and a little south of there, aligned along a major fault structure (the Breibogen Fault, see Figure 7.5). They are known collectively as the Bockfjorden Volcanic Complex.

Sverrefjellet

The volcanic cone-shape of Sverrefjellet somewhat resembles that of a present-day volcano (see Fig. 7.6), yet it has been partly reshaped by glacial activity. It rises up to 506m from the fjord and is the largest of the Quaternary volcanic units in Spitsbergen. It is made up predominantly of pyroclastic/hydroclastic basaltic

Figure 7.6 Photos from Sverrefjellet. **A**) The eroded volcanic centre of Sverrefjellet gives the impression of a fairly recent volcano. **B**) Close up of pillow lobes, some of which contain olivine-rich xenoliths. (Photos courtesy of Allan Treiman.)

rocks with some lava and also intrusions. Inspection of the exposed sections reveals spectacular pillows, some with many mantle xenoliths inside (e.g. Fig. 7.6b), suggesting that the eruptions occurred partly with ice cover present. The presence of mantle xenoliths points to a deep source for these volcanoes, which must have made their way up through the crustal weaknesses presented by the large faults (e.g. the Breibogen Fault). Much of the remains of the volcanic material is composed of frost-shattered scree, but some lava surfaces can be found amongst the predominance of basaltic pyroclasts. Dykes and pillar-like magma conduits are also seen cutting up through the sequence. Hydrothermal alteration is also found in the core of Sverrefjellet, suggesting, along with the magma conduits, a volcanic vent. Several hot springs occur near Sverrefjellet, some with temperatures up to 24 degrees (pretty hot for Spitsbergen), and travertine/carbonate terraces also attest to hydrothermal waters.

Halvdanpiggen

Along the peninsula between Bockfjorden and Woodfjorden, a number of volcanic pipes are found, the largest of which is Halvdanpiggen (some 200m across), forming a prominent raised plug on the landscape (elevated at 580–834m above sea level). This is made up mostly of volcanic breccia with basaltic pyroclastic rocks containing xenoliths from high pressure

regions beneath. There is a radially fanning columnar intruded basalt in its core, and surrounding pyroclastic remnants suggest that its original footprint may have been closer to 400m in diameter. A number of associated satellite vents/intrusions occur within a two-kilometre radius of the site. Again, the high-pressure xenoliths that are being brought up with the volcanics suggest a deep-seated origin, and features like Halvdanpiggen could represent deeply eroded vents such as diatremes.

Sigurdfjellet

Further south along the Breibogen Fault we find the Sigurdfjellet centre some 20km from Sverrefjellet. This forms a ridge of volcanic material some 200–250m wide more or less aligned normal to the Breibogen Fault over 4.5km, at an elevation of 910–1162m. It consists mainly of primary pyroclastic rocks and small explosion vents, with a small occurrence of pahoehoe lava. It is likely to have a similar origin to the Halvdanpiggen and Sverrefjellet examples, though little information is available.

Undersea volcanic activity – Loki's Castle

South of Svalbard between Norway and Greenland, in an area known as Loki's Castle (Lokes Slott in Norwegian), a discovery in 2005 added another location to the catalogue of volcanoes of Norway. Here is a series of active hydrothermal vents along the Mid-Atlantic Ridge (Fig. 7.7). The field lies in Norwegian waters and is located around 73°33′N 8°09′E, about 300km west of Bear Island and about 600km east of Jan Mayen Island (see Fig. 7.1).

The Loki vent field has been located at around 2400m, and since 2005 the researchers at the University of Bergen have discovered seven underwater vent fields hosted by volcanoes ranging in depths from 100m to 2500m. The southernmost, named 'Seven Sisters', was discovered at the northern part of the Kolbeinsey Ridge at 71N in 2013, and the northernmost was discovered at the Knipovich Ridge at around 77N in 2016. There is much research to do with this relatively new discovery. The hydrothermal vents are hosts to an array of marine life that cling to this undersea warm spot. The material that makes up the deposits is rich in rare metals, and there is the distant possibility that the area could one day emerge as a new volcanic island.

Figure 7.7 Submarine hydrothermal vents from Loki's Castle along the Arctic Mid-Ocean Ridge (images from the Centre for Geobiology, University of Bergen courtesy of Rolf Pedersen).

Chapter 8

France

Introduction

No volcano has erupted in France in historical times, but prehistoric people would have witnessed some of the latest activity. The most recent act in several million years of eruptions occurred about 7600 years ago when powerful explosions formed the crater now occupied by Lac Pavin in the Auvergne. Further eruptions could occur in that province, as seismic and petrologic data suggest that molten magma is still present, especially under the Chaîne des Puys.

During the past 20 million years, eruptions have been concentrated in the Massif Central, in relation to tectonic movements. As a result, a great variety of rifting and volcanic landforms dominate the scenery of Auvergne and Velay, in the centre and southeast of this vast upland. Thus, the youthful Chaîne de Puys has scores of cinder cones, lava flows, and majestic domes that stand high above the Limagne Rift fault. Also, the Mont-Dore and Cantal massifs are stratovolcanoes forming the highest peaks in the centre of the region. Widespread basaltic plateaux are formed by the Cézallier, Aubrac and Devès. In the Velay, lava plateaux predominate in the west and the extruded *sucs* (phonolite domes) in the east. More sporadic activity also occurred in Bourgogne and Forez in the north, in Ardèche in the southeast, and in a zone stretching across the Causses to the Escandorgue Chain in Languedoc. There are volcanic lakes, such as Lac d'Aydat, which have been impounded by lava flows, and maars formed by explosions, such as the Narse d'Espinasse, Gour de Tazenat and the Lac du Bouchet. A generalized map of the tectonic and volcanic areas of France is given in Fig. 8.1.

Older lava flows have been set in inverted relief by the removal of weaker sediments around them. The type example of this is the Montagne de la Serre on the Limagne fault, which preserves a 3.4 million-year-old lava filled valley over 10km from vent to lava toe. In the Velay and in the Limagne Rift, erosion has reduced all flows over 2 million years old to ridges, buttes or necks that protect ancient villages or form the plinths of monuments that make Le Puy-en-Velay, for instance, one of the most spectacular towns in France. The plateau formed by an ancient lava lake at Gergovie made an ideal site for a Gaulish defensive site and forms a natural viewpoint for the Montagne de la Serre (Fig. 8.2). In the far southeast, the Coiron lavas protected a spur, like a great paw, that juts out from the Massif Central to threaten the Rhône Valley.

Volcanic activity has not migrated clearly from one area to another and, for example, some of the oldest and youngest volcanic features in the Massif Central are both found within sight of the Puy de Dôme. The volcanoes all stand upon a rifted basement with horsts and grabens, separated by pronounced escarpments. This is part of the Western European Rift that extends from the Massif Central, through the Rhine Graben to the Egger in Bohemia. The rift started to form about 40 million years ago, during the height of the Alpine compression. At this time the volcanism was very sparse and the basins were filled by sedimentary successions near sea level. Only as the rift stretched further and thinned the lithosphere, did volcanism begin. Just after this phase, hot asthenosphere pushed out from the thickening and sinking Alps came under the area, and caused even greater lithospheric thinning, which provided the melting to produce the main volcanic phase of the Cantal, Mont-Dore volcanoes, and which continues today with the Chaîne des Puys.

Uplift due to the thermal erosion of the lithosphere has caused surface erosion. This has incised the area markedly during the past 2 million years, so that streams and glaciers eroded deep into the main volcanoes, both as they grew up and after they became extinct: the central core of Mont-Dore was gutted and deep gorges radiated from the Cantal stratovolcano. They developed a labyrinthine stratigraphy of inverted relief as eruptions followed erosion, filling up valleys, and then further erosion cut new valleys, that were in turn filled up. The most recent theatre of activity in the area is the Chaîne

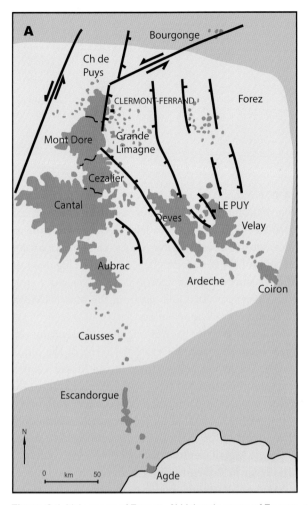

Figure 8.1 Volcanoes of France. **A)** Volcanic areas of France, along with major structural features (map with specific addition from Ben van Wyk de Vries). **B)** False colour topography map of broader area (NASA).

Figure 8.2 The Montagne de la Serre is an extraordinary ridge made of a 3.4 million years old lava flow, raised into inverted relief by the erosion of the Limagne rift sediments during the Quaternary doming of the Massif Central. The Serre ridge is bracketed by two young lava flows, notably the 'la Vache' flow, which is the youngest of the Chopine des Puys. **A)** Photograph: a view of the Serre from Gergovie plateau. **B)** Shaded relief map showing the Serre pinned between younger lavas (photo and interpreted topography map courtesy of Ben van Wyk de Vries).

des Puys, whose fresh and clear-cut features culminate in one of the most distinctive and iconic mountains of European volcanism, the Puy de Dôme.

The eruptive climaxes of the various areas of activity began with the Cantal, and the Aubrac and Cézallier plateaux, between 9 and 7 million years ago. They continued with those in Velay from about 8 to 6 million years ago. The most voluminous eruptions of the Mont-Dore

Figure 8.3 Puy de Sancy: at 1886m it is the second highest peak in the Massif Central in France (photo courtesy of Ben van Wyk de Vries).

occurred about 3 million years ago, with the Grande Nappe Ignimbrite, but eruptions continued until about 200,000 years ago on the Sancy volcano (Fig. 8.3), and the most recent eruptions have occurred on its very flanks with Lac Pavin. The Devès plateau erupted mainly about 2 million years ago, but the apogee of the Chaîne des Puys took place only 14,000 to 8000 years ago. The intervals of repose were undoubtedly much longer than the eruption durations. Thus, the total number of active years was small in comparison to the millions of years encompassed by the whole eruptive sequence.

Most of the lavas erupted in the Massif Central are basanites and alkaline trachybasalts. They occur in the older eruptions but are also very common in the *planèzes* of Mont-Dore and the Cantal, as well as the lava plateaux of Aubrac and the Cézallier. These lavas contain abundant olivine, and mantle xenoliths are common.

More differentiated lavas extend from trachybasalts to trachyandesites. They have a plethora of local names in the Mont-Dore such as doreites, ordanchites, and sancyites, which reflect their variable petrography. As the large volcanic centres began to form, trachytes and rhyolites eventually developed in the Mont-Dore, and trachyrhyolites in the Cantal. These formed domes and ignimbrites. Phonolites also developed widely in Velay and from small secondary reservoirs in Mont-Dore. The phonolite plugs of the Sanadoire and Tuillere on the Mont-Dore are good examples. The whole range of these rocks from basalt to trachyte is preserved in the Chaîne des Puys (Fig. 8.4), where individual volcanoes, like the Nugère or Pariou erupted the entire set in one eruption, leading to highly complex monogenetic volcanic landforms.

Aligned vents have produced by far the most numerous and widespread eruptions in the Massif Central. The vast majority are related to eruptive fissures from dykes, and the major faults have permitted the ascent of magma only very rarely. Thus, for example, in the area near Clermont-Ferrand, which is riddled by hundreds

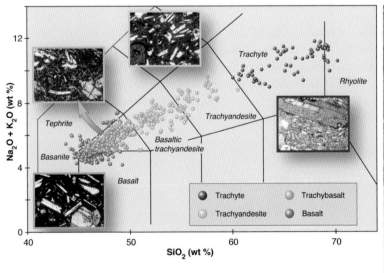

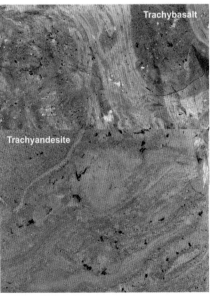

Figure 8.4 The principal rock types in the Chaîne des Puys and a compositional diagram of Silica vs. Alkaline elements, with thin section photographs of the corresponding rock types. The main rocks to see in hand specimen in the field are: basalt and trachybasalt, trachyandesite and trachyte. These are seen in the photos showing rocks from Lemptégy (trachybasalt), the Pariou (trachyandesite). Both seen in the field and in thin section, these go from dark, with low silica, to light-coloured with high silica. (Image – Pierre Boivin pers. comm. (see Boivin et al., 2009), photos courtesy of Ben van Wyk de Vries).

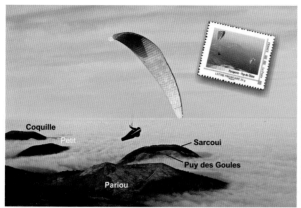

Figure 8.5 A view across the Chaîne des Puys reveals the variety of landforms developed by the protracted eruptions along the feature. The volcanoes in each image are labelled. The area is very popular with hang-gliders, paragliders and microlights. The inset stamp shows paragliding over the Chaîne des Puys (Shutterstock images/Pommeyrol Vincent/ MrCed; inset stamp courtesy of Josef Schalch, mountainstamp.com).

of vents, the major faults forming the border between the Limagne Rift and the plinth of the Chaîne des Puys have produced only two volcanoes. Most of the fractures and fissures trend from north to south, northwest to southeast, or north-northeast to south-southwest. An important feature of these alignments is that eruptions occurred upon them from time to time, but not all at once, and this also has contributed to the variety of landforms (Fig. 8.5).

Closely spaced eruptions along such fissures gave rise to the major basaltic plateaux of the Massif Central. The Cézallier, Aubrac, Devès and Coiron plateaux all trend from northwest to southeast, and their lavas show differentiation towards trachybasalts. These lavas often emerged in such quantities that they swamped the previous relief. But in Devès and Chaîne des Puys, in particular, the fissures have not only formed lava flows but also many maars, and line after line of cinder cones, known as 'Gardes' in the Devès. Aligned volcanic activity reached its climax in eastern Velay and especially in the Chaîne des Puys, where not only basaltic but also more evolved magmas were erupted. Thus, in eastern Velay, viscous phonolitic extrusions formed the steep-sided *sucs* that rise above the basalts; and, in the centre of the Chaîne des Puys, trachytic domes dominate the scenery among the lines of cinder cones (Fig. 8.6).

Figure 8.6 The classic dome of Gerbier de Jonc (photo by PRA – Creative Commons).

The centrally clustered vents of the Massif Central form the smallest category of volcanic features, but also two of its most prominent landforms: the stratovolcanoes of the Mont-Dore and Cantal. However, much smaller stratovolcanoes have been all but hidden in the Boutières of eastern Velay and the Luguet in the Cézallier plateau. These stratovolcanoes developed after an initial phase of more scattered basaltic eruptions. The spatial concentration was associated with magma evolution towards trachyandesitic, followed on the one hand by trachytes and phonolites and on the other by less common rhyolites. The central clusters of vents erupted repeatedly over several million years, piling up material into quite gently sloping stratovolcanoes perhaps 2000m high, and spreading more than 30km in diameter. Violent eruptions then ensued, which expelled vast blankets of trachytic or rhyolitic pumice and ignimbrites, which were followed by the collapse of calderas in both the Mont-Dore and the Cantal. Their last phases comprised caldera infilling and especially lower-flank basaltic eruptions forming the *planèzes* skirting both volcanic massifs. There was thus a late return to the first phase of predominantly effusive activity.

The Massif Central is one of the classic areas of volcanology. It was here that Guettard (1752) first identified the extinct volcanoes of France. It was here, too, that Desmarest (1806) and Poulett-Scrope (1827) demonstrated that rivers did indeed erode their valleys, by comparing the present valleys with the ancient, higher courses fossilized beneath lava flows on the edges of the Limagne Fault. Thus not only do the volcanoes of France represent a diverse range of eruptive styles and rock types; they have also played an important role in the history of Earth sciences.

The Chaîne des Puys

The Chaîne des Puys forms the beautiful western skyline to Clermont-Ferrand, the capital of Auvergne. It stretches from north to south for 40km along the granitic block of the Plateau des Dômes, which rises west of the Limagne Rift along the Limagne Fault (Fig. 8.7). The Chaîne des Puys has a small annexe on the fringes of the Mont-Dore stratovolcano. Most of the puys are 50–200m high, but the Puy de Dôme, rising 500m, provides a majestic exception in the centre of the Chaîne.

The puys, or peaks, arose from almost a hundred vents that provide a natural museum of cinder cones,

lava flows and domes. The eruptions began about about 95,000 years ago and the latest occurred no more than about 8000 years ago, so the puys are so well preserved that they make up for the absence of an active volcano among them. However, the Chaîne is probably not extinct, although there have been no credible accounts of activity during the 2000 years since Vercingetorix repulsed Julius Caesar at Gergovia in 52BC. Possible tephras collected in peat bogs give ages of 4–3000 years, but they may be related to erosive events in the area, as no vents have yet been located. Most eruptions took place along fissures running from north-northeast to south-southwest, but they rarely occurred simultaneously. Many vents gave off lava flows 10–20m thick, which often have rugged aa surfaces, known locally as *cheires*, or stony ground. Each eruption probably continued for a few weeks or a few years. Thus, the eruptions can scarcely have lasted more than a total of 500 years during the whole activity period, and there were long episodes of quiescence when the whole Chaîne would have looked just as calm as it does today. Evidence from the Beaunit Maar at the north end indicates that basaltic eruptions occurred rapidly, while the trachyte eruptions often happened after a protracted period of swelling as magma slowly rose to near the surface. Several swellings are still visible, such as the Petit Puy de Dome and the Grosmanaux. Magma swelling in the Puy de Gouttes volcano caused its collapse in a Mount St Helens type event that sent an avalanche and pyroclastic flows over most of the north of the area. Other swells or 'craters of elevation', as von Buch described them in the late eighteenth century, erupted vertically, such as the one at the Killian Crater.

The eruptions were probably fed by many superposed reservoirs, one about 25–30km deep, and the other lying 5–15km below the surface, where the magmas have undergone two periods of evolution. In spite of this evolution, 70 per cent of the lavas erupted are basaltic, 20 per cent are trachybasalts, and only 10 per cent are trachytes (Fig. 8.8).

The growth of the Chaîne des Puys

Basaltic eruptions had been occurring in the Limagne Rift and on its western shoulders for over 20 million years; however, around 100,000 years ago they began to concentrate on the Limagne Fault shoulder (see Fig. 8.7).

The first Chaîne des Puys *sensu strictu* eruptions began about 95,000 years ago. As the lower reservoir developed,

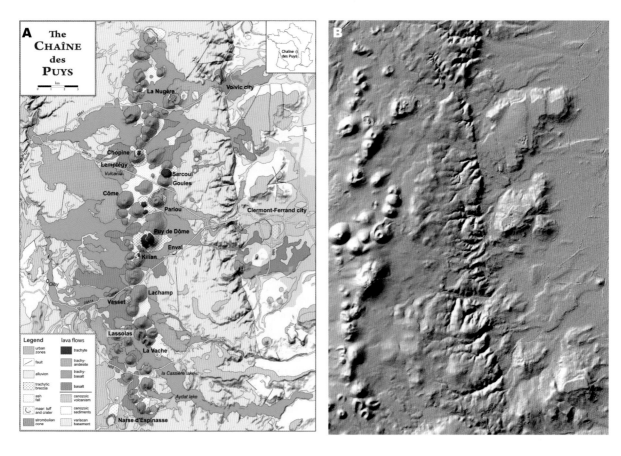

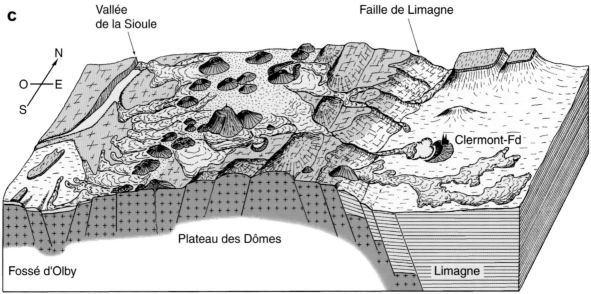

Figure 8.7 The Chaîne des Puys. **A**) Geological map of the Chaîne des Puys with the main units and volcanoes highlighted. **B**) Shaded high-resolution relief map where the landforms of the volcano and the main structures, e.g. Limagne Fault, can be realized (image courtesy of Ben van Wyk de Vries). **C**) Idealized cross-section through the Massif and the Limagne Fault (images A & C courtesy of Pierre Boivin pers comm, see Boivin *et al.*, 2009).

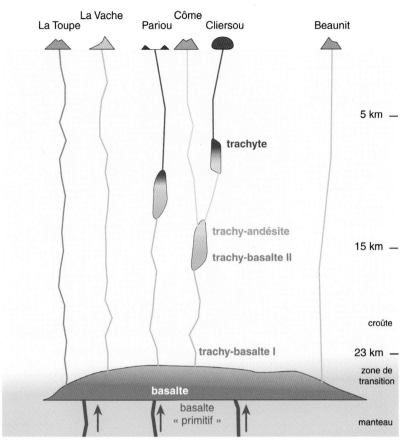

first basalts and then trachybasalts were expelled, and the earliest maars formed. These emissions formed, for instance, the lava flows from the Puy de Chanat, which date from about 90,000 years ago, and the flow in the River Auzon valley, which is about 60,000 years old. The Gravenoire volcano erupted at this time directly on the Limagne Fault Scarp, sending flows into the Clermont area. These are now inverted up to 60m above their present level.

A period of calm then ensued until about 45,000 years ago. The basalts stored in the shallow reservoir gradually evolved to trachybasalts and trachyandesite, which led to the first eruptive climax in the Chaîne about 30,000 years ago. The Puy de Lemptégy, the Puy de Laschamp, the Puy de la Taupe and the Puy des Goules, as well as the now covered or deformed cones around the Puy de Dome, formed. Most of the lava flows went westwards towards the River Sioule, but some from the central part flowed eastwards and entered the Limagne Rift through the rift flank ravines. Maar-forming explosions preceded the eruption of many cinder cones, possibly aided by the permafrost on the tundra during the glacial periods.

A second period of repose then followed, which was broken only by rare eruptions, but during this interval basaltic magma accumulated in the lower reservoir and started the second episode of evolution as trachybasalts began to develop again. From about 15,000 to about 12,000 years ago, these rose into upper reservoirs and occasionally erupted at the surface. During the next few thousand years, trachyandesites and trachytes developed in the upper reservoir. The stage was set for the latest episode in the growth of the Chaîne des Puys, which lasted from 12,000 until about 8000 years ago, and has contributed so much to their character. In the central part of the Chaîne, and for the first time in the area, trachytic magmas extruded the domes of the Puy de Dôme, Sarcoui, Clierzou, the Puy de Vasset and the Puy Chopine. At times the viscous trachyte magma collected under the older volcanics, to create laccoliths that raised up bulges like the Petit Puy de Dome and Grosmanaux. Strong vulcanian eruptions from these trachytic intrusions produced pyroclastic flows from the Puy Vasset, and from the Kilian crater, just south of the Puy de Dôme. The Puy Chopine grew in the collapsed amphitheatre of one such eruption, which destroyed the Puy de

Gouttes. Three marker beds can be followed over much of western Europe from these eruptions. Eruptions of trachyandesites were even less common than trachytes, but they were well represented by the Puy de Pariou and the lava flows from the Puy de la Nugère that provided the Volvic stone used to build the striking cathedral in Clermont-Ferrand.

About 8000 years ago, basaltic eruptions resumed. They came from the least evolved magmas in the history of the Chaîne and seem to have risen straight from the mantle, without halting in either reservoir; they might, therefore, signify the start of yet another period of magma evolution. Their finest expression is in the twin breached cones of the Puy de la Vache and the Puy de Lassolas, which together formed the dark grey flow of the Cheire d'Aydat. The basaltic Puy de Côme could also belong to an earlier episode of leaking of deep magmas. The distribution of some of the main volcanoes in the central part of the Chaîne des Puys is given in Fig. 8.9.

The latest eruptions broke out about 7000 years ago and were exceptional in two respects: they occurred in a small annexe area, due south of the Chaîne, on

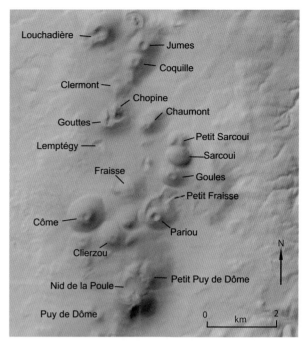

Figure 8.9 False-coloured topography map of the central part of the Chaîne des Puys with volcano locations labelled (image courtesy of Ben van Wyk de Vries).

Meet the scientist – Ben van Wyk de Vries

Ben lives and works in Clermont-Ferrand as a professor of Volcanology. He even has a lava flow in his back garden, and a bedroom view of the Limagne Fault and the Chaîne des Puys. He once lived in a crater in Managua, Nicaragua. He started his volcanology in that country, and France is just an 18-year stop-over on a journey around the world and other planets studying how volcanoes work. Ben works on many aspects of volcanology and is particularly passionate about communicating this to the general public.

Just what is it you like about volcanoes and faults?
The Nicaraguan years got me into the relationship between faults and volcanoes, and while trying to do other things, I keep on getting dragged back into that. This is because volcanoes are really just expressions of plate tectonics at work where there is too much juice (the liquid lava). So being a volcanologist is just being a geologist, but trying to look cool. Or perhaps better: volcanoes are the ambassadors for the rest of geology to the general public.

How long have you been working in the Massif Central?
About 18 years, now. I came to Clermont-Ferrand in 1998 – well actually, three years before on my honeymoon – but somehow missed the volcanoes on that trip. When I stayed, the volcanoes were a little forgotten, but bit by bit I began to realize that I was in a wonderland for a geologist.

What has been your most exciting discovery about it?
Craters of Elevation – big bulges in the volcanoes where magma has arched up the surface. It's actually an old observation from von Buch, a German naturalist who was responsible for some amazing eighteenth or nineteenth

century geological observations (Fig. 8.10). Unfortunately his theory of 'Hebungs Krateren' (craters of elevation) got taken too far; the idea was forgotten. I rediscovered his ideas on the Petit Puy de Dome, and the theory is now solidly back in place as a serious scientific feature.

What is the most important thing we still need to find out?
What the Chaîne des Puys is doing now. We are sure the volcanic field is active, but why are there no eruptions? When will they happen? There are really two main possibilities: one is that the Chaîne is dying down and eruptions have finished with the flare-up that produced the characteristic domes and the la Vache lava; the other is that the system is building up to produce more trachyte and will develop a stratovolcano, as the Puy de Sancy and Mont-Dore did to the south. That's a million year question to wait for, but there is a small probability that eruptions will resume during our civilization, and then we'd have to adapt.

What inspired you to study volcanoes?
They are the the the essence of geology, the most obvious phenomena in plate tectonics, and great fun to get to know! I wasn't so interested in them to start with, as geology in the UK does not really prepare you for the real thing. But going and seeing and experiencing active faults and volcanoes in Nicaragua hooked me. I tried the oil industry, but despite the money, it was never quite as much fun.

How important do you think the volcanoes of France have been in our understanding of volcanology?
They shaped many of the fundamental ideas in geology and volcanology in the eighteenth and nineteenth centuries and continue to do so, because they are small, highly diverse, and are expressed in a clear geological system, where you can see the tectonic and surface processes that shape much of our planet.

Any other comments about the volcanoes of Europe?
Europe has an amazing variety of volcanoes and a strong geoheritage structure that means they have a very important role in educating the public about Earth sciences. Those of the Chaîne des Puys attract hundreds of thousands of visitors each year, which, put together with those who visit the Eifel, Bohemian Paradise, Hungarian volcanoes and the many other sites, makes them a huge resource for informing the general public about Earth processes. We are trying to build a Pan European volcano route, to link all these and maximize on the interchanges.

Puy La Goutte from the East.   Puy Chopine.   Débris.

Figure 8.10 A classic sketch depicting craters of elevation by Scrope (1858) after von Buch. The intrusion of magma produces a raised fold with outer fractures forced up by the inflation of magma from underneath (image courtesy of Ben van Wyk de Vries).

The Puy de Dôme

The Puy de Dôme is, indeed, a volcano that stands out in a crowd. Although this striking individual reaches only 1465m above sea level, it soars over 200m above any rival in the Chaîne des Puys, is a landmark for over 100km around, and is the very symbol of Auvergne. Initially it was thought to be two volcanoes joined together side-by-side, which arose in succession from the same vent about 10,000 years ago. However, more recently paleomagnetic work has shown that there is only one main growth period, and that the volcano was tilted and split during a later intrusion, creating the Petit Puy de Dome. During the split a phreatic eruption left a thin deposit on top of the Puy de Dome.

The smoother eastern side of the Puy de Dôme contrasts clearly with its more rugged western flanks. This is also due to the 18 degrees of tilt that steepened the west side and caused parts of that flank to slip off (Fig. 8.11).

Figure 8.11 Aerial view across the Puy de Dôme. **A)** The classic outline of the Puy de Dôme. **B)** Close-up view of the Puy de Dôme and the Petit Puy de Dôme (photos courtesy of Ben van Wyk de Vries).

the eastern flanks of the Mont-Dore massif, and they resulted from a very rapid magma evolution that expelled materials ranging from basanite to trachyte in perhaps less than 200 years.

Domes

The domes form a small but distinctive group in the centre of the Chaîne, which comprises the Puy de Dôme itself, Sarcoui, Clierzou, the Puy Chopine and the Puy de Vasset. They are composed of pale-grey trachyte rich in silica, which resembles, and is almost as light as, pumice. They formed when the viscous lava extruded in a bulbous mass, although explosions and collapses sometimes destroyed or interrupted their growth. At least three of the domes – Sarcoui, the Puy Chopine and the Puy de Vasset – were extruded from vents that had previously undergone strong vulcanian or directed blast eruptions.

Sarcoui is the most perfect dome in the Chaîne des Puys, resembling an upturned cauldron that rises 250m above a base 800m across (Fig. 8.12). It is bald and stark, and its pale grey rocks show through its thin grass and forest cover. The remains of a tephra ring on its eastern base show that the eruption began with a strong explosion. The onion-skin layers, formed during the extrusion, can be seen in the quarries excavated in Gallo-Roman times. Small collapses, resulting in pyroclastic density currents from the flanks, have left arcuate niches. The largest of these collapses spread northeastwards for 600m, where secondary explosions formed a rash of rootless cones. Its youthful appearance reveals its age, for Sarcoui formed between 12,000 and 8500 years ago, but it is older than the Puy Chopine, because it is mantled by ash deposits from that eruption that contain distinctive carbonized wood.

The steep, darkly wooded pyramid of the Puy Chopine rises only 160m, but it has one of the most distinctive profiles in the Chaîne des Puys (Fig. 8.13). The Chopine has formed in a collapse crater, like that of Mt. Saint Helens in the USA. The Puy des Gouttes was intruded

Figure 8.12 View of the Petit Sarcoui, the Grand Sarcoui and the Puy des Goules from the air (photo courtesy of Ben van Wyk de Vries).

Figure 8.13 View of the Puy Chopine and Puy des Gouttes from the air (photos courtesy of Ben van Wyk de Vries).

by trachyte that deformed its flanks until it collapsed, uncorking the trachyte and creating a powerful northerly directed blast. The Chopine then formed as an almost solid piston of trachyte extruded from the vent. The lavas dragged blocks of granite and diorite, 15–20m across, from the walls of the vent and pushed them upwards, and these are now largely responsible for its distinctive pointed crest.

Monogenetic cones

More than three-quarters of the puys are scoria cones, and most of them are about 100m high and some 400m across, with craters less than 50m deep. They erupted from vents along en-echelon fissures trending from north-northeast to south-southwest. The younger cones have well-defined craters, and their straight, scarcely altered outer flanks usually lie at 30°, akin to the angle of rest of 32°–35°of their bedded fragments. Each cone was formed by a brisk phase of Strombolian eruptions lasting for a few weeks, a few months, or possibly a few

years. While most cones are still pristine, exceptionally fine cross-sections are available at the Puy de Lemptégy, which has been excavated right to its roots and is now a superb volcano museum. This one small cone has formed a major focus of research into monogenetic volcano formation in the last few years.

The simplest cones result from eruptions of similar intensity from a single vent. One such eruption built the Puy des Goules about 30,000 years ago. It is 150m high and has a crater 200m wide and 40m deep, partly filled by fragments ejected from its younger neighbours. Its sharp crater rim is some 10m higher on the northeastern side, where the southwesterly winds blew the fragments during the eruption. Its slopes are at an angle of 25°, low for a scoria cone, but the cone has no gullies, and is remarkably fresh for its age. When the Sarcoui erupted some 20,000 years later, its lower northern flanks were destroyed.

Multiple eruptions from a single vent often build up larger-than-average cones, but their forms depend upon when the main eruptions took place in the sequence. A large final eruption can bury the earlier cones: thus, the bulky form of the Puy de Mercoeur buries a smaller cone that erupted earlier from the same vent. On the other hand, when the later eruptions give off fewer fragments than their predecessors, a smaller cone then often nestles within the crater of the larger. This sequence happened at the Puy de Côme (Fig. 8.14), where the nested cone is much smaller than the puy as a whole. A circular atrium, about 10m deep, separates the two crater rims, which are both highest on their northeastern sides, where the southwesterly winds blew slightly more fragments as they erupted. The bulk of the Puy de Côme could have formed in several months, or possibly over several years, because it is the largest cinder cone

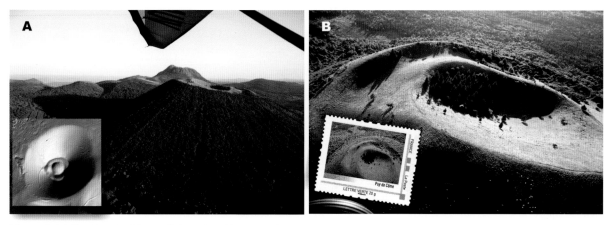

Figure 8.14 The Puy de Côme: **A**) from the side most of its slopes are forested (inset of high-resolution topography); **B**) the nested cone can be discerned on the grass-covered summit (inset of Puy de Côme stamp). (Photos and topography image courtesy of Ben van Wyk de Vries; inset stamp courtesy of Josef Schalch mountainstamp.com).

in the Chaîne, with a height of 250m. Then, as activity waned, a short eruption built the cone nested in the larger crater, perhaps in a matter of days.

While the Puy de Côme is an imposing feature, the bulk of its lavas were erupted from a small lava pond at the western base. This pond was discovered with the recent LiDAR images and has a rim of overflow lavas, lava terraces and a beautiful cliff section of pahoehoe lavas.

When vents are closely spaced along a fissure, the fragments often pile up into a single elongated cone, such as the Puy de Barme, which rises 4km southwest of the Puy de Dôme. The two small craters at its summit operated simultaneously to form an elliptical cone about 15,000–12,000 years ago. Subsequently, a third vent erupted on its southwestern flanks and extended the cone down the fissure. More widely spaced vents can sometimes give rise to twin cones if they function at the same time. The Puy de Jumes and the Puy de Coquille are twin cones, set 600m apart, on the same fissure. Both are composed of basaltic lapilli and cinders and are so close together that their beds intermingled, which confirms that they erupted simultaneously.

Clustered vents form bulky cones, crowned by a group of craters. Four vents built up the Puy de Monchier about 8540 years ago, and four separate craters of similar size are assembled near its summit. On the other hand, eruptions from the central vent built up most of the Puy de la Taupe, but its companion cones are scarcely more than satellites on its flanks, because their vents probably functioned for only a few weeks.

Breached cones

The Chaîne des Puys has over 60 cones that were breached by lava flows. Generally, a whole sector of the cone is missing down to its very base, but all variations are possible from this to just a nick at the top of the crater. For instance, after the Puy de Louchadière had formed, a lava flow welled up inside the crater, destroyed its upper western sector, and spread down beyond the base of the cone. The rest of the crater now forms an armchair-like hollow that probably gave Louchadière its name – 'the armchair'. After the Puy de Chalard had formed, the breaching was even more severe, because a flow not only removed a whole sector of the cone, but was powerful enough to push it bodily some 100m to the northwest, where it now remains as a ridge of cinders. This also happened to the Puy Pariou, where the first original tuff ring was breached, pushing the ring aside and letting out a flood of trachyandesite lava.

Most breaches develop while the cones are erupting, and they are almost always caused by lavas that leave the vent at the same time. This is how the Puy de la Vache and the Puy de Lassolas attained the characteristic shapes that have made them among the best known of all the cones in the Chaîne (Fig. 8.15). Their reddish cinders form two horseshoe masses rising 200m above the lavas that they both erupted to form the Cheire d'Aydat about 8000 years ago. The lavas carried away the scoria that fell upon them as if they were on an escalator, and the sectors where each flow emerged were therefore never constructed. The volcanoes have remained crescent-shaped, like a French croissant, ever since the eruption began.

Figure 8.15 The twin breached cones of the Puy de Lassolas and the Puy de la Vache (from the air with Puy de Dôme labelled behind), began by a powerful subPlinian explosive phase before settling down to make one of the longest lavas in the Chaîne. Their lava flows combine to form the now forested Cheire d'Aydat, which flowed away to the southeast (photo courtesy of Ben van Wyk de Vries).

Cones breached solely by explosions are rare in the Chain of Puys because moderate eruptions prevailed. However, the collapse of the northern sector of the Puy des Gouttes, for instance, was caused by an intrusion of trachyte. On collapse, the trachyte decompressed and exploded, then the rest of the magma rose up as a spine to create the Puy Chopine. In Lemptégy several basaltic intrusions also deformed the flanks, leading to partial collapse and generating small lava flows (Fig. 8.16).

Figure 8.16 Example of faulted parts of Lemptégy above a shallow basaltic intrusion (photo courtesy of Ben van Wyk de Vries).

Complex cones

The Puy de Pariou (Fig. 8.17) grew up in several stages. A first volcano formed a cinder cone about 9000 years ago, but it remains as only a ridge of basaltic cinders at the western base of the volcano, because it was largely destroyed by subsequent phreatomagmatic explosions that formed a tuff ring of the first phase of the Pariou at about 8500 years ago. The tuff ring has a trachyte composition and was filled by trachyandesitic lavas that formed a lava lake, pushed away the northern sector of the tuff ring, left it stranded near the Puy des Goules, and then flowed out eastwards in two long tongues that reached the suburbs of Clermont-Ferrand. The final eruptions then built the 160m-high uppermost cone of the Puy de Pariou. Its pristine crater, which is 90m deep and enclosed within the sharpest rim in the Chaîne, betrays its youth. It is about 8000 years old.

The Puy de Tartaret is composed of two superimposed cinder cones that stand about 90m above the Couze Chambon stream to the extreme south of the Chaîne, often pictured with the Château de Murol (Fig. 8.18). About 27,000 years ago, rising magma met the waters of the Couze, and the resulting phreatomagmatic explosions formed a tuff ring that now emerges from the southern base of the volcano. When the water could no

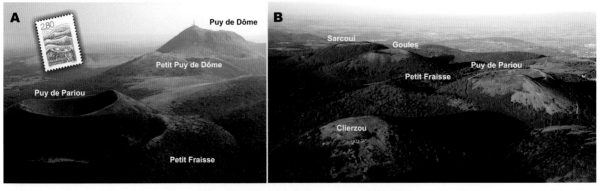

Figure 8.17 Area of the Puy de Pariou. A) The Puy de Pariou and the Puy de Dôme from the air. B) View looking NNE (photos courtesy of Ben van Wyk de Vries, inset stamp courtesy of Josef Schalch mountainstamp.com).

Figure 8.18 Classic view of Le Château de Murol with the Puy de Tartaret (photo by Pierre Andre Leclercq – Creative Commons).

The latest eruptions in France

The latest eruptions in France took place beyond the southern end of the Chaîne des Puys, on the edge of the Mont-Dore stratovolcano, about 10km southwest of Besse-en-Chandesse (Fig. 8.19). They all took place within the space of about a couple of centuries some 7000 years ago. The episode began with the eruption of the breached basanitic cinder cone of the Puy de Montcineyre that held back the Lac de Montcineyre. Thus, as the cone grew up, phreatomagmatic explosions interrupted the Strombolian eruptions whenever the water invaded the vent. Soon afterwards, a shallow hydrovolcanic explosion burst out 4km to the north and formed the maar of Estivadoux and its thin ring of tuffs and fragments of country rock. A little later, 1km to the west of Estivadoux, another Strombolian eruption built up the cinder cone of the Puy de Montchal and also expelled three trachybasaltic flows. The first two flows were small, but the third was 12km long. It filled the maar of Estivadoux and sent out a thin tongue, with a fine blocky surface, that invaded the upper reaches of the Couze Pavin valley. During the short interval that ensued, magmatic differentiation continued apace, for the next and latest eruption was a trachytic explosion on the lower flanks of the Puy de Montchal. Pyroclastic flows reached up to 15km to the southeast, and formed the hole that was then filled by

Lac Pavin – L'épouvantable ('the fearsome') – a name that would have been appropriate for the eruption itself, but in fact it describes the 92m-deep lake and the surrounding dark wooded hollow.

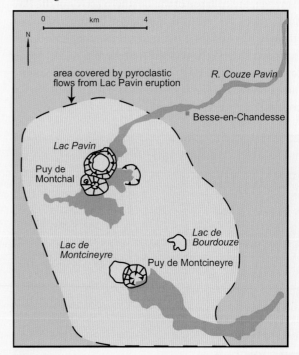

Figure 8.19 Location and features associated with the latest eruptions in France.

longer gain access to the vent, 'dry' Strombolian eruptions built a cinder cone, and a basaltic flow invaded the Couze Chambon valley. Together, they impounded the river and formed Lac Chambon. This Puy de Tartaret lava is one of the finest basaltic flows in Auvergne; it moulds the floor of the Couze Chambon valley for 22km, as far as the Limagne Basin. When the molten lavas flowed into the stream, about 30 rootless conelets, from 5–30m high, developed over the first 2km of the flow. The Puy de Tartaret and its lavas used to hold back the waters of Lac Chambon, but the cone does not now reach its shores. A great landslide later avalanched from the northern slopes of the lake, came to rest between the cone and the waters, and left behind a scar 200m high, called the Saut de la Pucelle ('the virgin's leap'). The cliff of the landslide scar exposes a large section of an old Miocene maar and tuff ring, a debris avalanche layer from the Mont-Dore, and a lava lake with frozen-in convection cells that look like huge onions.

Hydrovolcanic eruptions

Most of the hydrovolcanic eruptions in the Chaîne des Puys formed maars, which are locally known as '*gours*' or, when they are filled by sediments, as '*narses*'. Eruptions into periglacial permafrost probably encouraged much hydrovolcanic activity, and many older maars have now been identified. Many maars are encircled by tuff rings of exploded fragments of country rock and shattered magma. In several cases, also, such as at the Puy de Beaunit, hydrovolcanic eruptions were a prelude to the formation of cinder cones. Otherwise, more than one vent erupted 'dry' and 'wet' magma outcrops onto the Narse d'Espinasse, for instance, which reveal alternations of hydrovolcanic and Strombolian eruptions from a vent close to a stream and from one higher on the valley side.

The most famous maar in Auvergne is the Gour de Tazenat in the north, which is 66m deep and 900m across (Fig. 8.20). Its age is uncertain, but the land surface is remarkably uneroded, sometimes making it difficult to find outcrops. It has been suggested that it was formed between 30,000 and 90,000 years ago, but it may be much younger. It formed when water from the Rochegude stream was met by the rising basaltic magma. The consequent explosions cored out a deep crater but piled up only thin layers of fine fragments of lava and country rock around the northeastern edge of the Gour.

The Narse d'Espinasse (Fig. 8.21) in the south started life as a maar, when magma met the waters of the River

Figure 8.20 View of the Gour de Tazenat (photo by Romary – Creative Commons).

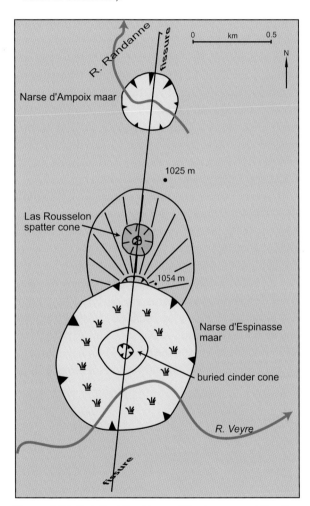

Figure 8.21 Detail of the Narse d'Espinasse and the Puy de l'Enfer.

Veyre and caused hydrovolcanic explosions at the same time as Strombolian eruptions were building the Puy de l'Enfer from a dry vent nearby. When the eruptions stopped, the Puy de l'Enfer formed a crescent of cinders around the maar. The River Veyre continued its course through the maar, filled it almost to the brim with sediments, and formed the flat-floored Narse d'Espinasse.

A smaller maar formed to the north of the same fissure at Ampoix. This has been dated at about 10,700 years. Cores taken from the Narse d'Espinasse have helped established a biostratigraphic chronology for the area.

The lava flows of the Chaîne des Puys

In spite of the tendencies towards magmatic differentiation in the Chaîne des Puys, nearly 90 per cent of the lavas in the flows are basaltic. The flows were fluid enough to enter the valleys draining to the River Sioule and down the Limagne fault. The lavas piled up more thickly in the west. The Cheire de Côme, for instance, has seven superposed flows with a total thickness of 135m, although not all of them necessarily came from the Puy de Côme. Also, the upper reaches of many flows are so thickly covered with ash that it is often hard to distinguish their sources. Thus, for instance, the fine trachyandesitic lava flow known as the Cheire de Mercoeur came, in fact, from the Puy de Mey. This is clearly shown in the new LiDAR data that is available (Fig. 8.22).

However, further away from the cones the lavas almost always develop the rugged maze of pits and pinnacles, flow channels, pressure ridges and lava levees typified by the aa surfaces of the Cheire de Côme, the Cheire d'Aydat, and the trachyandesite flow extending from the Puy de la Nugère to Volvic. The upper parts of many flows descending westwards are blistered by tumuli and occur most clearly on the lavas emitted from the Puy de Pourcharet and the Puy de Combegrasse. Near Nébouzat, the lavas from the Puy de Pourcharet invaded a marshy area, and the steam explosions developed some small rootless cones. The lavas of the Puy de Come were erupted, not from the main crater, but from a pond at its base. The lavas collected here and were then fed through a network of conduits or tubes into the main flow field, which slowly inflated to raise the flow many meters above the surrounding ridges, inverting the relief in one go (see Fig. 8.22).

Elsewhere, the streams have continued to flow underneath or alongside the invading flows. The streams often rapidly eroded the ribbon of lava. Thus, for example, the basaltic flow that threaded into the River Auzon valley some 60,000 years ago has already been cut into several pieces; and, above Royat, the River Tiretaine has already laid bare the central structures of the 40,000-year-old basalts from the Petit Puy de Dôme. Even the most recent flow of the Puy de la Vache has its base exposed at Saint Saturnin by stream action that has cut down 30m in 8000 years.

In some cases the lava flows were voluminous enough to impound the drainage of the valleys that they entered. The temporary damming of the River Sioule at Pontgibaud by the trachybasalts of the Cheire de Côme has been famous ever since a sketch by Scrope in 1827 (Fig. 8.23), and the eminent geomorphologist WM Davis honoured it with an illustration in 1912. On a smaller scale, 10km to the south, the Cheire d'Aumone blocked the Mazayes stream about 14,000 years ago.

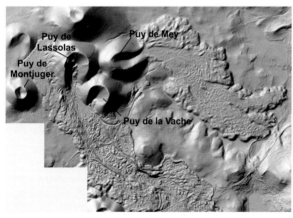

Figure 8.22 LiDAR topographic image with the volcanoes Puy de Montjuger, Puy de Lassolas, Puy de la Vache and Puy de Mey labelled. The ridged lava flow surfaces can be clearly seen in the LiDAR data (image from Ben van Wyk de Vries).

Figure 8.23 Scrope classic sketch of the prismatic lava flow on the banks of the Sioule, near Pontgibaud (image from Ben van Wyk de Vries).

Again, the lake drained and left behind the marshy zone that is called the Etang de Fung.

The most interesting of all the lava flows in the region is the Cheire d'Aydat. These fluid basalts swept in four quick pulsations from the Puy de la Vache and the Puy de Lassolas down into the River Veyre valley, and thence into the Limagne Rift. On its northern side, it impounded a small stream, surged 500m up its valley, and blocked the Lac de Cassière. On its southern side, it held back the Lac d'Aydat when it invaded the River Veyre valley. The river was forced to flow first alongside, and then underneath, the Cheire d'Aydat, before it emerged once again alongside the lavas. The River Veyre is now eroding the weaker sands of the Limagne at Saint Saturnin. This erosion has thus started volcanic inversion of relief, which is so well represented in the ridge of the three-million-year-old Montagne de la Serre, just to the north. In another three million years, the Cheire d'Aydat could be in the same state as its neighbour is at present.

Creating a UNESCO World Heritage site – the Chaîne des Puys–Limagne fault

In July 2018 The Chaîne des Puys–Limagne fault will hopefully become a UNESCO World Heritage site. This site is a geological scale model of continental drift in its first stages, where the Earth's crust begins to thin by faulting and associated volcanism. The Limagne Rift is part of the Western European Rift, which, with the Baikal Rift in Russia, forms one of the two main 'Orogenic Rifts' in the world, each thousands of kilometres long. Unlike the East African rift, which is related to hotspots and an old continent splitting apart, the Western European Rift is due to the suction of sinking lithosphere under the roots of the Alps, part of a convergence of continents. The Limagne Fault is a compact, clearly defined fault scarp, draped by lava flows and surmounted by the Chaîne des Puys. This conjunction of features allows visitors to get a clear feeling for continental separation from the landscape.

Because of the evident relationship between the volcanoes and the faults, linked by inverted relief, the site has been used since the eighteenth century to study landscape evolution, and it is still being used today. This means the site has high scientific value, and needs the highest level of protection. World Heritage is a means of protecting the site, as well as giving it

Figure 8.24 Panoramic view across the Chaîne des Puys and the groups behind the UNESCO World Heritage site bid. (Photo and logos courtesy of Ben van Wyk de Vries; inset stamp courtesy of Josef Schalch mountainstamp.com).

an internationally acknowledged label of quality. The World Heritage project began eight years ago with the decision of one local politician. Since then geologists and local administrators have joined forces with the international scientific community to propose the site.

The actors:

- Cecile Olive-Garcia – the head of the project –a political scientist by training.
- Christine Montoloy – 'The lady of the park' – runs the protection in the park and the interface with the local population, who are vital in keeping the landscape intact.
- Benjamin van Wyk de Vries – the international scientist, lost in the Auvergne for 18 years, has resurfaced to re-establish the Chaîne des Puys and the Fault in modern science.
- Pierre Boivin – the ancient scientist with years of knowledge of the area.
- Yves Michelin – not 'bibendum' but from the other Michelin family – an agricultural and landscape management specialist.
- Aurelie and Vincianne, the foot soldiers who make the link between the project and the general public.

How a Project gets together?

Protecting a volcanic site needs 1) interest, 2) political backing and 3) hard work.

Interest: the Chaîne des Puys has been a site of curiosity and scientific interest for over 250 years, and a tourist site for over 2000 years, since before the Romans made a holiday village at the foot of the Puy de Dome. The site also had religious importance to the Druids; neolithic burials face the volcanoes, and even now the Auvergne inhabitants are religiously attached to the Puy de Dome and its summit spike (an outcry ensued one 1 April, when the local paper suggested it was to be taken down).

Political backing comes from the need to raise the profile of the Auvergne, probably the least visited region of France; yet, as exemplified by the Chaîne des Puys, one of the richest in landscape and geology. Industries that are based in Clermont-Ferrand, like Michelin, need this profile and support the project.

Hard work is involved in organizing the huge proposal to UNESCO. All the geological features need to be described and inventoried, and the science done on the site has to be collected. For the volcanoes and faults this is a formidable exercise in assembling 250 years of studies. A new management plan has also to be worked out, integrating existing protection (Regional Park, National Monument, and even urban planning) with the new plan.

Finally, the site has to be proposed to the national government and then discussed at UNESCO. In 2014 the site reached this last critical stage; the site became a *cause célèbre* in a struggle to uphold the integrity and credibility of geology on the international stage. The forces of geological sciences assembled around the project in an epic struggle to defend the scientific case for the Chaîne des Puys–Limagne fault as a World Heritage site.

Education Sites: Puy de Dome, Gergovie, Vulcania, Lemptégy, Château Montlosier – all these existed before the UNESCO project, but are integrated into and support the project. These are the key educational sites when you visit the Chaîne des Puys, and each provides a different experience:

- Puy de Dome gives you the 360 degrees panorama of the Chaîne and fault;
- Gergovie, the view of the ensemble from a classic battle site;
- Vulcania is the geoscience theme park that takes you through the whole experience of geology;
- Lemptégy gives a view into the centre of a real volcano; and
- the Château Montlosier is the park headquarters.

Figure 8.25 Picturesque photo of the Puy de Dome and spike with snow on (photo courtesy of Ben van Wyk de Vries).

Chapter 9

Germany

Introduction

Famous as the home and birthplace of the maar volcanic crater, the Eifel region in Germany has seen volcanic activity from around 800,000 to 10,000 years ago. The volcanic eruptions in Germany were initially concentrated in a belt, about 600km long, following the valley of the River Rhine, which occurred when a rift developed across Europe as the Atlantic Ocean opened. Most of the early eruptions took place along major faults trending mainly from north to south, or sometimes from northwest to southeast. Successive episodes of rifting were reflected in the activity that reached its climax in areas such as the Odenwald, Spessart, Taunus and Westerwald, Swabia and the Kaiserstuhl more than 15 million years ago.

It was only after a long interval that eruptions resumed in the Eifel Massif, which forms an extensive plateau, about 500m high, in the area between the River Rhine and its tributary the River Moselle (Fig. 9.1), this most recent phase of activity being broadly distributed in two zones (the East and West Eifel). The West Eifel zone, near Daun, runs for 50km from northwest to southeast in a band 20km wide. The East Eifel zone, centred on the Laacher See, near Mayen, is an area with some 50 vents that stretches 35km from the River Rhine in a belt about 25km broad. The West Eifel massif has long been famous for its maars, the circular lakes reaching up to more than 2km in diameter, for which the region has become the type locality. But these beautiful relics of hydrovolcanic eruptions are also accompanied by tuff rings and many cinder cones, lava flows and domes, as well as extensive blankets of pumice. The younger flows, cinders and pumice have been widely quarried, whereas the older materials have weathered to provide a mixture of rich arable and meadow lands and woods surrounding prosperous rural towns.

Volcanic activity began in the Eifel region about 800,000 years ago and reached its distinct late climax between about 13,000 and 10,000 years ago. It was also some 13,000–12,900 years ago that the great eruptions

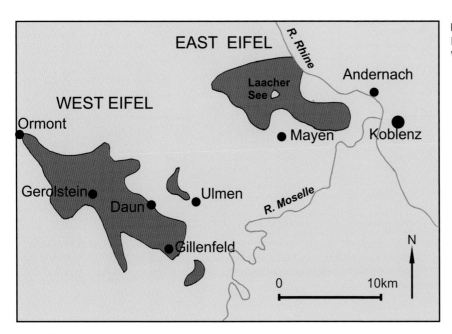

Figure 9.1 Location and Distribution map of East and West Eifel.

Constructing a maar

The Eifel region is the very founding place for the term 'maar' in a volcanic sense. It was Sebastian Münster, way back in 1544 in his book *Cosmographia*, who applied the term 'maar' (as *Marh*) to the Ulmener Maar (arguably amongst the most classic maars in the Eifel) and the Laacher See (which by many is considered rather to be a larger caldera structure, but will have gone through a maar phase during its eruption). The term was fully adopted into volcanology much later. A maar itself is a low-relief volcanic crater that is the result of a hydrovolcanic/phreatomagmatic eruption, where the eruption of a diatreme interacts with shallow groundwater. Recent examples of these eruptions have been witnessed; for example, in Alaska, where some of the largest diatremes/maars are found. The maar itself forms when a lake forms in the crater left after the eruption (e.g. Fig. 9.2). These circular to semi-circular lakes and sometimes barren (filled) craters are littered around the Eifel, and can be seen from the air and experienced by boat.

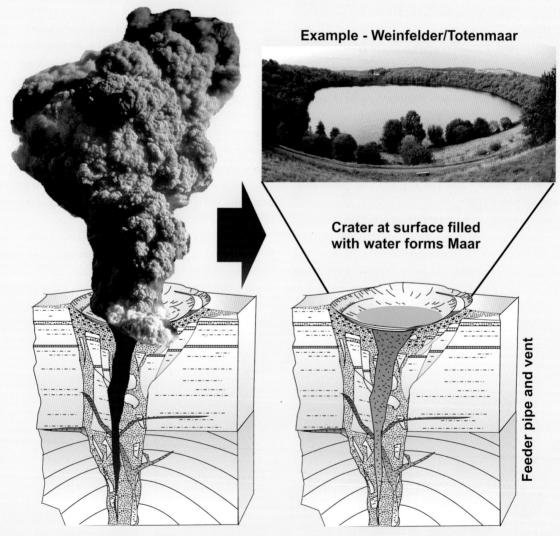

Example - Weinfelder/Totenmaar

Crater at surface filled with water forms Maar

Feeder pipe and vent

Figure 9.2 Formation of a maar. A) Feeder pipe transports magma to the surface, where it interacts with wet, unconsolidated substrate and erupts (photo of eruption cloud from phreatomagmatic eruption column rising from the east Ukinrek Maar crater in 1977 by R. Russell, USGS). B) After the eruptions, the depression fills with water, creating a maar (maar photo by Dietrich Krieger – Wikimedia Commons).

of Laacher See distributed indicator beds as far afield as Switzerland, northern Italy and southern Sweden. One or perhaps two magma reservoirs erupted alkaline basalts, basanites, tephrites, phonolites and some trachytes. Basalts, tephrites and basanites were responsible for most of the lava flows, cinder cones, maars and tuff rings, whereas the domes and most of the layers of pumice are chiefly phonolitic. Some domes and layers of tuff are also composed of selbergite, a local phonolite rich in the mineral leucite. Although the region has been quiet for several thousand years, emissions of carbon dioxide from Laacher See, for instance, indicate that the magma reservoir has cooled only to about 400°C.

East Eifel

The volcanic activity of the East Eifel region was helped by the development of a zone of fault troughs, running west-southwest across the plateau from the River Rhine near Andernach. The eruptions of alkali basalts, tephrites and basanites often formed cinder cones and lava flows, but violent widespread phonolitic explosions of pumice were sometimes followed by volcano-tectonic foundering or by extrusions of phonolitic domes. Although hydrovolcanic activity was both powerful and frequent, voluminous lavas were emitted fast enough to prevent the

formation of maars here; and the volcanotectonic depressions, such as those at Wehr, Rieden and the Laacher See, are at least twice the size of the maars further west (Fig. 9.3). Two dozen or so basanitic cinder cones predominate in the fault-trough zone east of Laacher See, and a similar number of leucitic or leucitic phonolite cones and domes mark the up-thrown blocks west of Laacher See. Many of the parent vents in both areas are aligned on regional fissures trending from northwest to southeast or northeast to southwest. Many of the cones have been quarried for their scoria. Examples of these quarried cones include Rothenberg (by Laacher See), Nastberg in Andernach/Eich, Karmelenberg between Ochtendung and Bassenheim, and the Eppelsberg, located directly by the Laacher See (e.g. Fig. 9.4).

The volcanic depressions of Wehr, Rieden and Laacher See seem too large to be maars and too small to be calderas. They are clearly associated with volcano-tectonic collapse as well as with violent explosions that often had a distinct hydrovolcanic component. These larger depressions were, perhaps, simply caused by more violent explosions of phonolitic magma in eruptions that reached Plinian proportions.

The Rieden depression is the largest in the East Eifel area and is 2km broad, stretching 4km from northwest to southeast. Its edges are sometimes scalloped, which

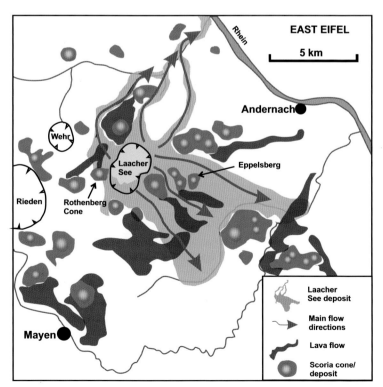

Figure 9.3 General outline of Wehr, Rieden and the Laacher See depressions, surrounding scoria cones, lava flows and the eruption flow directions and deposit outline of the Laacher See pyroclastic density currents.

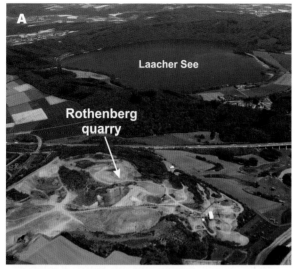

Figure 9.4 Examples of quarry workings. A) Rothenberg quarry and Laacher See from the air (photo courtesy of Walter Müller, Niederzissen). B) Quarry workings at the Eppelsberg scoria cone, East Eifel (photo © Kappest/ Vulkanpark - courtesy of Jörg Busch).

The growth of Rothenberg cone

Rothenberg, which rises on the low ridge separating the Rieden depression from Laacher See, is perhaps typical of the rather complex evolution of the basanitic–tephritic cinder cones in the East Eifel. It was emitted from six vents aligned on a north-northeast to south-southwest fissure, about 600m long. The present much-quarried summit of Rothenberg now lies 60m, and originally probably rose about 120m, above its base. The bulk of the volcano is composed of two coalescent cinder cones, although Rothenberg was constructed in six phases of activity. The eruption probably started about 12,000 years ago in an area blanketed by more than 20m of phonolitic pyroclastic fragments that had exploded from the Rieden depression. These deposits constitute a major water-holding layer, and were primarily responsible for the hydrovolcanic explosions from the Rothenberg vents. The first hydrovolcanic eruption formed the small tuff ring, 200m across, of thin and fine tephritic beds. However, a rapid increase in magma discharge soon excluded any groundwater from the vent and built the first tephritic cinder cone, 65m high, which eventually buried the tuff ring. Then, when the discharge decreased again, the magma could no longer rise up the main northern vent and could reach the surface only to the south. Here, abundant groundwater was still available in the Rieden layers, and hydrovolcanic explosions immediately generated another set of tuffs. The last two phases occurred when basanitic magma erupted. Activity was only weak on the northern cone, but a new vent formed further south along the fissure. Hydrovolcanic eruption again began the sequence on the new vent. But once again, as magma discharge increased and eliminated the groundwater, Strombolian eruptions expelled more cinders to form the southern basanitic cinder cone of Rothenberg, which soon coalesced with its northern predecessor. The Rothenberg cone has been extensively quarried and is also a favourite spot for mineral collectors, as it contains a number of exotic minerals.

suggests that the depression collapsed intermittently. It contains 15 volcanic vents, which were the likely sources of the thick blanket of pale phonolitic (selbergitic) fragments that filled the depression and radiated more than 5km from it. The fragments were expelled in pyroclastic density currents, which most probably accompanied the collapse of the Rieden depression, which began 470,000 years ago and climaxed about 410,000 years ago.

A smaller and simpler collapse then occurred at Wehr, some 2km to the northwest. The Wehr depression stretches almost 2km from northwest to southeast. The phonolitic-trachyte fragments erupted about 213,000 years ago and, no doubt, mark the initiation of the depression. It has a simple outline and encloses only one vent, although the Dachsbusch cone of alkali basalt also rises on its northern rim. The later eruptions from Wehr covered the Dachsbusch basalts and formed four white layers of phonolitic pumice, dated to 60,000, 52,000, 32,000 and 25,000–12,000 years ago, which were themselves interspersed with basaltic eruptions. The Wehr depression is not entirely extinct, for Welschmiesenmuhle, at its northern end, still manifests fumarole activity.

Laacher See

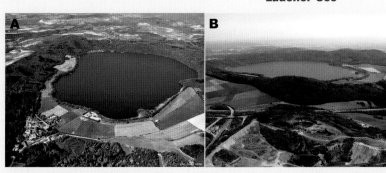

Figure 9.5 The Laacher See Maar. A) View of Laacher See with spa complex and monastery. B) Aerial view of Laacher See over Rothenberg quarry (photos courtesy of Walter Müller, Niederzissen).

Situated 40km south of Bonn, Laacher See is the largest lake and the most famous volcanic formation in the Eifel Massif (Fig. 9.5). It is 2.5km in diameter and up to 51m deep (average depth is 31m), and lies in the midst of a swarm of tephritic and basanitic cinder cones that rise from a plateau blanketed with thick layers of white pumice. Laacher See seems to be a complex ashflow caldera formed by a combination of vigorous explosions and volcanotectonic foundering. There is no doubt about the explosions. About 13,000–12,900 years ago, in the space of some tens of days to about four to five months, Laacher See was the site of a Plinian eruption, with a number of associated pyroclastic density currents. This produced over 16km³ of fragments and flows of phonolitic ash and pumice that are still over 50m thick near the vent. The debris covers many adjacent cinder cones and forms a clear indicator bed in deposits far beyond the confines of the region (e.g. Fig. 9.6).

The eruptions were generated by a phonolitic magma in a reservoir situated between 3km and 6km below the surface, which had probably evolved from an original basanitic magma that had already erupted the surrounding swarm of cinder cones about 27,000 years ago. The most differentiated, highly alkaline and gas-rich phonolites from the uppermost parts of the reservoir were ejected first, followed by less differentiated crystal-rich phonolites, at temperatures between 800°C and 880°C, as the eruptions concluded. The eruption may have been precipitated a few hours after fresh basanitic magma invaded the reservoir, although it is possible, on the other hand, that the new magma arose only when the old had been ejected from the reservoir.

The eruptions proceeded at a very rapid pace and most of them had marked hydrovolcanic components.

The breccias expelled during the initial vent clearing were soon superseded by the Plinian columns of phonolitic fragments. The wind winnowed pumice from the columns and spread it widely, and pyroclastic density currents surged outwards whenever the columns collapsed. The pyroclastic density currents rushed between the cinder cones and swamped the valleys with as much as 10m of pumice for up to 3km around the vent. The finer material spread in a blanket of better-sorted pumice that reached a depth of 1m over a wide area of the East Eifel. The finest materials covered some 700,000 km² and reached as far afield as Stockholm, Berlin and Turin. This eruption brought the activity of the East Eifel area to a spectacular, but perhaps temporary, conclusion. Indeed, as fumaroles are still present at Alte Burg, the Laacher See volcano should not be considered extinct.

Most of deposit accumulated in less than 5 months

Figure 9.6 Tourists view the Laacher See deposit with information displays.

West Eifel

The volcanic activity of the West Eifel area spread densely over two areas of the plateau centred on Daun, and most of the vents are aligned along en-echelon fissures. Examples of the location of some of these maars are given in Figure 9.7. The basaltic, tephritic and basanitic magmas were erupted at intervals that began about 400,000 years ago and ended about 12,000 years ago in the northwest, but lasted only from about 60,000 years until 10,000 years ago in the southeast. The major differences in both these volcanic areas are determined by the presence or absence of groundwater. In the northwestern part of the region, Strombolian eruptions took place along the fissures without water interference, and formed tephritic and basanitic cinder cones, such as the Dohm, Hoher List and Radersberg, and they were often accompanied by small lava flows. Most of these cones have already lost their steepness and sharp outlines after many millennia of weathering.

In the southeast, the eruptions not only happened generally more recently, but also were greatly altered by water interference. Although a dozen or so ordinary cinder cones have been formed, the area is dominated by the famous maars. These are rounded hollows of varying depth, and they derive their name from the lakes or maars that now occupy the enclosed depressions, simply because rainwater cannot completely drain away from them. Most maars are 500–1,000m across and 10–200m deep. Pulvermaar is perhaps the finest example, 900m across, 120m deep and encompassed by a regular basaltic tuff ring over 10m high. Some maars, such as the older half of Schalkenmehrener-maar, the Dürresmaar and the Oberwinkelermaar, have already been filled by sedimentary or volcanic accumulations (see Fig. 9.7), but others, like the Meerfeldermaar, are still over 200m deep. The maars were caused by shallow hydrovolcanic explosions that were accompanied by caldera-like subsidence down ring faults. The repeated explosions formed a low tuff ring of volcanic and lithic fragments. No fewer than 28 of the 29 maars in this area were formed in valleys where streams provided the water that infiltrated down fissures, met the rising magma, suddenly generated steam in a confined space, and triggered violent explosions. The Weinfeldermaar, where copious groundwater was available anyway, provides the only exception.

Views from the air highlight how closely clustered some of the maars are (e.g. Fig. 9.8). Most of the maars were formed during the waning phases of the most recent glaciation, when melting permafrost would also provide great quantities of additional water to generate further explosions. The maars have characteristically rounded outlines; many, like Pulvermaar, are circular; some have slightly scalloped edges, like Gemündenermaar and

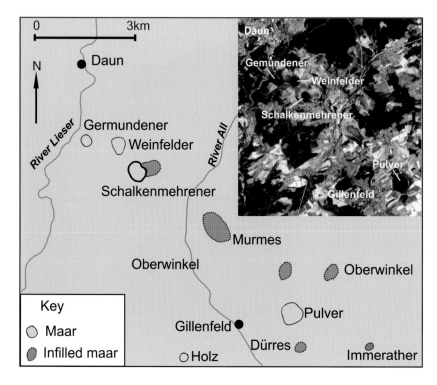

Figure 9.7 Distribution of some of the maars in the West Eifel. Both filled and unfilled maars are shown. Satellite inset (NASA World Wind) highlighting the three maars Gemündener, Weinfelder, and Schalkenmehrener (see also Fig. 9.8).

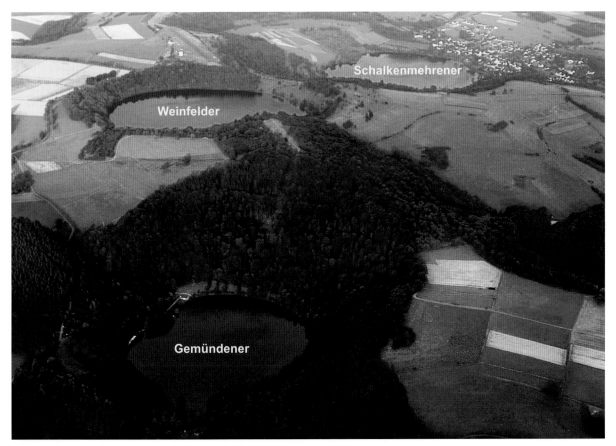

Figure 9.8 Three water-filled maars in the West Eifel, Germany (Gemündener Maar, Weinfelder Maar, Schalkenmehrener Maar) (photo by Martin Schildgen – Wikimedia Commons).

Weinfeldermaar; and others, such as Immerathermaar and Meerfeldermaar, are elliptical. The surrounding tuff rings often have steep inward-facing slopes of about 25° and gentler outer slopes that rarely exceed 12°. These tuff rings are often only between 10m and 30m high and contain many thin beds of fine basanitic or tephritic tuffs and varying amounts of fragments of the country rock. However, there are differences of detail in the maars, which are related, for example, to variations in the amount of magma supplied and to the depth of the explosion. The greater amount of magma erupted at Gemündenermaar, for instance, was probably responsible for the construction of its 100m-high tuff ring. The 200m-deep Meerfeldermaar, on the other hand, is surrounded by a ring that is chiefly made up of fragments of the country rock, which indicates that little magma participated in most of its eruptions. At Ulmenermaar, too, similar explosions dominated, because the ring is also composed largely of country-rock fragments.

It was during the period of the waning phases of the most recent glaciation, from 12,500 to 10,000 years ago,

that the present maars were formed. For example, one of the oldest, Holzmaar, formed 12,500 years ago, and it was followed about 11,000 years ago by Mosbruchmaar and the largest in the area, Meerfeldermaar, which is 1480m long and 1200m wide. At about the same time a pair of maars was formed that soon coalesced into the Schalkenmehrenermaar, where the products of the younger western eruption choked the eastern maar so completely that it is now dry. Between 11,000 and 10,000 years ago came Pulvermaar, Gemündenermaar and Weinfeldermaar. At that time, also, the explosion of Dürresmaar made a cavity in the side of the adjacent Römerberg cinder cone, and Sprinkermaar cut into the southern flanks of the Wartgesberg cinder cone. These are but examples of many that mark the most recent volcanic activity in the West Eifel region. They owe their survival to their relative youth, and it is likely that maars also exploded during earlier periods. But the older specimens have probably been eroded or weathered away, or masked beneath a cover of younger volcanic materials.

Volcano tourism in the Eifel

Given its rather hidden initial appearance amid the German countryside, the Eifel volcanic field has been developed into a modern example of volcanic tourism, providing a large number of opportunities to explore its volcanic past. This is in part due to its links with the old Roman workings of the volcanic materials, more recent workings of pumice in the area, and its wonderful maar lakes. The region boasts volcanic trails, information centres and underground caves, all with a specific flavour of volcanic or human/volcanic interaction.

The 'Vulkanpark' in East Eifel (which translates into 'Volcano Park') was established in 1996 from an initial idea that seeded in the 1980s. The extraction of pumice and other volcanic stone initially started with the Romans extracting lava and tuff, and later companies extracting pumice from around 1850s onwards. By the mid-80s there were as many as 110 companies. During the early '90s these numbers declined or merged into 10–15 companies with a large amount of disused quarries that exposed the volcanic rocks, and in places some of the Roman workings. It was decided to use some of these areas for tourism, and so the Vulkanpark was born.

This enterprise is funded by the district government and is also operated in collaboration with the Roman–Germanic Central Museum in Mainz. The various themes and exhibits that have been developed follow the volcanic and human history themes. There are information centres that provide an overview of the area from the eruptions to the Roman and later excavations. There is a lava dome centre, where you can see a working model of a volcano (Fig. 9.9) and experience a virtual eruption as seen by a news team, as though the Laacher See volcano was erupting today. You can go underground to where industrious locals excavated the natural columns of basalt to form underground cave systems, which proved ideal for brewing beer. You can see walls of ash that erupted from the famous Laacher See eruption; also a fantastic excavation of some of the original Roman workings reveals how they worked the tuff stone (ignimbrite) and used the basaltic lava rocks for millstones (see examples in Fig. 9.10).

In the West Eifel the project 'volcano tourism' in the Eifel was started in the mid-80s. At the same time in the East, another initiative, the 'Vulkanpark Brohltal/ Laacher See' was also started around the Laacher See area, which today is called the 'Ferienregion Laacher-See'. The West Eifel area is now recognized as the 'Vulkaneifel UNESCO Global Geopark' and is listed as one of the European Geoparks. It includes the municipalities of Daun, Hillesheim and Kelberg as well as parts of the municipalities of Gerolstein, Obere Kyll, Manderscheid and Ulmen, and it covers 980km². The region offers Geo-themed trails, nature guides and events programmes, and some museums (e.g. the Maar-Museum in Manderscheid, the Vulkanhaus in Strohn, and the Eifel-Vulkanmuseum in Daun).

All three Geoparks (Westeifel, Vulkanpark and Brohltal) are partners in the 'Deutsche Vulkanstraße', the 'German Volcano Route', a signposted holiday route linking 39 places of interest around the topic of volcanic activity in the Eifel, which runs through both the West and East Eifel (see Fig. 9.11). In all, the region has done a wonderful job of integrating its volcanic past into its modern tourism. Further information about the Vulcan Park (East) can be found at (http://www.vulkanpark.com), the Ferienregion Laacher-See (www.ferienregion-laacher-see.de), and the Natur- & Geopark Vulkaneifel (West) (http://www.geopark-vulkaneifel.de).

Figure 9.9 Volcano model at the Lava-Dome German Volcano Museum, part of the Vulkanpark, East Eifel (photo © Kappest/ Vulkanpark – courtesy of Jörg Busch).

Figure 9.10 Examples from the Vulkanpark 'Volcano Park' in East Eifel. **A**) Information centre. **B**, **C** & **D**) Roman Mine at Meurin. Tunnels 'B' were excavated by the Romans to work the ignimbrite 'tuff stone' used for building materials. The excavations took place under the thick blanket of pumice, as depicted in the cross-section in 'C'. Wood material trapped in the ignimbrite has been used to help date the eruption (e.g. burned log in 'D'). **E**) Underground excavation of columnar lava flows made caverns ideal for brewing beer (the 'Lava Cellar' 30 metres under the town of Mendig) (photos Dougal Jerram).

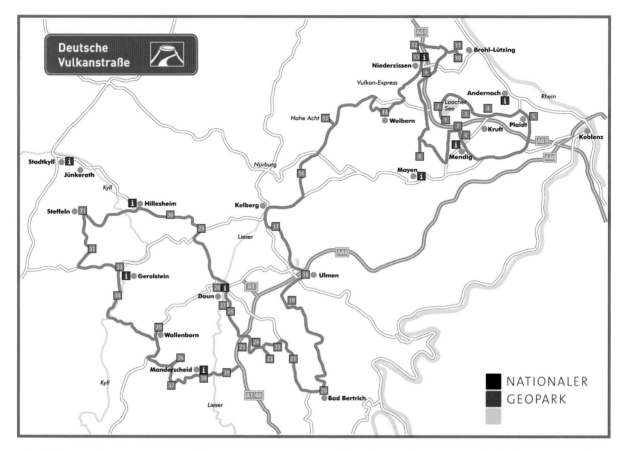

Federal highway 61 (COLOGNE - KOBLENZ) AS MENDIG

1. Parking at harvest Cross: starting point overlooking the Laacher See.

2. Benedictine Abbey at Laacher See.

3. Eppelsberg in Nickenich, digestion of a cinder cone.

4. Roman Mine Meurin at Kretz, antique technique park.

5. Volcano Park Information Centre Plaidt/Saffig.

6. German Volcano Museum and Museumslay in Mendig.

7. Wingertsbergwand, Diary of the Laacher See volcano.

8. Mayener pit filed/mine area, 7000 years' mining history.

9. Gleitfalte Dachsbusch and Wehrer Boiler (Caldera).

10. Kurklinik Bad Tönisstein, ash stream (Trass), mineral water.

11. Trass-mining caves in the Lower Brohltal at the inn 'Jägerheim'.

12. Bausenberg, cinder cones with horseshoe crater.

13. Info centre Vulkanpark Brohltal/Laacher See.

Federal highway 48 (KOBLENZ - TRIER) AS MANDERSCHEID

14. Tuff stone Centre, Weibern: tuff quarries and Museum Island.

15 Höhe Acht, the highest mountain in the Eifel.

16. 'Raustert' in Welcherath, old quarry.

17. Mosbrucher Weiher, maar volcano.

18. Ulmener Maar, Germany's youngest volcano.

19 'Steineberger Ley', tertiary temporal basalt dome with stone wall.

20 Bad Bertrich, cheese cave, hot springs.

21 Immerather Maar – young quaternary maar with lake.

22 Pulvermaar, Maarvulkan, deepest natural lake in Germany.

23 Strohn, huge volcanic bomb.

24. Holzmaar, young quaternary maar with lake.

25 Durres Maar (bog) and Hitsche (smallest maar in the Eifel).

26 View of the vineyards and Schalkenmehrener maar.

27 Gemündener Maar.

28 Eifel Volcano Museum.

29 Dreiser Weiher (pond), on the east side with cracks 'olivine bomb'.

30 Arensberg, tertiary temporal basalt dome with tephra.

31 Steffeln, well-preserved Quaternary palagonite tuff ring.

32. Ice Cave in Quaternary cinder Rother head.

33. Papenkaule and Hagelskaule, young Quaternary crater.

34. Hundsbachtal, cut into the Basaltic lava stream of Kalem.

35. Wallenborn, intermittent mineral spring.

36. Meerfelder maar, largest maar of Eifel.

37. Hinkelmaar, slag ring, with crater Windsborn.

38. Maarmuseum Manderscheid.

39. Fold in Lower Devonian rocks: the base of the volcanoes.

Figure 9.11 The 'German Volcano Route' ('Deutsche Vulkanstraße'), a signposted holiday route linking 39 places of interest around the topic of volcanic activity in the Eifel, running through both the West and East Eifel (copyright Nat. Geopark Laacher See).

Bibliography

Abbruzzese, D. (1936) Sulla catastrofica esplosione dello Stromboli dell'11 settembre 1930. *Atti Accademia Gioenia di Scienze Naturali in Catania* (6th series) 1 (memoria 4), 1–13.

Abdel-Monem, A. A. et al. (1975) K–Ar ages from the eastern Azores (Santa Maria, São Miguel and the Formigas Islands). *Lithos* 8, 247–54.

Ablay, G. J. et al. (1995) The ~2ka sub-Plinian eruption of Montaña Blanca, Tenerife. *Bulletin of Volcanology* 57, 337–55.

Albore Livadie, C. (ed.) (1986) *Tremblements de terre, éruptions volcaniques et vie des hommes dans la Campanie antique*. Naples: Institut Français de Naples, Publications du Centre Jean Bérard, 2ème série, VII, 1–233.

Allard, P. (1997) Endogenous magma degassing and storage at Mount Etna. *Geophysical Research Letters* 24, 2219–22.

Ancochea, E. et al. (1990) Volcanic evolution of the island of Tenerife (Canary Islands) in the light of new K–Ar data. *Journal of Volcanology and Geothermal Research* 44, 231–49.

Ancochea, E. et al. (1994) Constructive and destructive episodes in the building of a young oceanic island, La Palma, Canary Islands, and genesis of the Caldera de Taburiente. *Journal of Volcanology and Geothermal Research* 60, 243–62.

Ancochea, E. et al. (1996) Volcanic complexes in the eastern ridge of the Canary Islands: the Miocene activity of the island of Fuerteventura. *Journal of Volcanology and Geothermal Research* 70, 183–204.

Andronico, D. et al. (1998) Introduction to Somma–Vesuvius. *Cities on Volcanoes, International Meeting*, G. Orsi, M. Di Vito, R. Isaia (eds), 14–25. Naples: Osservatorio Vesuviano.

Anguita, F. and Hernán, F. (1975) A propagating fracture model versus a hotspot origin for the Canary Islands. *Earth and Planetary Science Letters* 27, 11–19.

Anguita, F. et al. (1991) Roque Nublo Caldera: a new strato-cone caldera in Gran Canaria, Canary Islands. *Journal of Volcanology and Geothermal Research* 47, 45–63.

Annertz, K. et al. (1985) *The post-glacial history of Dyngjufjoll*. Report 8503, Nordic Volcanological Institute, Reykjavík.

Araña, V. and Carracedo, J. C. (1978) *Los volcanes de las Islas Canarias, I: Tenerife*. Madrid: Rueda.

Araña, V. and Carracedo, J. C. (1979) *Los volcanes de las Islas Canarias, II: Lanzarote y Fuerteventura*. Madrid: Rueda.

Araña, V. and Coello, J. (eds) (1989) *Los volcanes y la caldera del Parque Nacional del Teide (Tenerife, Islas Canarias)*. Serie Tecnica 7, Instituto para la Conservación de la Naturaleza, Madrid.

Archambault, C. and J-C. Tanguy (1976) Comparative temperature measurements on Mount Etna lavas: problems and techniques. *Journal of Volcanology and Geothermal Research* 1, 113–25.

Baillie, M. G. L. and Munro, M. A. R. (1988) Irish tree rings, Santoríni and volcanic dust veils. *Nature* 332, 344–6.

Barberi, F. et al. (1973) Evolution of Eolian Arc volcanism (South Tyrrhenian Sea). *Earth and Planetary Science Letters* 21, 269–76.

Barberi, F. et al. (1974) Evolution of a section of the Africa/Europe plate boundary: paleomagnetic and volcanological evidence from Sicily. *Earth and Planetary Science Letters* 22, 123–32.

Barberi, F. et al. (1981) The Somma–Vesuvius magma chamber: a petrological and volcanological approach. *Bulletin Volcanologique* 44(3), 295–315.

Barberi, F. et al. (eds) (1984) *The bradyseismic crisis at Phlegraean Fields, Italy. Bulletin Volcanologique* 47 [special issue], 173–412.

Barberi, F. et al. (1993a) Volcanic hazard assessment at Stromboli based on review of historical data. *Acta Vulcanologica* 3, 173–87.

Barberi F. et al. (1993b) The control of lava flow during the 1991–1992 eruption of Mt Etna. *Journal of Volcanology and Geothermal Research* 56, 1–34.

Barberi, F. and Carapezza, M. L. (1996) The Campi Flegrei case history. In R. Scarpa and R. I. Tilling (eds), *Monitoring and mitigation of volcano hazards*, 771–86 Berlin: Springer.

Barker, A. K. et al. (2015) The magma plumbing system for the 1971 Teneguía eruption on La Palma, Canary Islands. *Contributions to Mineralogy and Petrology* 170:54. doi: 10.1007/s00410-015-1207-7.

Barton, M. and Huijsmans, J. P. P. (1986) Post-caldera dacites from the Santoríni volcanic complex, Aegean Sea, Greece: an example of the eruption of lavas of near-constant composition over a 2200 year period. *Contributions to Mineralogy and Petrology* 94, 472–95.

Beccaluva, L. et al. (1985) Petrology and K/Ar ages of volcanics dredged from the Eolian seamounts: implications for geodynamic evolution of the southern Tyrrhenian basin. *Earth and Planetary Science Letters* 74, 187–208.

Bertagnini, A. et al. (1999) Violent explosions yield new insights into dynamics of Stromboli volcano. *Eos* 80, 633–6.

Bigazzi, G. and Bonadonna, F. (1973) Fission track dating of the obsidian of Lìpari Island (Italy). *Nature* 242, 322–3.

Björnsson, A. et al. (1979) Rifting of the plate boundary in north Iceland, 1975–1978. *Journal of Geophysical Research* 84, 3029–3038.

Bogaard, P. van den and Schmincke, H-U. (1985) Laacher See tephra: a widespread isochronous Late Quaternary tephra layer in central and northern Europe. *Geological Society of America, Bulletin* 96, 1554–71.

Bogaard, P. van den et al. (1987) $^{40}Ar/^{39}Ar$ laser dating of single grains: ages of Quaternary tephra from the East Eifel volcanic field, FRG. *Geophysical Research Letters* 14(12), 1211–14.

Boivin, P. et al. (2009) Volcanologie de la Chaîne des Puys Massif Central Français, 5th Edition, scale 1:25 000, 1 sheet.

Boivin, P. and Thouret, J. C. (2013) The volcanic Chaîne des Puys: a unique collection of simple and compound monogenetic edifices. In M. Fort and M.F. André (eds), *Landscapes and Landforms of France* 9, 81–91, Heidelberg: Springer.

Booth, B. et al. (1978) A quantitative study of and five thousand years of volcanism on São Miguel, Azores. *Royal Society, Philosophical Transactions* A288, 271–319.

Borgia, A. et al. (1992) Importance of gravitational spreading in the tectonic and volcanic evolution of Mount Etna. *Nature* 357, 231–5.

Branca S. et al. (2011) Geological map of Etna volcano, 1:50,000 scale. *Italian Journal of Geosciences*, 130(3), 265–291, doi: 10.3301/IJG.2011.15.

Brown, R. J. and Branney, M.J. (2004) Event stratigraphy of a caldera-forming ignimbrite eruption on Tenerife: the 273ka Poris Formation. *Bulletin of Volcanology* 66, 392–416.

Brousse, R. et al. (1981) Fayal dans l'Atlantique et Rapa dans le Pacifique: deux series faiblement alcalines évoluant sous conditions anhydres. *Bulletin Volcanologique* 44(3), 393–410.

Buch, L. von (1825) *Physikalische Beschreibung der Kanarischen Inseln*. Berlin: Academie der Wissenschaften. [1836: published in French as *Description physique des Iles Canaries*, translated by C. Boulanger. Paris: Levrault].

Buch, L. von (1836) On volcanoes and craters of elevation. *The Edinburgh New Philosophical Journal* 21:189–206.

Buchner, G. (1986) Eruzioni vulcaniche e fenomeni vulcano-tettonici di età preistorica e storica nell'isola d'Ischia. In C. Albore Livadie (ed.) *Tremblements de terre, éruptions volcaniques et vie des hommes dans la Campanie antique*, 145–88. Naples: Institut français de Naples.

Buckland, P. C. et al. (1997) Bronze Age Myths? Volcanic activity and human response in the Mediterranean and North Atlantic regions. *Antiquity* 273, 581–93.

Bungum, H. and Husebye, E. S. (1977) Seismicity of the Norwegian Sea: the Jan Mayen fracture zone. *Tectonophysics* 40, 351–60.

Butler, G. W. (1892) The eruptions of Vulcano, August 3 1888 to March 22 1890. *Nature* 46, 117–9.

Cadogan, G. (1988) Dating of the Santoríni eruption. *Nature* 332, 401–2.

Calvari, S. et al. (1994) The 1991–1993 Etna eruption: chronology and flow-field evolution. *Acta Vulcanologica* 4, 1–14.

Calvari, S. and Pinkerton, H. (1999) Lava tube morphology and evidence for lava flow emplacement mechanisms. *Journal of Volcanology and Geothermal Research* 90, 263–80.

Cantagrel, J. M. et al. (1984) K–Ar chronology of the volcanic eruptions in the Canarian archipelago: island of Gomera. *Bulletin Volcanologique* 47, 597–609.

Canto, E. P. (1879, 1880, 1881, 1882, 1883, 1884, 1887) Vulcanismo nos Açores, (various reports), *Archivos dos Açores*.

Capaldi, G. et al. (1985) Geochronology of Plio-Pleistocene volcanic rocks from southern Italy. *Rendiconti della Societá Italiana di Mineralogia e Petrologia* 40, 25–44.

Carracedo, J. C. (1994) The Canary Islands: an example of structural control on the growth of large oceanic island volcanoes. *Journal of Volcanology and Geothermal Research* 60, 225–41.

Carracedo, J. C. (1996) Morphological and structural evolution of the western Canary Islands: hotspot-induced three-armed rifts or regional tectonic trends? *Journal of Volcanology and Geothermal Research* 72, 151–62.

Carracedo, J. C. (1999) Growth, structure, instability and collapse of Canarian volcanoes and comparisons with Hawaiian volcanoes. *Journal of Volcanology and Geothermal Research* 94, 1–19.

Carracedo, J. C. (2015) The 2011–2012 submarine eruption off El Hierro, Canary Islands: new lessons in oceanic island

growth and volcanic crisis management. *Earth-Science Reviews* 150: 168–200.

Carracedo, J. C. and Rodríguez Badiola, E. (1991) *Lanzarote, la erupción volcánica de 1730*. Las Palmas de Gran Canaria: Servicio de Publicaciones, Cabildo Insular de Lanzarote.

Carracedo, J. C. and V. R Troll (eds) (2013) Teide volcano – geology and eruptions of a highly differentiated oceanic stratovolcano. Berlin Heidelberg: Springer-Verlag 2013, *Active Volcanoes of the World XIV*, 279pp.

Carracedo, J. C. et al. (1992) The 1730–1736 eruption of Lanzarote, Canary Islands: a long, high-magnitude basaltic fissure eruption. *Journal of Volcanology and Geothermal Research* 53, 239–50.

Carracedo, J. C. et al. (1996) The 1677 eruption of La Palma, Canary Islands. *Estudios Geológicos* 52, 103–114.

Carracedo, J. C. et al. (1999) Later stages of volcanic evolution of La Palma, Canary Islands: rift evolution, giant landslides, and the genesis of the Caldera de Taburiente. *Geological Society of America, Bulletin* 111, 2–16.

Carracedo, J. C. et al. (1999) Giant Quaternary landslides in the evolution of La Palma and El Hierro, Canary Islands. *Journal of Volcanology and Geothermal Research* 94, 169–90.

Carta, S. et al. (1981) A statistical model for Vesuvius and its volcanological implications. *Bulletin Volcanologique* 44(2), 129–51.

Chester, D. K. et al. (1985) *Mount Etna: the anatomy of a volcano*. London: Chapman & Hall.

Chiodini, G. et al. (1997) Chemical and isotopic variations of Bocca Grande fumarole (Solfatara volcano, Phlegraean Fields). *Acta Vulcanologica* 8, 228–32.

Cioni, R. et al. (1995) Compositional layering and syn-eruptive mixing of a periodically refilled magma chamber: the AD79 Plinian eruption of Vesuvius. *Journal of Petrology* 36, 739–76.

Cioni, R. et al. (1999) Pyroclastic deposits as a guide for reconstructing the multi-stage evolution of the Somma–Vesuvius caldera. *Bulletin of Volcanology* 60, 207–22.

Civetta, L. and Santacroce, R. (1992) Steady-state magma supply in the last 3400 years of Vesuvius activity. *Acta Vulcanologica* 2, 147–60.

Civetta, L. et al. (1988) The eruptive history of Pantelleria (Sicily Channel) in the last 50ka. *Bulletin of Volcanology* 50, 47–57.

Civetta, L. et al. (1991) Sr and Nd isotope and trace-element constraints on the chemical evolution of the magmatic system of Ischia (Italy) in the last 55ka. *Journal of Volcanology and Geothermal Research* 47, 213–30.

Civetta, L. et al. (1997) Geochemical zoning, mingling, eruptive dynamics and depositional processes: the Campanian Ignimbrite, Campi Flegrei caldera, Italy. *Journal of Volcanology and Geothermal Research* 75, 183–219.

Civetta, L. et al. (1998) Volcanic, deformational and magmatic history of the Campi Flegrei caldera. In G. Orsi, M. Di Vito and R. Isaia (eds), *Cities on Volcanoes, International Meeting*, 26–71. Naples: Osservatorio Vesuviano.

Clocchiatti, R. et al. (1994) Assessment of a shallow magmatic system: the 1888–90 eruption, Vulcano Island, Italy. *Bulletin of Volcanology* 56, 466–86.

Clocchiatti, R. et al. (1998) Earlier alkaline and transitional magmatic pulsation of Mt Etna volcano. *Earth and Planetary Science Letters* 163, 399–407.

Clocchiatti, R. et al. (1999) Lava-flow temperature measurements on Etna, Italy. *Smithsonian Institution Global Volcanism Network Bulletin* 24(6), 6–7.

Cole, P. D. and Scarpati, C. (1993) A facies interpretation of the eruption and emplacement mechanisms of the upper part of the Neapolitan Yellow Tuff, Campi Flegrei, southern Italy. *Bulletin of Volcanology* 55, 311–26.

Cole, P. D. and Scarpati, C. (2010) The 1944 eruption of Vesuvius, Italy: combining contemporary accounts and field studies for a new volcanological reconstruction. *Geological Magazine* 147(3), 391–415.

Cole, P. D. et al. (1995) An historic sub-Plinian/phreatomagmatic eruption: the 1630 AD eruption of Furnas volcano, São Miguel, Azores. *Journal of Volcanology and Geothermal Research* 69, 117–35.

Cole P. D. et al. (2001) Capelinhos 1957–1958, Faial, Azores: deposits formed by an emergent Surtseyan eruption. *Bulletin of Volcanology* 63(2–3), 204–20.

Coltelli, M. et al. (1998) Discovery of a Plinian basaltic eruption of Roman age at Etna volcano, Italy. *Geology* 26(12), 1095–98.

Condomines, M. and Allègre, C. J. (1980) Age and magmatic evolution of Stromboli volcano from ^{230}Th/^{238}U disequilibrium data. *Nature* 288, 354–7.

Condomines, M., et al. (1982) Magmatic evolution of a volcano studied by ^{230}Th–^{238}U disequilibrium and trace elements systematics: the Etna case. *Geochimica et Cosmochimica Acta* 46, 1397–416.

Condomines, M. et al. (1995) Magma dynamics at Mt Etna: constraints from U–Th– Ra–Pb radioactive disequilibria and Sr isotopes in historical lavas. *Earth and Planetary Science Letters* 132, 25–41.

Cortese, M. et al. (1986) Volcanic history of Lìpari (Aeolian Islands, Italy) during the last 10,000 years. *Journal of Volcanology and Geothermal Research* 27, 117–33.

Crisci, G. M. et al. (1983) Age and petrology of the Late Pleistocene brown tuffs on Lìpari, Italy. *Bulletin Volcanologique* 46, 381–91.

Day, S. J. et al. (1999) Recent structural evolution of the Cumbre Vieja volcano, La Palma, Canary Islands: volcanic rift zone reconfiguration as a precursor to volcano flank instability? *Journal of Volcanology and Geothermal Research* 94, 135–67.

De Astis, G. et al. (1997) Eruptive and emplacement mechanisms of widespread fine-grained pyroclastic deposits on Vulcano Island (Italy). *Bulletin of Volcanology* 59, 87–102.

Delcamp, A. et al. (2014) The endogenous and exogenous growth of the monogenetic Lemptégy volcano, Chaine des Puys, France. *Geosphere* issue 5, vol.10.

Desmarest, N. (1806) *Sur la détermination de trois époques de la nature par les produits des volcans, et sur l'usage qu'on peut faire de ces époques dans l'étude de ces volcans.* Mémoires de l'Institut Tome VI, 1ère Prairial: an 12 (published in 1806).

Di Vito, M. et al. (1987) The 1538 Monte Nuovo eruption (Campi Flegrei, Italy). *Bulletin of Volcanology* 49, 608–15.

Di Vito, M. et al. (1999) Volcanism and deformation since 12,000 years at the Campi Flegrei caldera (Italy). *Journal of Volcanology and Geothermal Research* 91, 221–46.

Dollar, A. T. J. (1966) Genetic aspects of the Jan Mayen fissure volcano group on the mid-oceanic Mohns Ridge, Norwegian Sea. *Bulletin Volcanologique* 29, 25–6.

Doúmas, C. (1974) The Minoan eruption of the Santoríni volcano. *Antiquity* 48, 110–15.

Doúmas, C. (ed.) (1978) *Thera and the Aegean world I* [proceedings of the First International Scientific Congress on the Volcano Thera, Santoríni, Greece]. London: Thera Foundation.

Doúmas, C. and Papazoglou, L. (1980) Santoríni tephra from Rhodes. *Nature* 287, 322–4.

Driscoll, E. M. et al. (1965) The geology and geomorphology of Los Ajaches, Lanzarote. *Geological Journal* 4, 321–34.

Druitt, T. H. et al. (1999) *Santoríni volcano*. Memoir 19, Geological Society, London.

Dugmore, A. J. et al. (1996) Long-distance marker horizons from small-scale eruptions in Iceland and Scotland. *Journal of Quaternary Science* 11, 511–16.

Edgar, C.J. et al. (2007) The late Quaternary Diego Hernandez Formation, Tenerife: volcanology of a complex cycle of voluminous explosive phonolitic eruptions. *Journal of Volcanology and Geothermal Research* 160, 59–85.

Eldholm, O. (1991) Magmatic tectonic evolution of a volcanic rifted margin. *Marine Geology* 102(1–4), 43–61.

Ellam, R. M. et al. (1988) The transition from calc-alkaline to potassic orogenic magmatism in the Aeolian Islands, southern Italy. *Bulletin of Volcanology* 50, 386–98.

Elsworth, D. and Day, S. J. (1999) Flank collapse triggered by intrusion: the Canarian and Cape Verde archipelagoes. *Journal of Volcanology and Geothermal Research* 94, 323–40.

Evans, A. J. (1921–1936) *The palace of Minos at Knossos* [4 vols]. London: Macmillan.

Fazellus, T. (1558) *De rebus Siculis decades duae.* Panormi, Sicily: Maida & Carrara.

Féraud, G. et al. (1980) K–Ar ages and stress and pattern in the Azores: geodynamic implications. *Earth and Planetary Science Letters* 46, 275–86.

Féraud, G. et al. (1981) New K–Ar ages, chemical analyses and magnetic data of rocks from the islands of Santa Maria (Azores), Porto Santo and Madeira (Madeira Archipelago) and Gran Canaria (Canary Islands). *Bulletin Volcanologique* 44(3), 359–75.

Fisher, R. V. et al. (1983) Origin and emplacement of a pyroclastic flow and surge unit at Laacher See, Germany. *Journal of Volcanology and Geothermal Research* 17, 375–92.

Fisher, R. V. et al. (1993) Mobility of a large volume pyroclastic flow: emplacement of the Campanian Ignimbrite. *Journal of Volcanology and Geothermal Research* 59, 205–220.

Fitch, F. J. et al. (1965) Potassium–argon ages of rocks from Jan Mayen and an outline of its volcanic history. *Nature* 207, 1349–51.

Forjaz, V. H. (1997a) *Alguns vulcões da Ilha de São Miguel.* Ponta Delgada [São Miguel]: Observatório Vulcanológico Geotérmico dos Açores.

Forjaz, V. H. (ed.) (1997b) *Vulcão dos Capelinhos. Retrospectivas: vol. I.* Ponta Delgada [São Miguel]: Observatório Vulcanológico Geotérmico dos Açores.

Forjaz, V. H. and Fernandes, N. S. M. (1975) *Noticia explicativa das folhas "A" e "B": Ilha de São Jorge.* Lisbon: Serviços Geológicos, Carta Geológica de Portugal.

Forjaz, V. H. et al. (2000) *Noticias sobre o vulcão oceânico da Serreta, Ilha Terceira dos Açores.* Ponta Delgada [São Miguel]: Observatório Vulcanológico e Geotérmico dos Açores.

Fouqué, F. (1867) Sur les phénomènes volcaniques observés aux Açores. *Académie des Sciences à Paris, Comptes Rendus* 65, 1050–3.

Fouqué, F. (1873) Voyages géologiques aux Açores. *Revue des deux mondes* (2ème année, 2ème période) 103, 40–65.

Fouqué, F. (1879) *Santorin et ses éruptions.* Paris: Masson.

Frazzetta, G. et al. (1983) Evolution of the Fossa cone, Vulcano. *Journal of Volcanology and Geothermal Research* 50, 329–60.

Frazzetta, G. et al. (1984) Volcanic hazards at Fossa of Vulcano: data from the last 6000 years. *Bulletin Volcanologique* 47, 105–124.

Frepoli, A. et al. (1996) State of stress in the southern Tyrrhenian subduction zone from fault-plane solutions. *Geophysical Journal International* 125, 879–91.

Fructuoso, G. (1591) *Saudades da terra (livro IV)* [reprinted 1966]. Ponta Delgada [São Miguel]: Instituto Cultural de Ponta Delgada.

Gabbianelli, G. C. et al. (1993) Marine geology of the Panarea–Stromboli area. *Acta Vulcanologica* 3, 11–20.

García Cacho, L. et al. (1994) A large volcanic debris avalanche in the Pliocene Roque Nublo stratovolcano, Gran Canaria, Canary Islands. *Journal of Volcanology and Geothermal Research* 63, 217–29.

Gaspar, J. L. et al. (2003) Basaltic lava balloons produced during the 1998–2001 Serreta Submarine Ridge eruption (Azores). In J. White, D. Clague, and J. Smellie (eds), *Subaqueous Explosive Volcanism*, American Geophysical Union, Geophysical Monograph 140, 205–12.

Gasparini, P. and Musella, S. (1991) *Un viaggio al Vesuvio*. Naples: Liguori.

Gastesi, P. (1973) Is the Betancuria Massif, Fuerteventura, an uplifted piece of oceanic crust? *Nature* 246, 102–4.

Geldmacher, J. et al. (2005) New $^{40}Ar/^{39}Ar$ age and geochemical data from seamounts in the Canary and Madeira volcanic provinces: support for the mantle plume hypothesis. *Earth and Planetary Science Letters* 237, 85–101.

Gemmellaro, C. (1831) *Relazione dei fenomeni del nuovo vulcano sorto dal mare fra la costa di Sicilia e l'isola di Pantelleria nel mese di Luglio 1831*. Catania: Pastore Carmelo.

Georgalás, G. C. (1962) *Catalogue of the active volcanoes of the world including solfatara fields*, part XIII: *Greece*. Rome: International Association of Volcanology.

Gèze, B. (1979) *Languedoc, Méditerranéen, Montagne Noire*. Paris: Masson.

Gillot, P. Y. and Keller, J. (1993) Radio-chronological dating of Stromboli. *Acta Vulcanologica* 3, 69–77.

Gillot, P. Y. and Villari, L. (1980) *K–Ar geochronological data on the Aeolian arc volcanism: a preliminary report*. Report 3-80, Istituto Internazionale di Vulcanologia, Catania.

Gillot, P. Y. et al. (1994) The evolution of Mount Etna in the light of potassium–argon dating. *Acta Vulcanologica* 5, 81–7.

Goër de Hervé, A. de and Belin, J. M. (1985) *Le volcanisme en Auvergne*. Clermont-Ferrand: Centre de Recherches et Documentation Pédagogiques.

Goër de Hervé, A. de et al. (1995) *Volcanisme et volcans de l'Auvergne*. Aydat, Puy-de-Dôme: Parc Naturel Régional des Volcans d'Auvergne.

Gorceix, H. (1874) Phénomènes volcaniques de Nísyros. *Académie des Sciences à Paris, Comptes Rendus* 78, 444–6.

Goree, Father (1711) A relation of a new island which was raised from the bottom of the sea on the 23rd of May 1707 in the Bay of Santoríni, in the archipelago. Written by Father Goree (a Jesuit) an eye witness. *Royal Society of London, Philosophical Transactions* 2(332), 353–74.

Grönvold, K. (1984) *Mývatn Fires 1724–1729: chemical composition of the lava*. Report 8401, Nordic Volcanological Institute, Reykjavík.

Grönvold, K. et al. (1983) The Hekla eruption 1980–1981. *Bulletin Volcanologique* 46(4), 349–63.

Gudmundsson, A. (1987) Lateral magma flow, caldera collapse, and a mechanism of large eruptions in Iceland. *Journal of Volcanology and Geothermal Research* 34, 65–78.

Gudmundsson, A. (1995) Ocean ridge discontinuities in Iceland. *Geological Society of London, Quarterly Journal* 152, 1011–15.

Gudmundsson, A. et al. (1992) The 1991 eruption of Hekla, Iceland. *Bulletin of Volcanology* 54, 238–46.

Gudmundsson, A. et al. (1997) Ice–volcano interaction of the 1996 Gjálp sub-glacial eruption, Vatnajökull, Iceland. *Nature* 389, 954–7.

Guérin, G. et al. (1981) Age subactuel des dernières manifestations éruptives du Mont-Dore et du Cézallier. *Académie des Sciences à Paris, Comptes Rendus* D292, 855–7.

Guest, J. E. and Murray, J. B. (1979) An analysis of hazard from Mount Etna volcano. *Geological Society of London, Journal* 136, 347–54.

Guest, J. E. et al. (1984) Lava tubes, terraces and mega-tumuli on the 1614–24 pahoehoe lava flow field, Mount Etna, Sicily. *Bulletin Volcanologique* 47(3), 635–48.

Guest, J. E. et al. (1987) The evolution of lava flow fields: observations of the 1981 and 1983 eruptions of Mount Etna, Sicily. *Bulletin of Volcanology* 49, 527–40.

Guettard, J. E. (1752) *Mémoire sur quelques montagnes de la France qui ont été des volcans*. Mémoire 27 de l'Académie des Sciences de Paris pour 1752 (published in 1756).

Guillou, H. et al. (1996) K–Ar ages and magnetic stratigraphy of a hotspot-induced fast grown oceanic island: El Hierro, Canary Islands. *Journal of Volcanology and Geothermal Research* 73, 141–55.

Gvirtzman, Z. and Nur, A. (1999) The formation of Mount Etna as the consequence of slab rollback. *Nature* 401, 782–5.

Hamilton, Sir W. (1772) *Observations on Mount Vesuvius, Mount Etna, and other volcanoes of the two Sicilies*. London: Cadell.

Hamilton, Sir W. (1776/9) *Campi Phlegraei: observations on the volcanoes of the two Sicilies*. Naples: Fabris.

Hamilton, Sir W. (1795) An account of the late eruption of Mount Vesuvius in a letter to Sir Joseph Banks. *Royal Society, Philosophical Transactions* 85, 73–116.

Hammer, C. U. et al. (1987) The Minoan eruption of Santoríni in Greece dated to 1645 BC? *Nature* 332, 517–19.

Hardy, D. A. (ed.) (1990) *Thera and the Aegean world III* [3 vols]. London: Thera Foundation.

Havskov, J. and Atakan, K. (1991) Seismicity and volcanism of Jan Mayen island. *Terra Nova* 3, 517–26.

Hawkins, J. R. W. and Roberts, B. (1963) Agglutinate in north Jan Mayen. *Geological Magazine* 100, 156–63.

Heiken, G. and McCoy, F. (1984) Caldera development during the Minoan eruption, Thira, Cyclades, Greece. *Journal of Geophysical Research* 89, 8441–62.

Hirn, A. et al. (1997) Roots of Etna in faults of great earthquakes. *Earth and Planetary Science Letters* 148, 171–91.

Hornig-Kjarsgaard, I. et al. (1993) Geology, stratigraphy and volcanological evolution of the island of Stromboli, Aeolian arc, Italy. *Acta Vulcanologica* 3, 21–68.

Houghton, B. F. and Schminke, H-U. (1989) Rothenberg scoria cone, East Eifel: a complex Strombolian and phreatomagmatic volcano. *Bulletin of Volcanology* 51, 28–48.

Huijsmans, J. P. P. et al. (1988) Geochemistry and evolution of the calc-alkaline volcanic complex of Santoríni, Aegean Sea, Greece. *Journal of Volcanology and Geothermal Research* 34, 283–306.

Humboldt, A. von (1814) *Voyages aux régions équinoxiales du nouveau continent 1799–1809*, vol. I. Paris: Dufour.

Hutchinson, R. W. (1962) *Prehistoric Crete*. Harmondsworth, England: Penguin.

Imbò, G. (1949) L'attività eruttiva vesuviana e relative osservazioni nel corso dell'intervallo intereruttivo 1906–1944 ed in particolare del parossismo del marzo 1944. *Annali dell'Osservatorio Vesuviano*, 5, 185–380.

Imsland, P. (1978) The geology of the volcanic island Jan Mayen, Arctic Ocean. *Nordic Volcanological Institute Report* 7812, 1–74.

Imsland, P. (1986) The volcanic eruption of Jan Mayen, January 1985: interaction between a volcanic island and a fracture zone. *Journal of Volcanology and Geothermal Research* 28, 45–53.

Imsland, P. (1989) Study models for volcanic hazards in Iceland. In J. H. Latter (ed.), *Volcanic hazards: assessment and monitoring*, 36–56. Berlin: Springer.

Jackson, E. L. (1982) The Laki eruption of 1783: impacts on population and settlement in Iceland. *Geography* 67, 42–50.

Jakobsson, S. P. (1979) Petrology of recent basalts of the eastern volcanic zone, Iceland. *Acta Naturalia Islandica* 26, 1–103.

Jerram, D. A. (2011) *Introducing Volcanology: a Guide to Hot Rocks*. Edinburgh: Dunedin Academic press.

Jerram, D. A. and Martin, V. M. (2008) Understanding crystal populations and their significance through the magma plumbing system. In C. Annen and G. F. Zellmer (eds), *Dynamics of Crustal Magma Transfer, Storage and Differentiation*. Geological Society, London, Special Publications, 304, 133–48.

Jerram, D. A. and Petford, N. (2011) *The Field Description of Igneous Rocks*. Oxford: Wiley.

Johnson, G. L. and Heezen, B. C. (1967) The morphology and evolution of the Norwegian–Greenland Sea. *Deep-Sea Research* 14, 755–71.

Johnston-Lavis, H. J. (1890) The eruption of Vulcano Island. *Nature* 42, 78–9.

Jones, J. G. (1970) Interglacial volcanoes of the Laugarvatn region, southwest Iceland – II. *Journal of Geology* 78, 127–40.

Judd, J. W. (1875) Contributions to the study of volcanoes. The Lìpari Islands: Vulcano. *Geological Magazine* 2, 99–105.

Jung, J. (1946) *Géologie de l'Auvergne et de ses confins bourbonnais et limousins*. Mémoire 11, Service de la Carte Géologique de France, Paris.

Juvigné, E. and Gwelt, M. (1987) La narse d'Ampoix comme tephrostratotype dans la Chaîne des Puys méridionale (France). *Bulletin de l'Association française pour l'étude du Quaternaire* 1987(1), 37–49.

Keller, J. (1980) The island of Vulcano. *Rendiconti della Società Italiana di Mineralogia e Petrologia* 36, 29–75.

Kieffer, G. (1982) Les explosions phréatiques et phréatomagmatiques terminales à l'Etna. *Bulletin Volcanologique* 44, 655–60.

Kieffer, G. and Tanguy, J-C. (1993) L'Etna: évolution structurale, magmatique et dynamique d'un volcan polygénique. In R. Maury (ed.), *Pleins feux sur les volcans*, 253–71. Mémoire 163, Société Géologique de France, Paris.

Kilburn, C. R. J. and Lopes, R. M. C. (1988) The growth of aa flow fields on Mount Etna, Sicily. *Journal of Geophysical Research* 93, 14759–72.

Kokelaar, B. P. and Durant, G. P. (1983) The submarine eruption and erosion of Surtla (Surtsey), Iceland. *Journal of Volcanology and Geothermal Research* 19, 239–46.

Kokelaar, P. and Romagnoli, C. (1995) Sector collapse, sedimentation and clast evolution at an active island arc: Stromboli, Italy. *Bulletin of Volcanology* 57, 240–62.

Krafft, M. and Larouzière, F. D. de (1999) *Guide des volcans d'Europe et des Canaries*, 3rd edn. Lausanne: Delachaux et Niestlé.

Kuniholm, P. I. et al. (1996) Anatolian tree rings and the absolute chronology of the eastern Mediterranean, 2220–718BC.

Nature 381, 780–3.

Lacroix, A. (1908) *La Montagne Pelée après ses éruptions, avec observations sur les éruptions du Vésuve en 79 et en 1906*. Paris: Masson.

Lanza, R. and Zanella, E. (1991) Palaeomagnetic directions (223–1.4ka) recorded in the volcanites of Lìpari, Aeolian Islands. *Geophysical Journal International* 107, 191–6.

Lanzafame, G. and Bousquet, J. C. (1997) The Maltese escarpment and its extension from Mt Etna to the Aeolian Islands (Sicily): importance and evolution of a lithosphere discontinuity. *Acta Vulcanologica* 9, 113–20.

Larsen, G. (1984) Recent volcanic history of the Veidivötn fissure swarm, southern Iceland: an approach to volcanic risk assessment. *Journal of Volcanology and Geothermal Research* 22, 33–58.

Larsen, G. and Thórarinsson, S. (1977) H4 and other acid Hekla tephra layers. *Jökull* 27, 28–45.

Larsen, G. et al. (1979) Volcanic eruption through a geothermal borehole at Námafjall, Iceland. *Nature* 278, 707–10.

Larsen, G. et al. (1999) Geochemistry of historical-age silicic tephras in Iceland. *The Holocene* 9, 463–71.

Limburg, E. and Varekamp, J. C. (1991) Young pumice deposits on Nísyros, Greece. *Bulletin of Volcanology* 54, 68–77.

Lorenz, V. (1973) On the formation of maars. *Bulletin Volcanologique* 37, 183–203.

Lorenz, V. (1986) On the growth of maars and diatremes and its relevance on the formation of tuff rings. *Bulletin of Volcanology* 48, 265–74.

Luce, J. V. (1969) *The end of Atlantis*. London: Thames & Hudson.

Luongo, G. and Scandone, R. (eds) (1991) Campi Flegrei. *Journal of Volcanology and Geothermal Research* (special issue) 48, 1–227.

Luongo, G. et al. (1991) A physical model for the origin of volcanism of the Tyrrhenian margin: the case of the Neapolitan area. *Journal of Volcanology and Geothermal Research* 48, 173–86.

Lyell, Sir C. (1855) *A manual of elementary geology*. London: John Murray.

Maccaferri, F. et al. (2014) Off-rift volcanism in rift zones determined by crustal unloading. *Nature Geosciences* 7(4), 297–300.

Machado, F. et al. (1962) Capelinhos eruption of Faial volcano, Azores, 1957–58. *Journal of Geophysical Research* 67, 3519–29.

Mahood, G. A. and Hildreth, W. (1986) Geology of the peralkaline volcano at Pantelleria, Strait of Sicily. *Bulletin of Volcanology* 48, 143–72.

Manning, S. W. (1988) Dating of the Santoríni eruption. *Nature* 332, 401.

Marinátos, S. (1939) The volcanic destruction of Minoan Crete. *Antiquity* 13, 425–39.

Marini, L. et al. (1993) Hydrothermal eruptions of Nísyros (Dodecanese, Greece): past events and present hazard. *Journal of Volcanology and Geothermal Research* 56, 71–94.

Marquart, G. and Jacoby, W. (1985) On the mechanism of magma injection and plate divergence during the Krafla rifting episode in NE Iceland. *Journal of Geophysical Research* 90, 10178–92.

Martel, C. et al. (2013) Trachyte phase relations and implications for magma storage conditions in the Chaîne des Puys (French Massif Central). *Journal of Petrology* 54, 1071–1107.

Martí, J. et al. (1994) Stratigraphy, structure, age and origin of the Cañadas Caldera (Tenerife, Canary Islands). *Geological Magazine* 131, 715–27.

Martin, V. M. et al. (2010) Using the Sr isotope compositions of feldspars and glass to distinguish magma system components and dynamics. *Geology* 38(6) 539–42.

Masson, D. G. (1996) Catastrophic collapse of the volcanic island of El Hierro 15ka ago and the history of landslides in the Canary Islands. *Geology* 24, 231–4.

Maury, R. C. et al. (1980) Cristallisation fraction-née d'un magma basaltique alcalin: la série de la Chaîne des Puys (Massif Central France) I: pétrologie. *Bulletin Minéralogique* 103(2), 250–66.

McClelland, E. A. and Druitt, T. H. (1989) Paleomagnetic estimates of emplacement temperatures of pyroclastic deposits on Santoríni, Greece. *Bulletin of Volcanology* 51, 16–27.

McDougall, I. and Schmincke, H-U. (1976) Geochronology of Gran Canaria, Canary Islands: age of shield building, volcanism and other magmatic phases. *Bulletin Volcanologique* 40, 57–77.

McGarvie, D. W. (1984) Torfajökull: a volcano dominated by magma mixing. *Geology* 12, 685–8.

McGarvie, D. W. et al. (1990) Petrogenic evolution of the Torfajökull volcanic complex, Iceland II: the role of magma mixing. *Journal of Petrology* 31, 461–81.

McGuire, W. J. and Pullen, A. D. (1989) Location and orientation of eruptive fissures and feeder-dykes at Mount Etna: influence of gravitational and regional tectonic stress regimes. *Journal of Volcanology and Geothermal Research* 38, 325–44.

McKenzie, D. P. (1972) Active tectonics in the Mediterranean region. *Royal Astronomical Society, Geophysical Journal* 30, 109–85.

McKenzie, D. P. (1972) Active tectonics of the Mediterranean region. *Royal Astronomical Society, Geophysical Journal* 55, 217–54.

Mehl, K. W. and Schmincke, H-U. (1999) Structure and emplacement of the Pliocene Roque Nublo debris avalanche deposit, Gran Canaria, Spain. *Journal of Volcanology and Geothermal Research* 94, 105–34.

Mercalli, G. (1907) *I vulcani attivi della terra.* Milan: Ulrico Hoepli.

Métrich, N. and Clocchiatti, R. (1989) Melt inclusions investigation of volatiles behavior in historic alkaline magmas of Etna. *Bulletin of Volcanology* 51, 185–98.

Miller, J. (1989) *The 10th-century eruption of Eldgjá, southern Iceland.* Report 8903, Nordic Volcanological Institute, Reykjavík.

Mitchell-Thomé, R. C. (1981) Volcanicity of historic times in the middle Atlantic islands. *Bulletin Volcanologique* 44, 57–69.

Monaco, C. et al. (1997) Late Quaternary slip rates on the Acireale–Piedimonte normal faults and tectonic origin of Mt Etna (Sicily). *Earth and Planetary Science Letters* 147, 125–39.

Montalto, A. (1996) Signs of potential renewal of eruptive activity at La Fossa (Vulcano, Aeolian Islands). *Bulletin of Volcanology* 57, 483–92.

Montlosier, F. de (1788) *Essai sur la théorie des volcans d'Auvergne.* Riom.

Moore, J. G. (1985) Structure and eruptive mechanisms at Surtsey Volcano, Iceland. *Geological Magazine* 122, 649–61.

Moore, J. G. and Calk, L. C. (1991) Degassing and differentiation in subglacial volcanoes, Iceland. *Journal of Volcanology and Geothermal Research* 46, 157–80.

Moore, R. B. (1991) Geology of three late Quaternary stratovolcanoes on São Miguel, Azores. *United States Geological Survey, Bulletin* 1900, 1–46.

Moore, R. B. and Rubin, M. (1991) Radiocarbon dates for lava flows and pyroclastic deposits on São Miguel, Azores.

Radiocarbon 33, 151–64.

Morgan, D.J. et al. (2007) Combining CSD and isotopic microanalysis: magma supply and mixing processes at Stromboli Volcano, Aeolian Islands, Italy. *Earth and Planetary Science Letters* 260(3–4): 419–31.

Moss, J. L. et al. (1999) Ground deformation of a potential landslide at La Palma, Canary Islands. *Journal of Volcanology and Geothermal Research* 94, 251–65.

Murray, J. B. (1990) High-level magma transport at Mount Etna volcano, as deduced from ground deformation measurements. In M. P. Ryan (ed.), *Magma transport and storage*, 357–83. Chichester, England: John Wiley.

Nazzaro, A. (1997) *Il Vesuvio: storia eruttiva e teorie vulcanologiche.* Naples: Liguori.

Noe-Nygaard, A. (1974) Cenozoic to Recent volcanism in and around the North Atlantic Basin. In A. E. M. Nairn and F. G. Stehli (eds), *The ocean basins and margins*, vol. 2: *the North Atlantic*, 428–30. New York: Plenum.

Nomikou, P. et al. (2014) The emergence and growth of a submarine volcano: the Kameni islands, Santorini, Greece. GeoResJ 1–2, 8–18 (online). Available from URL: www.sciencedirect.com/science/article/pii/S2214242814000047 (accessed 6 August 2016).

Nomikou, P. et al. (2013) Submarine volcanoes along the Aegean volcanic arc. Tectonophysics, 597–598, 123–146 (online). Available from URL: www.sciencedirect.com/science/journal/00401951/597-598 (accessed 6 August 2016).

Orsi, G. et al.1991. The recent explosive volcanism at Pantelleria. *Geologische Rundschau* 80, 187–200.

Orsi, G., et al. (1996a) The restless, resurgent Campi Flegrei nested caldera: constraints on its evaluation and configuration. *Journal of Volcanology and Geothermal Research* 74, 179–214.

Orsi, G. et al. (1996b) ^{14}C geochronological constraints for the volcanic history of the island of Ischia (Italy) over the last 5000 years. *Journal of Volcanology and Geothermal Research* 71, 249–57.

Orsi, G. et al. (1998) The volcanic island of Ischia. In G. Orsi, M. Di Vito and R. Isaia (eds), *Field excursion guide: Cities on volcanoes, International Meeting*, 72–8. Naples: Osservatorio Vesuviano.

Orsi, G. et al. (1999) Short-term ground deformations and seismicity in the resurgent Campi Flegrei caldera (Italy): an example of active block resurgence in a densely-populated area. *Journal of Volcanology and Geothermal Research* 91, 415–51.

Ortiz, R. et al. (1986) Magnetotelluric study of the Teide (Tenerife) and Timanfaya (Lanzarote) volcanic areas. *Journal of Volcanology and Geothermal Research* 30, 351–77.

Óskarsson, N. (1984) Monitoring of fumarole discharge during the 1975–1982 rifting in Krafla volcanic center, north Iceland. *Journal of Volcanology and Geothermal Research* 22, 97–121.

Óskarsson, N. et al. (1982) A dynamic model of rift zone petrogenesis and the regional petrology of Iceland. *Journal of Petrology* 23, 28–74.

Palmieri, L. (1872) *Incendio Vesuviano del 26 Aprile 1872.* Torino: Fratelli Bocca.

Parks, M. M. et al. (2015) From quiescence to unrest – 20 years of satellite geodetic measurements at Santorini volcano, Greece. *Journal of Geophysical Research (Solid Earth)* 120, 1309–28, doi:10.1002/2014JB011540.

Patanè, G. et al. (1984) Seismic activity preceding the 1983 eruption of Mt Etna. *Bulletin of Volcanology* 47, 941–52.

Patanè, G. et al. (1996) A model of the onset of the 1991–1993 eruption of Mt Etna, Italy. *Physics of the Earth and Planetary*

Interiors 97, 231–45.

Paterne, M. et al. (1988) Explosive activity of the south Italian volcanoes during the past 80,000 years as determined by marine tephrochronology. *Journal of Volcanology and Geothermal Research* 34, 153–72.

Pé, G. G. (1974) Volcanic rocks of Methana, South Aegean arc, Greece. *Bulletin Volcanologique* 38, 270–90.

Pedersen, R. B. et al. (2010) Discovery of a black smoker vent field and vent fauna at the Arctic Mid-Ocean Ridge. *Nature Communications*. 1:126 doi: 10.1038/ncomms1124.

Pelletier, H. et al. (1959) Mesure de l'âge de l'une des coulées volcaniques issues du Puy de la Vache (Puy-de-Dôme) par la méthode du carbone 14. *Académie des Sciences à Paris, Comptes Rendus* D214, 2221.

Perret, F. A. (1916) The lava eruption of Stromboli summer–autumn, 1915. *American Journal of Science* 42, 443–63.

Perret, F. A. (1924) *The Vesuvius eruption of 1906: study of a volcanic cycle*. Publication 339, Carnegie Institution, Washington DC.

Peterlongo, J. M. (1978) *Massif Central*. Paris: Masson.

Pinkerton, H. and Sparks, R. S. J. (1976) The 1975 sub-terminal lavas, Mount Etna: a case history of the formation of a compound lava field. *Journal of Volcanology and Geothermal Research* 1, 167–82.

Pinkerton, H. and Wilson, L. (1994) Factors controlling the lengths of channel-fed lava flows. *Bulletin of Volcanology* 56, 108–20.

Pitman, W. C. and Talwani, M. (1972. Sea-floor spreading in the North Atlantic. *Geological Society of America, Bulletin* 83, 619–46.

Pliny [Gaius Plinius Secundus] 1969) *Letters and Panegyricus*, vol. I: letters 16 and 20 [translated by B. Radice]. London: Heinemann.

Poli, S. et al. (1989) Time dimension in the geochemical approach and hazard estimates of a volcanic area: the Isle of Ischia case (Italy). *Journal of Volcanology and Geothermal Research* 36, 327–35.

Prévost, C. (1835) Notes sur l'île de Julia, pour servir à l'histoire de la formation des montagnes volcaniques. *Mémoires de la Société Géologique de France* 2(5), 91–124.

Principe, C. (1998) The 1631 eruption of Vesuvius: volcanological concepts in Italy at the beginning of the XVIIth century. In N. Morello (ed.), *Volcanoes and History*, 525–42. *Proceedings of the 20th International Commission on the History of the Geological Sciences Symposium Napoli–Eolie–Catania (Italy)*, Genoa: Brigati.

Principe, C. et al. (1998) Archéomagnétisme et chronologie des éruptions du Vésuve. *Bulletin de la Section de Volcanologie de la Société Géologique de France* 44, 25.

Principe, C. et al. (2004) Chronology of Vesuvius activity from AD 79 to 1631 based on archeomagnetism of lavas and historical sources. Bulletin of Volcanology 66, 703–24.

Pyle, D. M. (1989) Ice-core acidity peaks, retarded tree growth and putative eruptions. *Archaeometry* 31, 88–91.

Pyle, D. M. and Elliott, J.R. (2006) Quantitative morphology, recent evolution and future activity of the Kameni islands volcano, Santorini, Greece. *Geosphere* 2(5), 253–68.

Real Audiencia de Canarias (1731) *Copia de las Ordones y providencias dadas para el alivio de los Vezinos de la Isla de Lanzarote en su dilatado padezer a causa del prodigioso volcán, que en ella rebentó el primer dia de Septiembre del año immediato pasado de 1730, y continúa hasta el dia de la fecha . . . Canaria y Abril 4 de 1731*. Legajo 89, Gracia y Justicia. Simancas [Spain]: Archivo General de Simancas.

Reck, H. (ed.) (1936) *Santorin: der Werdegang eines Inselvulkans und sein Ausbruch 1925–1928* [3 vols]. Berlin: Reimer.

Rehak, P. and Younger, J. G. (1998) Review of Aegean Prehistory VII: Neopalatial, Final Palatial, and Post-palatial Crete. *American Journal of Archaeology* 102, 91–173.

Riccò, A. (1892) Terremoti, sollevamento ed eruzione sotto-marina a Pantelleria nell seconda metà dell'Octobre 1891. *Bolletino della Società Geographica Italiana* 29, 130–56.

Riccò, A. and Arcidiacono, S. 1902–1904. L'eruzione etnea del 1892. *Atti Accademia Gioenia di Scienze Naturali in Catania* (series 4) 15–17.

Ridley, W. I. (1970a) The petrology of the las Cañadas volcanics, Tenerife, Canary Islands. *Contributions to Mineralogy and Petrology* 26, 124–60.

Ridley, W. I. (1970b) Abundance of rock types on Tenerife. *Bulletin of Volcanology* 34, 196–204.

Rittmann, A. (1938) Die Vulkane am Mývatn in nordost Island. *Bulletin Volcanologique* 4, 3–38.

Romano, R. (ed.) (1982) *Mount Etna volcano: a review of the recent Earth science studies*. Memorie 23, Società Geologica Italiana.

Romano, R. and Sturiale, C. (1982) The historical eruptions of Mt Etna (volcanological data). *Memorie della Società Geologica Italiana* 23, 75–97.

Rosi, M. (1980) The island of Stromboli. *Rendiconti della Società Italiana di Mineralogia e Petrologia* 36, 345–68.

Rosi, M. and Santacroce, R. (1983) The AD 472 'Pollena' eruption: volcanological and petrological data for this poorly known Plinian-type event at Vesuvius. *Journal of Volcanology and Geothermal Research* 17, 249–71.

Rosi, M. et al. (1993) The 1631 Vesuvius eruption: a reconstruction based on historical and stratigraphical data. *Journal of Volcanology and Geothermal Research* 58, 151–82.

Rosi, M. et al. (1996) Interaction between caldera collapse and eruptive dynamics during the Campanian Ignimbrite eruption, Phlegraean Fields, Italy. *Bulletin of Volcanology* 57, 541–54.

Rubin, A. M. (1990) A comparison of rift-zone tectonics in Iceland and Hawaii. *Bulletin of Volcanology* 52, 302–19.

Rymer, H. et al. (1998) Mount Etna: monitoring in the past, present and future. In D. J. Blundell and A. C. Scott (eds), *Lyell: the past is the key to the present*, 335–47. Special Publication 143, Geological Society, London.

Saemundsson, K. (1979) Outline of the geology of Iceland. *Jökull* 29, 7–28.

Saemundsson, K. (1986) Subaerial volcanism in the western North Atlantic. In P. R. Vogt and B. E. Tucholke (eds), *The geology of North America*, vol. M: *the western North Atlantic region*, 69–86. Boulder, Colorado: Geological Society of America.

Santacroce, R. (1983) A general model for the behaviour of the Somma–Vesuvius volcanic complex. *Journal of Volcanology and Geothermal Research* 17, 237–48.

Santo, A. P. et al. (1995) $^{40}Ar/^{39}Ar$ ages of the Filicudi Island volcanics: implications for the volcanological history of the Aeolian Arc, Italy. *Acta Vulcanologica* 7, 13–18.

Scandone, R. et al. (1991) The structure of the Campanian plain and the activity of the Neapolitan volcanoes. *Journal of Volcanology and Geothermal Research* 48, 1–31.

Scandone, R. et al. (1993) Mount Vesuvius: 2000 years of volcanological observations. *Journal of Volcanology and Geothermal Research* 58, 5–25.

Scarpati, C. et al. (1993) The Neapolitan Yellow Tuff: a large volume multiphase eruption from Campi Flegrei, southern Italy. *Bulletin of Volcanology* 55, 343–56.

Scarth, A. (1966) The physiography of the fault scarp between the Grande Limagne and the Plateaux des Dômes, Massif Central. *Institute of British Geographers, Transactions* 38, 25–40.

Scarth, A. (1967) The Montagne de la Serre. *Geographical Journal* 133(1), 42–8.

Scarth, A. (1983) Nísyros volcano. *Geography* 68(2), 133–9.

Scarth, A. (1989) Volcanic origins of the Polyphemus story in the Odyssey: a non-classicist's interpretation. *Classical World* 83(2), 89–96.

Scarth, A. (1994) *Volcanoes*. London: UCL Press.

Scarth, A. (1999) *Vulcan's fury: man against the volcano*. London: Yale University Press.

Scarth, A. (2000) The volcanic inspiration of some images in the Aeneid. *Classical World* 93, 591–605.

Schilling, J. G. et al. (1982) Evolution of the Icelandic hotspot. *Nature* 296, 313–20.

Schmincke, H-U. (1976) The geology of the Canary Islands. In G. Kunkel (ed.), *Biogeography and ecology of the Canary Islands*, 67–184. The Hague: Junk.

Schmincke, H-U. and Mertes, H. (1979) Pliocene and Quaternary volcanic phases in the Eifel volcanic fields. *Naturwissenschaften* 65, 614–15.

Schmincke, H-U. and Mertes, H. (1982) Volcanic and chemical evolution of the Canary Islands. In V. von Rad, K. Hinz, M. Sarnthein and E. Seibold (eds), *Geology of the northwest African continental margin*, 273–306. New York: Springer.

Schumacher, R. and Schmincke, H-U. (1990) The lateral facies of ignimbrites at Laacher See volcano. *Bulletin of Volcanology* 52, 271–85.

Scrope, G. J. P. [later Poulett-Scrope] (1825) *Considerations on volcanoes, the probable causes of their phenomena and their connection with the present state and past history of the globe: leading to the establishment of a new theory of the Earth*. London: W. Phillips.

Scrope, G. P. (1827) *Memoir on the geology of Central France: including the volcanic formations of Auvergne, the Velay and the Vivarais, with a volume of maps and plates*. London: Longman.

Scrope, G. P. (1858) *The geology and extinct volcanoes of Central France*. London: John Murray.

Self, S. (1976) The recent volcanology of Terceira, Azores. *Geological Society of London, Quarterly Journal* 132, 645–66.

Self, S. and Gunn, B. M. (1976) Petrology, volume and age relations of alkaline and saturated peralkaline volcanics from Terceira, Azores. *Contributions to Mineralogy and Petrology* 54, 293–313.

Self, S. and Sparks, R. S. J. (eds) (1981) *Tephra studies*. Dordrecht: Reidel.

Self, S. et al. (1974) The Heimaey Strombolian scoria deposit, Iceland. *Geological Magazine* 111, 539–48.

Selvaggi, G. and Chiarabba, C. (1995) Seismicity and *P*-wave velocity image of the southern Tyrrhenian subduction zone. *Geophysical Journal International* 121, 818–26.

Sheridan, M. F. and Malin, M. C. (1983) Applications of computer-assisted mapping to volcanic hazard evaluation of surge eruptions: Vulcano, Lìpari and Vesuvius. *Journal of Volcanology and Geothermal Research* 17, 187–202.

Sheridan, M. F. et al. (1981) A model for Plinian eruptions of Vesuvius. *Nature* 289, 282–5.

Siebert, L. (1984) Large volcanic debris avalanches: characteristics of sources, areas, deposits and associated eruptions. *Journal of Volcanology and Geothermal Research* 22, 163–97.

Siggerud, T. (1972) The volcanic eruption on Jan Mayen, 1970. *Yearbook: 1970* (pp. 5–18), Norsk Polarinstitutt, Oslo.

Sigmundsson, F. et al. (2010) Intrusion triggering of the 2010 Eyjafjallajökull explosive eruption. *Nature* 468, 426–30.

Sigmundsson, F. et al. (2015) Segmented lateral dyke growth in a rifting event at Bárðarbunga volcanic system, Iceland. *Nature* 517, 191–5.

Sigurdsson, H. S. and Loebner, B. (1981) Deep-sea record of Cenozoic explosive volcanism in the North Atlantic. In S. Self and R. S. J. Sparks (eds), *Tephra studies*, 289–316. Dordrecht: Reidel.

Sigurdsson, H. S. et al. (1982) The eruption of Vesuvius in AD 79: reconstruction from historical and volcanological evidence. *American Journal of Archaeology* 86, 39–51.

Sigurdsson, H. S. et al. (1985) The eruption of Vesuvius in AD 79. *National Geographic Research* 1(3), 332–87.

Sigvaldason, G. E. (1983) Volcanic prediction in Iceland. In H. Tazieff and J. C. Sabroux (eds), *Forecasting volcanic events*, 193–213. Amsterdam: Elsevier.

Sigvaldason, G. E. et al. (1992) Effect of glacier loading/unloading on volcanism: post-glacial volcanic production rate of the Dyngjufjöll area, central Iceland. *Bulletin of Volcanology* 54, 385–92.

Simkin, T. and Siebert, L. (1994) *Volcanoes of the world*, 2nd edn. Tucson: Geoscience Press.

Soler, V. and Carracedo, J. C. (1984) Geomagnetic secular variation in historical lavas in the Canary Islands. *Royal Astronomical Society, Geophysical Journal* 78, 313–18.

Spallanzani, L. (1792) *Viaggio alle Due Sicilie e in alcune parti dell'Appennino* (vol. 2). Pavia: Stamperia Baldassare Comini.

Sparks, R. S. J. et al. (1978) The Thera eruption and Late Minoan Ib destruction on Crete. *Nature* 271, 91.

Stillman, C. J. (1999) Giant Miocene landslides and the evolution of Fuerteventura, Canary Islands. *Journal of Volcanology and Geothermal Research* 94, 89–104.

Stothers, R. B. and Rampino, M. R. (1983) Volcanic eruptions in the Mediterranean before AD 630 from written and archaeological sources. *Journal of Geophysical Research* 88, 6357–71.

Sylvester, A. G. (1975) History and surveillance of volcanic activity on Jan Mayen island. *Bulletin Volcanologique* 39, 313–35.

Tanguy, J-C. (1978) Tholeiitic basalt magmatism of Mount Etna and its relations with the alkaline series. *Contributions to Mineralogy and Petrology* 66, 51–67.

Tanguy, J-C. (1980) *L'Etna: étude pétrologique et paléomagnétique; implications volcanologiques*. Thèse d'état, Université de Paris VI, France.

Tanguy, J-C. (1981) Les éruptions historiques de l'Etna: chronologie et localisation. *Bulletin Volcanologique* 44, 586–640.

Tanguy, J-C. and Kieffer, G. (1993) Les éruptions de l'Etna et leurs mécanismes. In R. Maury (ed.), *Pleins feux sur les volcans*, 239–52. Mémoire 163, Société Géologique de France, Paris.

Tanguy, J-C. and Patanè, G. (1996) *L'Etna et le monde des volcans*. Paris: Diderot Editeur.

Tanguy, J-C. et al. (1985) Geomagnetic secular variation in Sicily and revised ages of historic lavas from Mount Etna. *Nature* 318, 453–5.

Tanguy, J-C. et al. (1996) Dynamics, lava volume and effusion rate during the 1991–1993 eruption of Mount Etna. *Journal of*

Volcanology and Geothermal Research 71, 259–65.

Tanguy, J-C. et al. (1997) Evolution of the Mount Etna magma: constraints on the present feeding system and eruptive mechanism. *Journal of Volcanology and Geothermal Research* 75, 221–50.

Tanguy, J-C. et al. (1999) Variation séculaire de la direction du champ géomagnétique enregistrée par les laves de l'Etna et du Vésuve pendant les deux derniers millénaires. *Académie des Sciences de Paris, Comptes Rendus* 329, 557–64.

Tanguy, J-C. et al. (2012) New archeomagnetic and ^{226}Ra-^{230}Th dating of recent lavas for the geological map of Etna volcano. *Italian Journal of Geosciences* 131(2), 241–57.

Tazieff, H. (1958 and 1959) L'éruption de 1957–58 et la tectonique de Faial (Açores). *Bulletin de la Société Belge de Géologie* 67, 13–49 and *Serviços Geológicos de Portugal* 4 (1959), 71–8.

Thórarinsson, S. (1958) The Öraefajokull eruption of 1362. *Acta Naturalia Islandica* II(2), 1–98.

Thórarinsson, S. (1966) The Surtsey eruption: course of events and the development of Surtsey and other new islands. *Surtsey Research Progress Report* II, 117–123. Reykjavík: Nordic Volcanological Institute.

Thórarinsson, S. (1967a) Some problems of volcanism in Iceland. *Geologische Rundschau* 57, 1–20.

Thórarinsson, S. (1967b) The eruptions of Hekla in historical times: a tephrochronological study. In *The eruption of Hekla 1947–48*. Reykjavík: Societa Scientifica Islandica.

Thórarinsson, S. (1970) The Lakagígar eruption of 1783. *Bulletin Volcanologique* 33, 910–29.

Thórarinsson, S. (1972) The Hekla eruption of 1970. *Bulletin Volcanologique* 36, 270–88.

Thórarinsson, S. (1981) The application of tephrochronology in Iceland. See Self and Sparks (1981: 109–134).

Thórarinsson, S. and Sigvaldason, G. E. (1962) The eruption of Askja 1961. *American Journal of Science* 260, 641–51.

Thórarinsson, S. et al. (1973) The eruption on Heimaey, Iceland. *Nature* 241, 372–5.

Thordarson, T. and Self, S. (1993) The Laki (Skaftár Fires) and Grímsvötn eruptions in 1783–1785. *Bulletin of Volcanology* 55, 233–63.

Thordarson, T. and Self, S. (2003) Atmospheric and environmental effects of the 1783–84 Laki eruption. *Journal of Geophysical Research: Atmosphere* 108(D1).

Thordarson, T. and Sigmarsson, O. (2009) Effusive activity in the 1963–67 Surtsey eruption, Iceland: flow emplacement and growth of small lava shields. In Thordarson et al. (eds), *Studies in Volcanology: The Legacy of George Walker*. Special Publication IAVCEI No 3, 53–84.

Thordarson, T. et al. (1996) Sulfur, chlorine and fluorine degassing and atmospheric loading by the AD 1783–1784 Laki (Skaftár Fires) eruption in Iceland. *Bulletin of Volcanology* 58, 205–25.

Thordarson, T. et al. (2001) New estimates of sulfur degassing and atmospheric mass-loading by the AD ~935 Eldgjá eruption, Iceland. *Journal of Volcanology and Geothermal Research*. Volume 108, Issues 1–4, 15, 33–54.

Torriani, L. (1592) *Descripción de las Islas Canarias* [translated by A. Cioranescu, 1978]. Santa Cruz de Tenerife: Goya.

Treiman, A. H. (2012) Eruption age of the Sverrefjellet volcano, Spitsbergen Island, Norway. *Polar Research* 2012, 31, 17320. doi: 10.3402/polar.v31i0.17320.

Troll, V.R. et al. (2012) Floating stones off El Hierro, Canary Islands: xenoliths of pre-island sedimentary origin in the early products of the October 2011 eruption. *Solid Earth* 3, 97–110.

Troll, V.R. et al. (2015) Nannofossils: the smoking gun for the Canarian hotspot. *Geology Today* 31, No. 4, 137–45.

Tryggvason, E. (1984) Widening of the Krafla fissure swarm during the 1975–1981 volcanic–tectonic episode. *Bulletin Volcanologique* 47, 47–69.

Tryggvason, E. (1986) Multiple magma reservoirs in a rift zone volcano: ground deformation and magma transport during the September 1984 eruption of Krafla, Iceland. *Journal of Volcanology and Geothermal Research* 28, 1–44.

Tryggvason, E. (1989) Ground deformation in Askja, Iceland: its source and possible relation to flow in the mantle plume. *Journal of Volcanology and Geothermal Research* 39, 61–71.

Valentin, A. et al. (1990) Geochemical and geothermal constraints on magma bodies associated with historic activity, Tenerife (Canary Islands). *Journal of Volcanology and Geothermal Research* 44, 251–64.

Van Wyk de Vries, B. et al. (2014) Craters of Elevation revisited: forced folds, bulges and uplift of volcanoes. *Bulletin of Volcanology* doi: 10.1007/s00445-014-0875-x.

Ventris, M. and Chadwick, J. (1973) *Documents in Mycenean Greek*, 2nd edn. Cambridge: Cambridge University Press.

Villari, L. (1980a) The island of Alicudi. *Rendiconti della Società Italiana di Mineralogia e Petrologia* 36, 441–6.

Villari, L. (1980b) The island of Filicudi. *Rendiconti della Società Italiana di Mineralogia e Petrologia* 36, 467–88.

Walker, G. P. L. (1963) The Breiddalur central volcano, eastern Iceland. *Geological Society of London, Quarterly Journal* 119, 29–63.

Walker, G. P. L. (1973) Lengths of lava flows. *Royal Society, Philosophical Transactions* A274, 107–118.

Walker, G. P. L. (1981) Plinian eruptions and their products. *Bulletin Volcanologique* 44(2), 223–40.

Washington, H. S. (1909) The submarine eruptions of 1831 and 1891 near Pantelleria. *American Journal of Science* 27, 131–50.

Watts, A. B. and Masson, D. G. (1995) A giant landslide on the north flank of Tenerife, Canary Islands. *Journal of Geophysical Research* 100, 24487–98.

Weijermars, R. (1987) A revision of the Eurasian/African plate boundary in the western Mediterranean. *Geologische Rundschau* 76(3), 667–76.

Weston, F. S. (1964) List of recorded volcanic eruptions in the Azores with brief reports. *Boletim do Museu e Laboratório Mineralógico e Geológico da Faculdade de Ciências, Universidade de Lisboa* 10, 3–18.

White, J. D. L. and Schminke, H-U. (1999) Phreatomagmatic eruptive and depositional processes during the 1949 eruption on La Palma (Canary Islands). *Journal of Volcanology and Geothermal Research* 94, 283–304.

Wolfe, C. J. et al. (1997) Seismic structure of the Icelandic mantle plume. *Nature* 385, 245–7.

Wolff, J. A. (1985) Zonation mixing and eruptions of silica-undersaturated alkaline magma: a case study from Tenerife, Canary Islands. *Geological Magazine* 122, 623–40.

Woodhall, D. (1974) Geology and volcanic history of Pico Island Volcano, Azores. *Nature* 248, 663–5.

Wordie, J. M. (1922) Jan Mayen island. *Geographical Journal* 59, 180–85.

Wordie, J. M. (1926) The geology of Jan Mayen. *Royal Society of Edinburgh, Transactions* 104, 742–5.

Wörner, G. and Wright, T. L. (1984) Evidence for magma mixing within the Laacher See magma chamber (East Eifel, Germany). *Journal of Volcanology and Geothermal Research* 22, 301–27.

Zaczek, K. et al. (2015) Nannofossils in 2011 El Hierro eruptive products reinstate plume model for Canary Islands. *Scientific Reports* 5, 7945 doi: 10.1038/srep07945.

Zielinski, G. A. and Germani, M. S. (1998) New ice-core evidence challenges the 1620s BC age for the Santoríni (Minoan) eruption. *Journal of Archaeological Science* 25, 279–89.

Zielinski, G. A. et al. (1994) Record of volcanism since 7000 BC from the GISP2 Greenland ice core and implications for the volcano–climate system. *Science* 264, 948–52.

Zielinski, G. A. et al. (1995) Evidence of the Eldgjá (Iceland) eruption in the GISP2 Greenland ice core: relationship to eruption processes and climatic conditions in the tenth century. *The Holocene* 5, 129–40.

Zbyszewski, G. (1980) Géologie des Iles Atlantiques. In *Géologie des pays Européens*, 157–71. Paris: Dunod.

Zbyszewski, G. et al. (1958) *Noticia explicativa da folha B, São Miguel (Açores)*. Lisbon: Serviços Geológicos, Carta Geológica de Portugal.

Zbyszewski, G. and Ferreira, O. da V. (1959a) Le volcanisme de l'Isle Faial et l'éruption du volcan Capelinhos: rapport de la deuxième mission géologique. *Serviços Geológicos de Portugal* 4, 339–45. Lisbon.

Zbyszewski, G. et. al. (1959b, 1961, 1971) *Noticia explicative da folha Ilha Terceira (Açores)*. Lisbon: Serviços Geológicos, Carta Geológica de Portugal.

Zollo, A. et al. (1998) An image of Mt Vesuvius obtained by 2-D seismic tomography. *Journal of Volcanology and Geothermal Research* 82, 161–73.

Glossary

A

aa (lava): a type of lava flow, from the Hawaiian for 'stony rough lava', its top surface being made of a rubbly mixture of broken pieces of lava, some with very sharp spines, often termed clinker.

alkali basalt: basalt that is richer in alkalis and lower in silica than the more common tholeiitic basalt.

andesite: a greyish volcanic rock, intermediate in composition between basalt and rhyolite.

ankaramite: a type of basalt that is especially rich in large pyroxene and olivine crystals.

ash: the smallest component of explosive volcanoes, particles less than 2 mm in diameter, composed of broken rocks and volcanic glass.

B

barranco: a deep gully or ravine incised by a stream into any steep slope, and often particularly well developed where unconsolidated fragments clothe a volcano.

basalt: a dark, pasty-grey volcanic rock, containing only 40–52 per cent of silica, but relatively rich in iron, calcium and magnesium. When molten it can exceed 1000°C. It is by far the most common volcanic rock, forming the bulk of the ocean floors, and on land it occurs in many lava flows, cinder cones, shield volcanoes and volcanic plateaux.

basanite: a type of very alkaline basalt that is rich in ferromagnesian minerals.

black smoker: submarine chimneys formed from deposits of iron sulphide that occur at hydrothermal vents, usually occurring at mid-ocean ridges on the sea floor.

block: a large, solid angular fragment, often of old lava material, expelled during an eruption.

block and ash flow: deposit from a pyroclastic flow that lacks pumice fragments, often formed from the collapse of a dome.

blocky lava: a lava flow that has solidified with a surface of angular blocks.

bomb (volcanic): large ejected magmatic material from an explosive eruption. Volcanic bombs differ from volcanic blocks in that their shape records fluidal surfaces (bombs were juvenile liquid magma when erupted).

bradyseismic: small, oft-repeated upward and downward movements of the Earth's crust, which bring about slow and prolonged vertical displacements. They are often caused by oscillations in the levels of subterranean magma.

C

caldera (collapse): the collapse structure that occurs when a volcano has evacuated a large volume of magma during an eruption. The ground above the void left by the removal of the magma collapses into it, leaving a negative valley-like structure at the surface. The term is derived from the Spanish word for cauldron. The Portuguese term 'caldeira' is sometimes used.

caliche: a buff or pale ochre calcrete in the Canary Islands, which was formed by chemical precipitation after the capillary rise of water rich in calcium carbonate during the wetter episodes of the Glacial Period.

cinders: – see Scoria.

cinder cone/scoria cone: a steep conical hill, usually less than 250m high, with straight slopes that are initially at the angle of rest of the loose materials composing the cone. Formed above a vent when moderate explosions accumulate layers of scoria, lapilli and ash.

crust: the solid outer layers of the Earth, forming both the continents and the ocean floors.

D

dacite: a pale volcanic rock, rich in silica (63–68 per cent), which is emitted at temperatures usually about 800°C to 900°C and is viscous and slow-moving. A common constituent of domes, it is often also involved in violently explosive eruptions.

debris avalanche – see sector collapse.

dense rock equivalent (DRE): the fragments expelled during an eruption are riddled with holes that can double or triple their volume compared with that of a compact rock such as a lava flow. Thus, in order to compare the sizes of different eruptions, the measurements of erupted volumes taken in the field are changed into their equivalent in dense rock. These dense rock volumes correspond broadly with the volumes of magma expelled.

dome: a rounded, convex-sided mass of volcanic rock, which is usually silicic and too viscous to flow far from the vent. Often formed on stratovolcanoes towards the end of an eruption. Frequently composed of dacite, phonolite, trachyte or rhyolite.

dyke (US spelling – dike): a vertical/sub-vertical planar sheet of frozen magma that has intruded pre-existing rocks, commonly cutting across the bedding in sedimentary rocks, for example. In reality, dykes are rarely planar structures and can be intricately associated with sills and sill complexes as part of the plumbing system of volcanoes.

dyke swarm: a symmetrical or radial set of closely spaced vertical to sub-vertical sheet intrusions, sometimes, but not exclusively, associated with igneous centres.

E

effusive eruption: a volcanic eruption that is not explosive but involves lava flowing out from fissures or eruptive centres.

eruption: the way in which gases, liquids and solids are expelled onto the Earth's surface by volcanic action, ranging from violently explosive outbursts to effusive or hydrothermal outflow.

eruption column: the column of hot ash and volcanic gases that rise up from an explosive volcanic eruption above the vent, and can extend many kilometres into the atmosphere.

eustatic: the term used to describe worldwide changes in sea level, which are themselves caused by changes in the volume of water in the oceans.

F

fajã: a relatively isolated area in the Azores, which has been formed at the base of cliffs, either when rubble has fallen from them, or where lava has flowed into the sea in a broad delta.

fire fountain: a natural fountain of molten magma that is jetted into the sky through pressure caused by escaping gases as the magma reaches the surface.

fissure: a crack, fault or cluster of joints, cutting deep into the Earth's crust, which may allow magma to reach the surface. A fissure usually gives rise to effusive emissions, which may be accompanied by rather more explosive eruptions forming cones of cinders or spatter known as 'fissure vents'.

fragments: ash, bombs, cinders, lapilli or pumice shattered by explosions during an eruption. They are the main constituents of cinder cones and many stratovolcanoes. Also called pyroclasts and tephra.

fumarole: a small vent giving off gases or steam, and often surrounded by fragile precipitated crystals. They are a major aspect of hydrothermal activity, along with solfataras and geysers.

G

geological time: the whole history of the Earth, extending back about 4.6 thousand million years.

H

historical time: the timespan during which events have been recorded, in however fragmentary a fashion, by observers. In the Mediterranean area it may reach back 3000 years, whereas in the New World it can be less than 200 years.

hornito: a small cone or mound of spatter, usually less than 10m, but occasionally 100m in height. They are notably rough and steep-sided, whatever their size. The name is derived from the Spanish word for little oven, which smaller hornitos resemble.

hotspot (plume): term used for the area above an anomalously hot area of the mantle manifested by an increased amount of melting and volcanoes. A hotspot generates chains of volcanoes when the tectonic plates move over it.

hyaloclastite: a volcaniclastic rock made up of fragments of rock from a millimetre to a few centimetres, containing much fresh, glassy material formed by the interaction of lava with large water bodies, e.g. lakes and the sea, or ice.

hydrothermal: a term used to describe processes involving heated groundwater, usually due to hot igneous rocks near the surface (from Greek – 'hydros' meaning water and 'thermos' meaning heat).

hydrothermal eruption: a term used to describe eruptions of gases, steam, and hot water, without magma.

hydrovolcanic eruption: a term used to describe violently explosive eruptions in which both water or ice and magma play a significant role. Such eruptions can be termed Surtseyan in shallow sea water and lakes, and hydrovolcanic or hydromagmatic on land (sometimes termed phreatomagmatic eruptions).

I

Icelandic eruption: eruptions that are usually basaltic in character, which take place notably along fissures caused basically by plate divergence in Iceland. They form long rows of relatively small cones but often give out vast lava flows. Sometimes they also develop into large volcanic systems.

Iceland plume: the anomalously hot mantle upwelling, hot spot, that is situated under Iceland.

ignimbrite: the deposit from a pyroclastic density current (pyroclastic flow). They can be voluminous deposits, often erupting more than one cubic kilometre, of pumice, broken crystals, and elongated pieces of glass ('fiamme') in a matrix of ash. When they are deposited at high temperatures they may become welded.

island arc: a gently curving chain of volcanic islands rising above sea level from the ocean floor, formed when an oceanic plate is subducted beneath another. Volcanic chains are their equivalent on land.

J

jökulhlaup: large flood caused by meltwater from ice caps during the eruption of a subglacial volcano.

L

lahar: debris flows rich in volcanic particles and water, often called mudflows, which are very hazardous, as they can move very quickly and destroy almost anything in their path. Lahars can occur during eruptions, particularly where ice on a volcano is rapidly melted to provide an abundant source of water, or from heavy rains after an eruption, which mobilize loose volcanic debris.

lapilli: pyroclastic particles that range in size from 2–64mm in diameter.

latite: a silicic lava in which potassium and sodium occur in similar volumes. It is a term that is now little used, and is considered to be a potassic variety of trachyandesite.

lava: molten rock or magma that reaches the surface and solidifies on cooling. Lava occurs as flows, domes, fragments within cones, and as pillows formed on the ocean floors.

lava tube (tunnel): a partially filled or empty underground tube where lava is travelling along, or had travelled along in the past.

M

Maar: a German word used to describe an almost circular crater, often about 1km across, formed mainly by hydrovolcanic eruptions. They may or may not be bordered by a ring or crescent of fine fragments, sloping gently outwards from a low crest overlooking the crater. The crater is usually filled with a small lake, from which the German name is derived.

magma: hot mobile rock material, mainly formed by partial melting of the mantle, commonly at depths of between 70km and 200km. It is composed of hot, viscous liquid material also often containing crystals or rock fragments and small proportions of included gases. It is less dense than the materials surrounding it, and is thus able to rise slowly towards the Earth's surface by buoyancy. If it overcomes the pressure and resistance of the rocks of the Earth's crust, it erupts in a fluid state, releasing its contained gases with varying degrees of explosive violence, and emits lava in flows or fragments.

magma reservoir/chamber: a large zone of ill-defined fissures and cavities beneath a volcano, where rising magma halts for varying lengths of time. Reservoirs are most often a few cubic kilometres in volume and are situated usually between 2km and 50km in depth.

mantle (Earth): the viscous differentiated layer between the Earth's core and crust, which is around 2900km thick, and is so viscous as to be almost solid, and is composed mainly of silicates rich in iron and magnesium.

mid-ocean ridge: a ridge on the ocean floor where volcanic eruptions generate new oceanic crust and where two adjacent plates diverge.

mudflow – see lahar.

O

obsidian: volcanic glass. A dense, shiny black or brown, glassy and a rare form of rhyolite, which rises and cools rapidly, and is usually too viscous to flow far from the vent. It forms domes, mounds and short rugged lava flows.

P

pahoehoe: the term 'pahoehoe' is Hawaiian, meaning 'smooth, unbroken lava', and is one of the most common forms of basaltic lava flow, formed from fluidal, low viscosity lava. The surface can often have a ropy and bulbous surface appearance, and thick pahoehoe flows can develop through a process of lava flow inflation.

palagonite: yellowish-buff brown mineral formed by the breakdown of glassy fragments in hyaloclastites and pillow lavas.

pantellerite: rhyolite that is relatively rich in alkalis, and thus also known as peralkaline rhyolite.

Peléan eruption: a violent eruption that gives rise to blasts and pyroclastic density currents like those expelled by Montagne Pelée on 8 May 1902.

peralkaline: a volcanic rock in which the combined molecular proportions of sodium and potassium exceed those of aluminium oxides.

phonolite: a pale volcanic rock rich in sodium and potassium but relatively poor in silica (55–60 per cent), which is usually emitted at temperatures of less than 1000°C. It is derived from very alkaline basalts such as basanite or nephelinite. It is viscous and it occurs in domes and rugged lava flows. It is so named because it often breaks into plates that give a characteristic ringing sound when struck.

picón: the cover of black ash and lapilli expelled during the historical eruptions on Lanzarote.

pillow lava: lava that erupts at depth under water in the form of piles of pillows or cushions.

plate: the rigid upper slabs of the Earth that move slowly around the convecting and cooling mantle. Their edges constantly diverge or converge and plunge beneath each other. All are composed of oceanic crust and some also carry continental crust. Between 10 and 15 major plates are generally recognized, and a similar number of microplates.

plate tectonics: the process by which the rigid tectonic plates of Earth's crust and uppermost mantle move slowly around above the convecting mantle beneath. This causes the plates to separate, collide or rub past each other, resulting in magma generation, earthquakes, and much of the topographic variation we see on the Earth's surface and ocean floor.

Plinian cloud: the classic mushroom-like cloud formed by a Plinian eruption. Described by Pliny the Younger as being a shape like that of an umbrella pine (Mediterranean pine), a tree that can be found growing around Pompeii today.

Plinian eruption: named from the famous AD79 eruption of Mt. Vesuvius, described by Pliny the Younger. These eruptions are very powerful and result in a large column and cloud of ash and ejecta that travels high up into the atmosphere (up to 55km) and stretches out into a mushroom-like shape where it reaches the stratosphere. They are generated mainly by silicic magmas, although they can develop from hydrovolcanic eruptions of basalt. The gas and dust expelled to the stratosphere often form an acidic aerosol that can modify the weather over large tracts of the Earth.

pumice: very pale volcanic fragments riddled with gas holes, formed by the expansion of contained gases as the magma reaches the surface and explodes very violently over vast areas during an eruption. Most pumice floats on water and sometimes forms ephemeral floating islands after eruptions at sea. It varies from small fibrous chips to knobbly lumps and often resembles solidified foam. It is commonly expelled in eruptions of rhyolite, dacite, trachyte or phonolite.

pyroclastic density current (pyroclastic flow): term for a turbulent current of hot particles, rock fragments and gas, it is commonly termed 'pyroclastic flow'. When the current has a low concentration it is sometimes referred to as a 'pyroclastic surge'. There is a complete continuum between high- and low-concentration pyroclastic density currents, and they can mix between the two during an eruptive phase. Other old terminology includes the French word 'nuée ardente' for fiery cloud.

R

rhyolite: a pale volcanic rock very rich in silica (69–75 per cent), which is usually emitted at temperatures about 700–800°C and commonly forms extensive pumice and ashflows when expelled as fragments, but can also give rise to viscous lavas forming domes and stubby flows.

rootless cone: when molten lavas invade boggy land or very shallow surface water, their heat can suddenly convert the water into steam. The resultant explosions shatter the lava into fine fragments, which accumulate in small cones on the flow. They are termed 'rootless' because, unlike cinder cones, they have no vents extending down into the terrain beneath the lava flows.

S

satellite cone: a small cinder cone erupted on the flanks of a stratovolcano or shield. They can occur alone, in swarms, or along fissures radiating from the summit. They are also called lateral, adventive or parasite cones.

scoria: primary pyroclastic material made up of vesicular basalt or andesite, formed by explosive eruptions of more basic magmas.

seamount: a volcanic mountain found below sea level, rising from the ocean floor. Active seamounts could eventually form new volcanic islands. They may also be extinct submerged remains of old volcanoes.

sector collapse (debris avalanche/gravitational collapse): a landslide formed at the start of a volcanic eruption, where one side of the volcano collapses. This can be triggered by over-steepening of the volcano, and/or by earthquakes, and can cause an asymmetric eruption known as a lateral blast.

shield volcano: a large, gently sloping (shield-like) volcano, composed mainly of fluid basaltic lava flows with relatively few fragmented layers, emitted from clustered vents.

shoshonitic lava: a lava with a relatively high potassium content that occurs in subduction zones.

silica: the molecule, formed of silicon and oxygen (SiO_2), that is a fundamental component of volcanic rocks, and is the most important factor controlling the viscosity (fluidity) of magma. Other things being equal, the higher the silica content of a magma, the greater its viscosity.

silicic magma: magma rich in silica (60–75 per cent). Also called acid or felsic magma.

sill: tabular sheet intrusion that is concordant with the bedding within the host rock. In reality, sills are rarely simple planar structures and can be found as linked sill complexes with several levels of intrusion, as saucer-shaped intrusions, and in close association with dykes.

solfatara: an Italian word used to describe the emission of sulphurous gases from a fumarole.

spatter: lava fragments of cinder size, often emitted as clots in lava fountains. They are still molten when they return to the ground, and thus flatten out and form 'cowpats'. Spatter often welds together to form steep-sided cones and ramparts, as well as hornitos.

stratovolcano: a type of volcano found mainly at destructive plate boundaries that rises up from shallow slopes at the base to form steep-sided tops of the volcano, resulting in a cone-shaped mountain, often with a surprisingly small crater at the top. Eruptions from such volcanoes can be violent and explosive with Plinian type eruptions, and are often constructed mainly from more silicic/felsic magma types.

Strombolian eruption: a characteristic eruption style (named after the volcano Stromboli, where the eruptions are common) consisting of small amounts of hot basic magma that are erupted explosively. Continued eruptions build up a scoria cone, and when viewed in low light/night-time, a fountain-like (similar to roman candle fireworks) trajectory of the hot glowing scoria can be seen.

subduction zone: a zone where two plates converge and one plunges beneath the other into the mantle. The subducted slab releases volatiles that stimulate melting in the wedge of mantle above it, which helps form volcanoes. The plunging action of the slab also generates deep-seated violent earthquakes.

subglacial eruption: any eruption that takes place under ice, but most commonly associated with activity beneath the ice caps of Iceland.

sub-Plinian eruption: a less violent form of Plinian eruption.

Surtseyan eruption: an eruption that takes place in shallow lakes or seas where water can enter the vent, mix with the rising magma, and repeatedly form steam that shatters the magma into fine fragments, often called tuffs. Without such water interference, most of these eruptions would probably be Strombolian in nature. One of the major types of hydrovolcanic eruption.

T

tephrite: a volcanic rock that is relatively poor in silica (45–50 per cent) and rich in alkalis. After crystallization, some of the latter help to form feldspathoids, which are like feldspars but with less silica. Nepheline is a sodic feldspathoid; leucite is a potassic feldspathoid. The leucite tephrite of Vesuvius often contains large, rounded white crystals of leucite.

tholeiitic basalt: basalt containing relatively high amounts of silica, with relatively small proportions of sodium and potassium.

trachyandesite: a composition between trachyte and andesite, often containing alkali feldspar and sodic plagioclase.

trachybasalt: a grey alkaline volcanic rock, with about 50 per cent silica, that forms fragments and fairly fluid lava flows (more or less synonymous with hawaiite).

trachyte: a pale, greyish acidic volcanic rock relatively rich in silica (60–65 per cent) and in sodium and potassium, which is usually emitted at temperatures of about 1000°C. It is viscous, and can be involved in violent explosions. It also forms rugged lava flows and domes.

tsunami: a Japanese term used to describe huge, rapidly moving sea waves generated by violent eruptions or earthquakes. They increase in size and speed as they reach shallow water and often cause much damage and death on nearby coasts.

tuff cone: a steep, squat conical hill, usually less than 300m high, composed of innumerable thin layers of fine fragments with a deep wide crater, formed above a vent by Surtseyan eruptions.

tuff ring: a broad, circular accumulation of fine fragments, often 1km or more in diameter, surrounding a broad, shallow crater. Both the outer and crater-ward-facing slopes are relatively gentle compared with those of a cinder cone, and the crater is much wider. Commonly found where the crater is formed from an eruption involving water-saturated sediment or groundwater.

V

vent: the usually vertical conduit or pipe up which volcanic material travels from the magma source to the Earth's surface.

viscosity: a measure of the resistance of a fluid that is being deformed. High viscosity implies a sticky fluid and low viscosity a runny fluid.

volcanic gas: the volatile component of magma, mainly including steam, carbon dioxide, sulphur dioxide and smaller amounts of chlorine and fluorine. As the magma approaches the Earth's surface, the gases are exsolved and can become the chief factor in the violence of eruptions.

volcano: a hill or mountain formed around and above a vent by accumulations of erupted materials such as ash, pumice, cinders or lava flows. The term refers both to the vent itself and to the often cone-shaped accumulation above it.

Vulcanian eruption: first used to describe the 1888–1890 eruptions on the island of Vulcano. These eruptions form explosions like cannon fire at irregular intervals, and contain a large component of non-juvenile material. Vulcanian eruptions have been attributed to the interaction of the magma with groundwater, known as hydrovolcanic eruptions.

Vocabulary

This vocabulary is not intended to be exhaustive, but the selection of descriptive terms may help the reader to visualize the appearance of many volcanoes presented in the text.

F – French, G – Greek, Ger – German, It – Italian, Ice – Icelandic, P – Portuguese, S – Spanish

achada	P	abutting onto	eldgjá	Ice	fire fissure
ajuda	P, S	help	eldur	Ice	fire, eruption
alto	P	high	enfer	F	hell
ancien	F	old	enxofre	P	sulphur, fumarole
antico	It	old	fjal	Ice	mountain
antigua	S	old	fljöt	Ice	river
apalhraun	Ice	aa lava flow	fogo	P	fire
areia	P	sand, ash	fossa	It	crater
arena	It, S	sand, ash	fuego	S	fire
askja	Ice	box	fuencaliente	S	hot spring
áspros	G	white	fuoco	It	fire
azufrado	S	sulphur	furna	P	oven, cavern, hot spring
bagacina	P	cinders	gemelos	S	twins
baixo	P	low, lower	geysir	Ice	hot water fountain
bajo	S	low, lower	gjá	Ice	fissure
barranco	S	deep gully, ravine	gordo	P, S	large, fat
bermejo	S	red	gour	F	round lake or maar
bianco	It	white	hághios	G	holy
biscoitos	P	rugged lava flow	helluhraun	Ice	smooth pahoehoe lava flow
blanc	F	white			
blanco	S	white	hnúk	Ice	peak
boca	S	vent or mouth	hornito	S	small oven
bocca	It	vent or mouth	hryggur	Ice	ridge
borg	Ice	rocky hill	hver	Ice	hot spring
branco	P	white	jameo	S	lava tunnel in Canary Islands
butte	F	small mesa			
cabeço	P	head, cone	jökulhlaup	Ice	glacier burst
caldeira	P	cauldron, large crater	lagoa	P	lake
caldera	S	cauldron, large crater	lajes	P	stony, stones, flagstones
cancela	P	sheltered place	lombo	P, S	ridge
carvão	P	burnt, charcoal	loutrá	G	hot springs
cendres	F	ash, cinders	Maar	Ger	round lake or maar
couze	F	River in Auvergne	malpaís	S	area of rugged lava flows
cuchillo	S	narrow ridge	mávros	G	black
cumbre	S	ridge	megálos	G	big
dyngja	Ice	lava shield	mesa	S	tableland, or flat-topped hill
echeide	S	inferno (a Guanche word)			
égueulé	F	breached	micrós	G	small
eldfjell	Ice	volcano	mistério	P	lava flow in the Azores

259

mont	F	mountain
montaña	S	mountain
monte	It, S	mountain
narices	S	nostrils
narse	F	infilled maar in Auvergne
negro	P, S	black
neós	G	new
nero	It	black
nuée ardente	F	incandescent cloud
paleós	G	old
partido	S	cloven
pavin	F	fearsome
pic	F	peak
pico	P, S	peak
pietre	It	stones
puy	F	hill in Auvergne
queimado	P	burnt
rajada	S	split
raudur	Ice	red
redondo	P	round
reventado	S	breached
revento	S	breached
rhaun	Ice	lava flows

ribeira	P	river
rio	S	river
rivière	F	river
rojo	S	red
rosso	It	red
sarcoui	F	cauldron
sciara	It	rugged lava flow
secco	It	dry
seco	P, S	dry
serra	It, P	ridge
serre	F	ridge
solfatara	It	sulphur, sulphurous fumarole
soufrière	F	sulphurous place
stéfanos	G	crown
suc	F	narrow pointed lava dome
timão	P	arched yoke
vatn	Ice	water, lake
vecchio	It	old
vermelho	P	vermilion, red
vermelo	S	vermilion, red
vieux	F	old
víti	Ice	Hell, explosion crater

Eruptions in Europe in historical times

A question mark indicates doubt about:

- after the date: the date of the eruption;
- before the date: the occurrence of the eruption itself;
- after the place name: the location of the eruption.

Many eruptions cited without question marks are not necessarily certain to have taken place, but the reference often represents what seems to be the most likely location and date of the events. Many of the eruptions listed are discussed more fully in the text and in the relevant articles to which reference is made. Data come mainly from historical written documents; some information, however, is inferred as follows from dating systems: A archaeological, AM archaeomagnetic, ^{14}C radiocarbon, Ra radium disequilibria, K–Ar: potassium–argon.

Italy

The volcanoes in Italy provide one of the most extensive data bases of historical eruptions and this is somewhat reflected in the detail of the following notes.

Vesuvius

A	persistent mild activity
IE	intermediate eruption
FE	final eruption
SPE	sub-Plinian eruption
PE	Plinian eruption, see text

NB: the distinction between IE and FE, or SP and P, is often not so clear as it may seem in this list.

1944	18–29 March, strong FE, lava flows to the NW at Massa di Somma and San Sebastiano, lava fountains more than 2km in height, hydrovolcanic explosions and small nuées ardentes, 28 deaths.
1929	13 July to 17 March 1944, A including 4 IE.
1929	3–8 June, strong IE with lava fountains 500m high and lava overflows to the E base of the cone.
1913	5 July to 2 June 1929, increasing A fills the 1906 crater.
1906	4–22 April, very strong FE with lava fountains 3km high, flows to the SE destroying Boscotrecase,

	ballistic projectiles causing heavy damage at Ottaviano, 218 deaths.
1875	18 December to 3 April 1906, A including 6 IE, of which that of Colle Margherita and Colle Umberto.
1872	26–30 April, FE with lava fountains and flows to the NW, at least 12 deaths.
1870	2 November to 25 April 1872, A including one IE.
1868	15–26 November, FE.
1865	10 February to 14 November 1868, A with one IE.
1861	8–10 December, FE through a fissure extending to the sea on the SW side.
1855	19 December to 7 December 1861, A including one IE.
1855	1–27 May, FE (?) with lava flows to the NW reaching Cercola.
1850	6–16 February, FE.
1841	20 September to 5 February 1850, A including one IE.
1839	1–5 January, FE.
1835	13 March to 31 December 1838, A.
1834	23 August to 2 September, FE.
1824	2 July to 22 August 1834, A including 3 IE.
1822	21–25? October, strongly explosive FE.
1806	27 January to 20 October 1822, A including 4 IE.
1805	14–16 October?, FE.
1796	15 January to 13 October 1805, A including 3 IE.
1794	15 June to 7? July, large FE, fissure on the SW flank and lava flow to the sea destroying Torre del Greco, at least 18 deaths.
1783	18 August to 14 June 1794, A including 2 IE.
1779	3–15 August, strong FE, lava fountains 4km high.
1770	16 February to 2 August 1779, A including 2 IE.
1767	19–27 October, FE
1764	1 July to 18 October 1767, A with one IE.
1760	23 December to 5 January 1761, FE with fissure low on the S flank.
1744	1 July to 22 December 1760, A including 3 IE.
1737	20–30 May, strong FE.
1732	25 December to 19 May 1737, A including one IE.
1730	17–23 March, FE(?)
1712	5 February to 16 March 1730, A including 10 IE.
1707	28 July to 15? August, strongly explosive FE.
1699	22 April to 27 July 1707, A.
1698	20–31 May? FE (?).
1638?–98	A including 5 (?) IE.
1631	16–19 December, strong SPE with nuées ardentes

	(AM, ^{14}C), most of the villages S and W of the volcano destroyed, more than 4000 deaths.
*c.*1500	'fire' in the crater (hot gases?).
?1347–50	Strombolian activity?
1139	29 May to 5 June, strongly explosive eruption and flows (AM, ^{14}C).
1037	January–February, extensive lava flows to the sea (AM).
1006–1007	explosive eruption, blocks hurled '3 miles from the crater'.
999	cinders and lava flows (AM)
991	'flames and ashes'
968	lava fountains and flows to the sea (AM).
787	October–December, lava fountains and flows (AM, ^{14}C).
685	February–March, huge eruption column, lava or pyroclastic flows to the sea (^{14}C).
536	Strombolian activity.
?512	probable confusion with 472.
472	5–6 November, large SPE with nuées ardentes to the NW (AM, ^{14}C), fallout to the NE.
379–95	Strombolian activity and probable lava flows.
222–35	persistent Strombolian activity.
203	strong explosions and ash deposits (SPE?, ^{14}C).
172	Strombolian activity
AD 79	24–26 August, large PE burying Herculaneum (AM), Pompeii, Stabiae.
?217 BC	possible SPE
*c.*700 BC	(A), SPE
*c.*1000 BC	(A), SPE
?...	at least 3 undated explosive eruptions (SPE?)
c.1800 BC	(^{14}C), 'Avellino', PE
?...	at least 2 undated SPE
c.6000 BC	(^{14}C), 'Mercato' PE

Phlegraean Fields and Ischia

1538	29 September to 9 October, Monte Nuovo, Phlegraean Fields.
1302	18 January to March, Arso, Ischia.
*c.*670–890	Fiaiano, Ischia
*c.*430	Molara, Vateliero and Cava Nocelle, Ischia.
*c.*AD 60	Montagnone–Maschiatta, Ischia.
*c.*19 BC?	Rotaro II?
*c.*350 BC	Porto d'Ischia, Ischia?
*c.*470 BC	Porto d'Ischia, Ischia?
*c.*600 BC	Rotaro I, Ischia.

Aeolian Islands, Stromboli

Almost continuously active since at least 1788; only major or unusual events reported here.

2014	lava flows from side vent at 650m travel to the sea.
2010–2011	short-lived lava flow emissions from northern and southern vent areas.
2007	27 February, flank eruption begins with lava flows reaching the sea.
2003	flank lava eruptions.

2002	24 July, strong explosion; landslides on 30 December cause two tsunamis; damage in nearby villages.
2001	20 October, major explosion at Stromboli, tourist killed.
1999	26 August, strong explosion, tourists injured.
1998	16 January, 23 August, 8 September, stronger explosions.
1996	1 and 6 June, 4 September, incandescent material on vegetation.
1994	21–22 August, continuous lava fountaining.
1993	10 February, 16 and 23 October, strong explosions, tourists injured.
1990	15 April, lithic fallout on village.
1985	December to April 1986, lava flow to the sea.
1975	4–24 November, ashfall, lava flow to the sea.
1972	December, explosions, lithic fallout on village.
1967	19 April to 13 August, lava flow to the sea.
1959	19 May, 11 July, strong explosions.
1956	January–March, lava flows.
1955	28 February to 22 March, flank flow at the foot of Sciara del Fuoco.
1954	1 February, paroxysm, ash fall, hot avalanche, tsunami.
1952	7–22 June, explosion, intermittent lava flows.
1949	6–9 June, ash and block fall, fire on vegetation, lava flow.
1944	20 August, 2000m plume, hot avalanche, one house destroyed.
1943	December to October 1944, intermittent flows to the sea.
1943	3 December, ash and block fall, fires, houses damaged.
1938–9	lava fountains 1km high, several flows to the sea.
1934	2 February, block fall near village, air shocks damage houses.
1932	3 June to 1 February 1934, repose period.
1931	7 July to May 1932, repose period.
1931	23 April, 7 July, strong explosions.
1930	22 October to 2 December, intermittent lava fountains and flows.
1930	11 September, major paroxysm, block fall on villages, pyroclastic flow in the Vallonazzo through, tsunami, lava flows, 6 deaths.
1930	3 and 6 February, ash fall, lava flows.
1919	22 May, major paroxysm, bombs on villages, ash fall up to Sicily, tsunami, 4 deaths.
1916	30 June, major explosions and flows.
1915	18 June to 20 December, explosions and flows.
1912	June–November, intermittent lava fountains, block falls, ash to Calabria and Sicily.
1907	June to May 1910, repose period.
1907	January–April, intermittent ash flows and lava flows to the sea.
1905	February–December, repeated lava flows and ash flows.

1900	April–October, frequent explosions and ash falls, one hot avalanche.
1891	24 July and 31 August, two paroxysms, block falls, fire on vegetation.
1889	June to 1890, repose period.
1888	October to June 1889, lava fountains and flows.
1887	31 March and 18 November, pumice emissions.
1882	17–30 November, paroxysm, opening of new vents, block falls, landslides.
1879	February and August, lava fountains, ash falls, fire on vegetation.
1874	June, block fall on village.
1857	repose period
1855	3–4 October, lava fountains and scoria fallout
1850	ash and block falls, houses damaged.
1822	22 October, ash fall, hot avalanche, destruction of land.
1768–70	explosions, lava flows, submarine eruption.
1558?	destruction of land…

Lipari

c.1200	(AM) Rocche Rosse obsidian flow
AD 729	Mt Pilato Plinian eruption?

Vulcano

1888	3 August to 22 March 1890, series of powerful explosions.
1886	11 January to 31 March, intermittent explosions, new vents.
1881–3	strong fumaroles, incandescent gases.
1879	6–13 January, detonations heard at Lipari, 'flames'.
1877	September–October, strong fumaroles and (?) weak explosions.
1876	29 July (?), eruption with ashfall to Salina.
1873	22 July to June 1875, fumaroles, intermittent explosions.
1822–5	greater fumarole activity
1786	12 January to February, explosive eruption.
1783	February, explosions heard in Calabria.
1776–86	fumaroles larger than usual.
1731–39	eruptive period ending with Pietre Cotte obsidian flow.
1727	Forgia Vecchia satellite crater ?
1688	June, explosions
1626	March–April, greater activity?
1540–50?	erupted material links Vulcano to Vulcanello.
1444	5 February, powerful explosions with incandescent material.
c.1300–40	'great fire', detonations
c.1250	eruptions from both Vulcano and Vulcanello?
1184	December, possible confusion with Lipari.
?900–50	stronger activity?
?729–87	possible confusion with Lipari
c.500–80?	explosions?
c.200–250	activity

?AD 144	probably no eruption
c.AD 25–100	activity?
c.29–19 BC	activity
91 BC	both Vulcano and Vulcanello active?
126 BC	submarine eruption at or near Vulcanello.
?186–183 BC	doubtful reference to Vulcano or Vulcanello.
c.350 BC	activity?
c.425 BC	'fire' from Vulcano Island.
?475 BC	probable confusion with 425 BC.
c.1000–800 BC	to (AM) lava flow from the W basement of Vulcanello.
c.2600 BC	(K–Ar), major eruptive cycle
c.3500 BC	(K–Ar), beginning of La Fossa activity.

Etna

CC	Central Crater
V	Voragine
BN	Bocca Nuova
NEC	Northeast Crater
SEC	Southeast Crater
NSEC	New South East Crater

Note: in recent years Etna has continued to be very active with all types of eruptions being more actively recorded in the digital age. Most major events are summarized below (from Global Volcanism Program data), with examples from the known records that date back to 3300 BC. 'Paroxysm' is a shorthand for Etna's often intense Strombolian discharges that frequently include lava fountaining, lava flow emission, and tephra columns, which erupted at the summit craters (GVP).

2016	May, intense Strombolian and associated activity around NEC and V, with lava flows into Valle del Bove.
2015	2–8 December paroxysm V and associated eruptions (V, BN & NSEC) occurred. Paroxysm climaxed between 03:20 and 04:10 local time on 2 December with lava fountain that reached 1km (3300ft) in height, with ash plume up to 3km (9800ft) in height, largest ash plumes reached 7km (23,000ft).
2014–2015	continued activity with E- and SW-directed lava flows from NSEC flanks, some minor activity also recorded at V, BN and NEC.
2011–2013	birth of the NSEC, which formed at a large pyroclastic cone alongside the SEC. 18 paroxysms in 2011, 7 in 2012, and 21 in 2013.
2008	13 May, flank eruption E of summit craters, eruption continued for 417 days, until 6 July 2009.
2006–2007	East flank fissure eruptions and lava flows, many eruptive episodes and activity around SEC.
2004–2005	230m long N110° E-trending eruptive fissure opened at the base of the SEC with Lava flows into Valle del Bove.
2002–2003	27 October–28 January 2003, flank eruptions from N and S sides of volcano. Strong earthquakes on

	29 October with main shock recorded by Jean-Claude Tanguy in the SE region of the volcano (Trecastagni) at 17 seconds after 1102 (± 5 sec); 1000 people were left homeless.
2001	July–August, largest flank eruption since 1993 from around SEC and south flank of volcano.
2000	26 January. SEC eruptions (lava fountains and flows, more than 60 events as of 24 June).
1999	October, explosions and overflows from BN.
1999	4 September, lava fountains from V, explosions at BN.
1999	4 February to 6 November, fissure and flows at the SE base of SEC.
1998	September to January 1999, SEC eruptions (explosions and overflows).
1998	22 July, 5 August, lava fountains at V.
1998	27 March, lava fountains at NEC.
1996	November to July 1998, explosions BN, V, SEC, NEC, lava overflows at SEC.
1996	July–August, explosions and flows at NEC.
1995–6	intermittent lava fountains at NEC.
1991	14 December to 31 March 1993, SE flank eruption (231x10^{6}m^3 of lava).
1990	January, intermittent lava fountains and flows at SEC.
1989	10 September to 9 October, lava fountains at V and SEC, then NE flank eruption (~20x10^{6}m^3).
1986	30 October to 25 February 1987, SEC and NE flank eruption (~50x10^{6}m^3), Monte Rittmann.
1986	24 September, lava fountain NEC.
1985	25–31 December, E flank eruption.
1985	10 March to 13 July, S flank eruption.
1984	29 April to 16 October, SEC explosions and lava flows.
1983	28 March to 6 August, S flank eruption (79x10^{6}m^3).
1981	17–23 March, NNW flank eruption (18x10^{6}m^3).
1980	September, and February 1981, NEC lava fountains.
1979	3–9 August, SEC and SE, W, ENE flank eruptions.
1978	23–29 November, SEC and SE flank eruption.
1978	23–29 August, SEC and SE, ENE flank eruption.
1978	29 April to 5 June, SEC and SE flank eruption.
1977	July to March 1978, NEC intermittent lava fountains and flows.
1975	24 February to 8 January 1977, NEC and N flank.
1974	10 October…NEC explosions and N flank.
1974	30 January to 17 February and 11–29 March, W flank eruptions, Monte De Fiore.
1971	5 April to 12 June, S, SE and ENE flank eruptions (40x10^{6}m^3).
1966–71	March, NEC continuous explosions and flows.
1964	February–July, intermittent explosions and flows from CC.
1960	17 July, powerful lava fountain at V, minor explosions on following weeks.
1955–64	continuous explosions and flows at NEC

	(250x10^{6}m^3), various episodes at V.
1950	25 November to 2 December 1951, E flank eruption (124x10^{6}m^3).
1949	1–4 December, S, N and NW flank (4x10^{6}m^3).
1947	24 February to 10 march, N flank (7x10^{6}m^3).
1942	30 June, SSW flank and CC (3x10^{6}m^3).
1940	16 March, powerful lava fountain at CC.
1928	2–20 November, NEC and ENE flank (26x10^{6}m^3), Mascali overwhelmed.
1923	17 June to 18 July, NNE flank (48x10^{6}m^3).
1918	29 November?, NW flank (11x10^{6}m^3).
1917	24 June, lava fountain at NEC.
1911	10–21 September, NNE flank (27x10^{6}m^3).
1910	23 March to 18 April, S flank (37x10^{6}m^3).
1908	28 December, Messina tectonic earthquake.
1908	29–30 April, SE flank (2x10^{6}m^3).
1899	19 July and 5 August, powerful explosions at CC.
1892	9 July to 28 December, S flank (145x10^{6}m^3), Mt Silvestri.
1886	19 May to 7 June, S flank (51x10^{6}m^3), Mt Gemmellaro.
1883	22–24 March, S flank, Mt Leone.
1879	26 May to 7 June, SSW and NNE flanks (45x10^{6}m^3), Mt Umberto–Margherita.
1874	29 August, NNE flank (2x10^{6}m^3).
1869	26 September, E base of central cone (3x10^{6}m^3).
1868	27 November, 8 December, CC lava fountains 2km high.
1865	30 January to 28 June, NE flank, Mt Sartorius.
1852	20 August to 27 may 1853, ESE flank, Mt Centenari.
1843	17–28 November, W flank.
1838/39/42	CC explosions and overflows.
1832	1–22 November, W flank, Mt D'Ognissanti or Nunziata.
1819	27 May to 1 August, SE flank, 'La Padella' crater.
1811	27 October to 24 April 1812, E flank, Mt Simone.
1809	27 March to 9 April, NNE flank.
1802	15–16 November, E flank.
1798–1809	CC strong intermittent explosions.
1792	12 May to May? 1793, SSE flank, 'La Cisternazza' collapse pit.
1787	17, 18, 19 July, CC lava fountains 3km high and overflows.
1780	18–31 May, SSW flank.
1766	28 April to 7 November, S flank, Mt Calcarazzi.
1763	19 June to 10 September, S flank, 'La Montagnola' cone.
1763	6 February to 15 March, W flank, Mt Nuovo.
1755	10–15 March, E flank.
1732–65	CC recorded persistent activity
1723–24	CC recorded persistent activity
1702	8 march to 8 May, SE flank
1693	9 and 11 January, powerful tectonic earthquakes
1689	March, E flank

1682	September? upper E flank
1669	11 March to 11 July, S flank large eruption (600–800x10^{6}m^3), about 15 villages and part of Catania overwhelmed, strong explosions at CC on 25 March.
1651?	…(AM), lava flow towards Macchia di Giarre on the E flank.
1651	17 January to 1653, W flank large eruption destroying part of Bronte.
1646	20 November to 17 January 1647, NNE flank, northern Mt Nero.
1634	18 December to June? 1636, SSE flank.
1614	1 July to 1624, N flank voluminous eruption, 'Due Pizzi' or 'Fratelli Pii' hornitos.
1610	6 February to 15 August, SW flank, Grotta degli Archi crater row.
1607	June, NW flank, Pomiciaro cinder cone.
1603–1610	CC strong, persistent activity with frequent overflows.
1579	September, SE flank
1566	November–December? N flank
1537	10 May to July, S flank
1536–37	CC strong persistent activity
1536	22 March to 8 April? CC overflows, then S and possibly SW flanks.
1493–1500c.	CC persistent activity
1446	September, E flank
?1444…	SSE flank? Possible confusion with 1408
1408	8–25 November, SSE flank, partial destruction of Pedara village (AM, Ra).
?1381	5 August, SSE flank, possible confusion with 1329.
?1334	confusion with 1329
1329	28 June to August…, E and lower SE flanks, Mt Rosso di Fleri (AM, Ra), flows near Acireale.
1284–5	E or SE flank.
c.1250	(AM), S flank, flow near Serra La Nave (GM: indicated as 1536)
c.1200	(AM), SW flank, upper Gallo Bianco flow (GM: indicated as 1595)
c.1180	(AM, Ra), NE flank, flow to Linguaglossa village (GM: indicated as 1566)
1169	4 February, destructive tectonic earthquake, doubtful Etna eruption (see below)
c.1160	(AM), SSE flank, flow to the sea (GM: indicated as 1381)
1062–64	W flank, lower Gallo Bianco flow? (see below)
c.1060	(AM), WSW flank, lower Gallo Bianco flow (GM: indicated as 1595)
c.1050	(AM), E flank, Mt Ilice cone and extensive flow to the sea (GM: indicated as 1329)
c.1020	(AM), NE flank, Scorcia Vacca flow (GM: indicated as 1651)
c.1000	(AM), SSW flank, Mt Sona and flow to Paterno town (GM: indicated as 812/1169)
c.970	(AM), N flank, Mt Pizzillo cone
c.950	(AM, Ra), NW flank, lower Pomiciaro flow (GM: indicated as 1537)

c.800	(AM), SE flank, flow north and east of Trecastagni (GM: indicated as 1408)
c.700	(AM), SE flank, Mt Solfizio spatter rampart and flow.
c.650	(AM), W flank, Mt Lepre
c.500	(AM), S flank, Mt Ciacca lava flow
c.450	(AM), W flank, flow south of Bronte
c.300	(AM), S flank, Mompeloso cinder cone and flow
252	1–9 February, SSE flank, Cibali flow in Catania (AM)
c.200	(AM), SSE flank, Carvana flow in Catania
c.100	(AM), SSE flank, Mt Pizzuta Calvarina
AD 38–40	large explosive eruption heard in Messina
c.AD 0–20	persistent activity within summit caldera
32 BC	lava flow
36 BC	N or NW flank
44 BC	tephra fall to Reggio Calabria
49 BC	large summit plume, then flow on W flank
122 BC	(^{14}C) large Plinian eruption, flows near Catania
126 BC	earthquakes, summit and (?) flank eruptions
135 BC	gas plume and lava flows
140 BC	'fires'
396 BC	spring, E or SE flank, lava flow to the sea
425 BC	March–April, S flank, lava flow near Catania
?475 BC	probable confusion with 479
479 BC	August, SE flank, lava flow to the sea
?695 BC	…?...
c.1100 BC	(AM, Ra), SW flank Mt Arso and flow toward Licodia
c.1400 BC	large eruptions force the Sicanians to emigrate
c.3300 BC	(AM, Ra), Fortino Vecchio flow in Catania (GM: indicated as 693 BC)

Straits of Sicily

?2000	Graham Bank
1891	Foerstner Bank
?1863	
?1845	
1831	Graham Bank
?1707	
?1632	

Greece

1950	Neá Kaméni, Santoríni
1939–41	Neá Kaméni, Santoríni
1928	Neá Kaméni, Santoríni
1925–6	Neá Kaméni, Santoríni
1887	Caldera, Nísyros
1873	Caldera, Nísyros
1871	Caldera, Nísyros
1866–70	Neá Kaméni, Santoríni
1707–11	Neá Kaméni, Santoríni
1650	Kolómbos Bank, off Santoríni
1570	Mikrá Kaméni, Santoríni

726	Paleá Kaméni, Santoríni
AD 46	Paleá Kaméni, Santoríni
197 BC	Hierá (Bankos Bank), Santoríni
c.250 BC	Kaméni Xorió, Méthana

Spain, Canary Islands

2011–2012	El Hierro. October–March eruptions offshore along the southern submarine rift zone.
1971	Teneguía, La Palma
1949	San Juan, La Palma
1909	Chinyero, Tenerife
1824	Tao–Nuevo–Tinguatón, Lanzarote
1798	Narices de Teide, Tenerife
1793?	Lomo Negro, El Hierro?
1730–36	Montañas del Fuego, Lanzarote
1712	El Charco, La Palma
1706	Garachico, Tenerife
1705	Arafo, Tenerife
1705	Fasnia, Tenerife
1704–5	Siete Fuentes, Tenerife
1677–8	'San Antonio' – Fuencaliente, La Palma
1646	Martin, La Palma
1585	Tahuya, La Palma
1492	Teide, Tenerife
1478–80?	Teide, Tenerife
1470–90 or 1430–40	Montana Quemada, La Palma
1455?	Teide? Tenerife
1430?	Orotava Valley, Tenerife
1393? or 1399?	Tenerife

Portugal, Azores

1998–9	off Serreta, Terceira
1981	Monaco Bank, S of São Miguel
1964	off Rosais, São Jorge
1911	Monaco Bank, S of São Miguel
1907	Monaco Bank, S of São Miguel
1902	off Pico, Terceira
1867	off Serreta, Terceira
1811	Sabrina, off Ferraria, São Miguel
1808	Urzelina, São Jorge
1800	off Serreta, Terceira
1761	Negro & Fogo, Terceira
1720	Don João de Castro Bank
1682	off Ferraria, São Miguel
1652	Fogo I & II, São Miguel
1638	off Ferraria, São Miguel
1630	Furnas, São Miguel
1580	Queimada, São Jorge
1564	Lagoa do Fogo, São Miguel
1563	Sapateiro–Quiemado, São Miguel
1563	Lagoa do Fogo, São Miguel

1562–4	Praínha, Pico
1445	Furnas, São Miguel
1439	Sete Cidades, São Miguel

Iceland

2014–15	Holuhraun (Bárdarbunga)
2011	May, Grímsvötn
2010	March–June, Eyjafjallajökull
2004	November, Grímsvötn
2000	Hekla
1998	Grímsvötn
1996	Gjálp– Grímsvötn
1991	Hekla
1983	Grímsvötn
1981	Hekla
1980	Hekla
1975–84	Krafla
1973	Heimaey
1970	Hekla
1963	Surtsey
1961	Askja
1947	Hekla
1934	Grímsvötn
1922	Grímsvötn
1918	Katla
1903	Grímsvötn
1892	Grímsvötn
1883	Grímsvötn
1875	Askja
1862–4	Bárdarbunga–Veidivötn
1860	Katla
1845	Hekla
1838	Grímsvötn
1823	Grímsvötn
1821–3	Eyjafjallajökull
1783–5	Grímsvötn
1783–4	Laki
1774	Grímsvötn
1766–8	Hekla
1766	Bárdarbunga–Veidivötn
1755	Katla
1739	Bárdarbunga–Veidivötn
1724–9	Krafla
1727	Öraefajökull
1721	Katla
1717	Bárdarbunga–Veidivötn
1711	Bárdarbunga–Veidivötn
1706	Bárdarbunga–Veidivötn
1697	Bárdarbunga–Veidivötn
1693	Hekla
1660	Katla
1659	Grímsvötn
1636	Hekla

1625	Katla
1619	Grímsvötn
1612	Katla
1597	Hekla
1580	Katla
1510	Hekla
c.1500	Katla
1477	Bárdarbunga–Veidivötn
1416	Katla
1410	Bárdarbunga–Veidivötn
1389	Hekla
1362	Öraefajökull
1357	Katla
1354	Grímsvötn
1341	Hekla
c.1340	Bárdarbunga–Veidivötn
1300	Hekla
1262	Katla
1245	Katla
1231	Reykjanes
1226	Reykjanes
1222	Hekla

1206	Hekla
1179	Katla
1159	Bárdarbunga–Veidivötn
1158	Hekla
1104	Hekla
c.935	Eldgjá–Katla
c.920	Katla
870	Bárdarbunga–Veidivötn–Torfajökull

Norway, Jan Mayen

1985	Nordkapp
1984–5	Beerenberg?
1975	Nordkapp
1973	Nordkapp
1970–71	Nordkapp
1851?	Koksletta
1850?	Nordkapp?
1818	Dagnyhaugen–Eggoya?
1732	Dagnyhaugen–Eggoya?

Index

Page numbers in *italic* denote illustrations. Page numbers in **bold** denote major treatment of subject.